Modern Aerospace Engineering

Modern Aerospace Engineering

Edited by
Stephen Baggins

WILLFORD PRESS
www.willfordpress.com

Published by Willford Press,
118-35 Queens Blvd., Suite 400,
Forest Hills, NY 11375, USA

ISBN: 978-1-68285-633-8

Cataloging-in-Publication Data

Modern aerospace engineering / edited by Stephen Baggins.
    p. cm.
Includes bibliographical references and index.
ISBN 978-1-68285-633-8
1. Aerospace engineering. 2. Aeronautics. I. Baggins, Stephen.
TL545 .M63 2019
629.1--dc23

For information on all Willford Press publications
visit our website at www.willfordpress.com

WILLFORD PRESS

Contents

Preface

Over the recent decade, advancements and applications have progressed exponentially. This has led to the increased interest in this field and projects are being conducted to enhance knowledge. The main objective of this book is to present some of the critical challenges and provide insights into possible solutions. This book will answer the varied questions that arise in the field and also provide an increased scope for furthering studies.

Aerospace engineering is a branch of engineering that studies the design and development of aircraft and spacecraft. It branches into the two major disciplines of aeronautical engineering and astronautical engineering. The principles of propulsion are of utmost importance in aerospace engineering. An aircraft moves due to energy provided by jet engines, internal combustion engines and turbomachinery. Electric propulsion and ion propulsion are recent propulsion techniques. Modern aerospace engineering also uses computational fluid dynamics to simulate the behavior of fluids, reduce time and expenses. Further, the integration of software such as ground control software, flight software and test & evaluation software has resulted in the advancement of this field. This book attempts to understand the multiple branches that fall under the discipline of aerospace engineering and how such concepts have practical applications. The various studies that are constantly contributing towards advancing technologies and evolution of this field are examined in detail. Those with an interest in aerospace engineering would find this book helpful.

I hope that this book, with its visionary approach, will be a valuable addition and will promote interest among readers. Each of the authors has provided their extraordinary competence in their specific fields by providing different perspectives as they come from diverse nations and regions. I thank them for their contributions.

Editor

Quality Evaluation of Chromatic Interpolation Algorithms for Image Acquisition System

Diana Carolina Morón Hernández[1], Freddy Alexander Díaz González[1], Juan Sebastian Triana Correa[1], Pablo Roberto Pinzón Cabrera[1]

ABSTRACT: The main goal of the Libertad 2 mission is to take images of the Earth's surface in the visible spectrum with a multispectral sensor and send them as a Bayer array. To carry out the reconstruction in land of these images and to view them in a RGB color model, it is necessary to use a color interpolation algorithm to determine the values of the different channels of color in each pixel. This document presents an analysis of five chromatic interpolation algorithms (Nearest Neighbor, Adaptive Color Plane, Bilinear, Smooth Hue Transition and Median Based). The analysis also includes the identification of image evaluation algorithms without reference and the artifacts (False Color Measure, Blur Metric and Chromatic and Achromatic Zipper). Finally, the analysis results of the chromatic interpolation algorithms and the selection of the most suitable algorithm for the Libertad 2 satellite mission data are presented.

KEYWORDS: Bayer array, RGB images, Chromatic interpolation quality metrics.

INTRODUCTION

CubeSat satellites have gained recognition as possible platforms for future scientific missions (CubeSat Organization 2014), and previous studies determined the capabilities of these satellites and their functionality in the field of Earth observation. Daniel Selva, from the Massachusetts Institute of Technology (MIT), conducted a study about the feasibility of "Remote Sensing" technologies on CubeSat systems (Woellert *et al.* 2011). In this study, it was concluded that the type of supported technologies can be implemented in satellite systems under the CubeSat standard restrictions. Additionally, Daniel Selva proposes technologies to perform spectral estimations of the state of vegetation and its biomass, particularly to determine the Normalized Differential Vegetation Index (NDVI) (Greenland 2010).

Taking into account studies such as Selva and Krejci (2012) or Sandau (2008), the Universidad Sergio Arboleda is carrying out its second satellite mission called Libertad 2, which consists of the development and implementation of a 3U nanosatellite designed under the CubeSat standard. This mission's main goal is to validate the development of a Remote Sensing System (RSS) prototype, which will open the field for academic researches in Colombia based on the obtained images; these researches may focus on the analysis of land use for the agricultural sector, water resource analysis, planning of urban growth, among others.

The RSS of the Libertad 2 satellite mission will be based on a multispectral Complementary Metal-Oxide Semiconductor (CMOS) sensor, which will deliver the RAW data on a Bayer array. In order to one observe RAW data as a high-quality Red-Green-Blue (RGB) color model image, it is required the implementation

---

1.Universidad Sergio Arboleda – Escuela de Ciencias Exactas e Ingeniería – Bogotá – Colombia.

Author for correspondence: Freddy Alexander Díaz González | Universidad Sergio Arboleda – Escuela de Ciencias Exactas e Ingeniería | Calle 74 #14-14 | Bogotá – Colombia | Email: freddy.diaz@correo.usa.edu.co

of a chromatic interpolation algorithm that allows to rebuild the original image from the Color Filter Array (CFA) RAW data, calculating each pixel depending on its intensity within a bounded region of the arrangement (Maschal Jr *et al.* 2010). This bounded region of neighboring pixels can vary both the size and the gain of each channel; for this reason, a study should be made to determine the most appropriate algorithm to process the data captured by the Libertad 2 mission.

To perform a chromatic interpolation algorithms analysis, it is required to establish some metrics that allow to evaluate each algorithm's efficiency, since the image reconstruction procedure is usually carried out in the ground station of the satellite mission, where the computing cost is not relevant. The analysis cannot be based on a reference pattern because the input data of interpolation methods are only in Bayer array format and, in any case, in RGB model (Malvar *et al.* 2004). For this reason, the algorithms analysis should focus on the quality of the images without an original image reference to compare pixel by pixel. The interpolated image quality will be measured by the number of defects. These defects are often called "artifacts" and correspond to pixels with inconsistent values in an interpolated image. Some of these artifacts are Blur, Zipper and False Color and decrease the image quality. Thus, only the algorithms with the least amount of defects will be considered for the implementation of the image reconstruction system in the ground station of the mission.

This document describes the evaluation process of five chromatic interpolation algorithms; the first section describes the structure of an image sensor CMOS (which allows the capture of color images) and a RBG color model. Then, the chromatic interpolation algorithms and quality metrics that correspond to the selection criteria of the most appropriate algorithm, identifying the evaluation methods of images without reference, are explained in detail. Finally, the results are presented with some analysis, determining the optimal algorithm for the implementation of the Libertad 2 mission in the Earth observation system.

ACQUISITION OF SATELLITE IMAGES

A satellite image can be defined as the visual representation of the information captured by an installed sensor on-board an artificial satellite. A satellite image is composed of a set of elements with the same size, called "pixels", which are organized in rows and columns. Pixels contain a numeric value or digital number, obtained from the sensors when they capture the amount of energy reflected by the objects in the Earth's surface. When the satellite image is obtained by multispectral sensors, it is contained in a matrix of various dimensions, where each pixel's digital number is located in a row, a column and a band (Pérez *et al.* 2003). An image obtained from a four-band multispectral sensor is shown in Fig. 1.

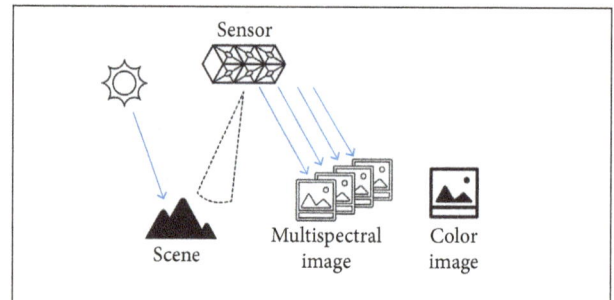

Figure 1. Image obtained from a four-band multispectral sensor.

Multispectral satellite images are those which obtain information in several bands. When the combination of bands is performed only in the visible spectrum, RGB-format images are obtained, often called "color images", because their representation is similar to the images perceived by the human eye. Sensors designed to obtain this type of images capture the energy in the corresponding bands to red, green and blue colors, as well as part of the near-infrared region, which gives more definition to the dynamic element of the captured image.

An image sensor is physically composed of three layers: the first, the outermost, consists of a microlenses array that focuses the incoming light towards each of the photosites; the second layer filters the light in the desired spectrum; the third one consists of an array of light-sensitive photodiodes which determine a value of voltage proportional to the current perceived. After the substrates, CMOS sensors incorporate a Digital Signal Processor (DSP) system with amplifiers and analog-to-digital (A/D) converters to process the image in different delivery formats and control the automatic values of gain for each channel's color and other digital functions available according to the manufacturer's ability (Lillesand *et al.* 2014).

In order to get color images, a filter has to be overlapped in each one of the cavities of an image sensor — CMOS or Charge-Coupled Device (CCD) —, which only allows the passage of required light intensity of primary colors (red, green, and blue) or complementary colors (cyan, magenta, and yellow), depending on how the sensor is built. The data delivered by the sensor form a matrix known as CFA (Li *et al.* 2008). There are different types of CFAs, such as Yamanaka CFA, Diagonal Stripe,

Vertical Stripe, RGBE, CYYM, Bayer (Fig. 2), among others, that vary depending on the manufacturer. CFA data are known as RAW data. To form a color image of the visible spectrum, the intensities of each array of the RGB format are calculated from the RAW data provided by the sensor (Wheeler 2009).

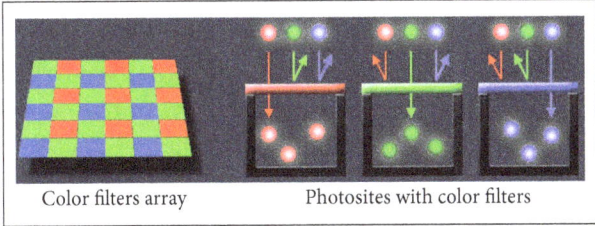

Figure 2. Bayer array (Cambridge in Colour 2013).

Colors can be represented by mathematical methods, known as "color models", for example, RGB or Lab (Hurkman 2010). The RGB color model obtains the different existing colors by combining the three primary ones of the light — red, green, and blue —, which are known as additive colors, because, by adding light intensity, new colors are created (Adobe Creative Team 2007). This color model is located at the RGB space represented in the trichromatic cube, as shown in Fig. 3. Each coordinate in the cube represents a color (Martínez and Díaz 2005).

The data from the RGB model are three stacked arrays which contain the image information in blue, red and green components, as shown in Fig. 4. Through the combination of the information of these components, a color image is obtained.

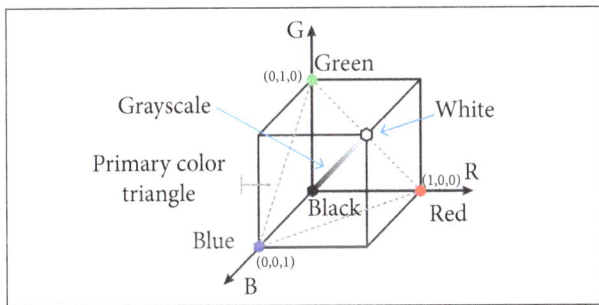

Figure 3. Trichromatic cube (Muñoz 2012).

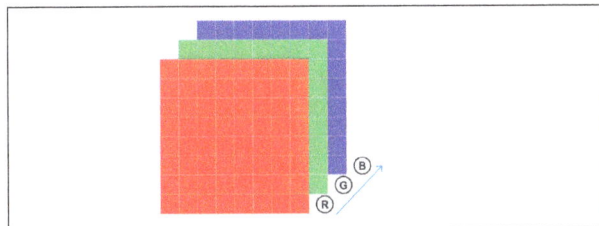

Figure 4. RGB matrix.

COLOR INTERPOLATION ALGORITHMS

Color interpolation is a process used on images to convert a Bayer array into a RGB color model (Losson *et al.* 2009). There are several mathematical methods to calculate RGB values from the data delivered by multispectral sensor CMOS. The most commonly used are: Nearest Neighbor, Bilinear, Adaptive Color Plane, Smooth Hue Transition and Median Based. The main purposes pursued by the color interpolation methods are to:

- Get an RGB image from a Bayer array.
- Rebuild a color image avoiding possible artifacts.
- Lose the fewest possible details in the color image reconstruction.
- Get low computational complexity to accelerate the processing when it is implemented on an embedded system or a mobile device with reduced computational resources.

Most color interpolation algorithms base the calculation of the RGB color model on the intensity of each element, inside a bounded region of the Bayer array, known as "neighboring elements". Generally, the algorithms vary their interpolation method depending on the quantity of elements in the region specified as neighboring elements (Elizondo and Maestre 2005). Color interpolation algorithms are divided into adaptive and non-adaptive algorithms.

Non-adaptive color interpolation algorithms calculate the pixels values of the missing colors taking into account the neighboring pixels. They use few computing resources and are easy to implement. Among these types of algorithms are the Nearest Neighbor, Bilinear and Smooth Hue Transition, which are explained next.

Nearest Neighbor algorithm calculates RGB color model of the image taking the nearest neighbors values and dividing the Bayer array into 2×2 subarrays, as in the following example (Ramanath *et al.* 2002).

If the pixel in analysis is the Bayer $(1,1)$, where i is the row and j is the column,

$$RGB(i, j, 1) = Bayer(i + 1, j)$$
$$RGB(i, j, 2) = Bayer(i, j) \qquad (1)$$
$$RGB(i, j, 3) = Bayer(i, j + 1)$$

Then $RGB(i,j,1)$ is the component in red of the pixel in analysis, $RGB(i,j,2)$ is the component in green and $RGB(i,j,3)$ is the component in blue. If the pixel in analysis is the Bayer (1,2),

$$RGB(i,j,1) = Bayer(i+1,j-1)$$
$$RGB(i,j,2) = Bayer(i,j-1) \qquad (2)$$
$$RGB(i,j,3) = Bayer(i,j)$$

If the pixel in analysis is the Bayer (2,1),

$$RGB(i,j,1) = Bayer(i,j)$$
$$RGB(i,j,2) = Bayer(i,j+1) \qquad (3)$$
$$RGB(i,j,3) = Bayer(i-1,j+1)$$

If the pixel in analysis is the Bayer (2,2),

$$RGB(i,j,1) = Bayer(i,j-1)$$
$$RGB(i,j,2) = Bayer(i,j)$$
$$RGB(i,j,3) = Bayer(i-1,j) \qquad (4)$$

Bilinear chromatic interpolation algorithm calculates the value of each pixel in the RGB image computing the average of the pixels of each missing colors in a 3 × 3 submatrix (Ramanath *et al.* 2002).

If the pixel in analysis is the Bayer (2,2),

$$RGB(i,j,1) = \frac{Bayer(i,j-1) + Bayer(i,j+1)}{2}$$
$$RGB(i,j,2) = Bayer(i,j) \qquad (5)$$
$$RGB(i,j,3) = \frac{Bayer(i-1,j) + Bayer(i+1,j)}{2}$$

If the pixel in analysis is the Bayer (2,3),

$$RGB(i,j,1) = Bayer(i,j)$$
$$RGB(i,j,2) = [Bayer(i-1,j) + Bayer(i,j-1) +$$
$$\qquad\qquad + Bayer(i+1,j) + Bayer(i,j+1)]/4 \qquad (6)$$
$$RGB(i,j,3) = [Bayer(i-1,j-1) + Bayer(i-1,j+1) +$$
$$\qquad\qquad + Bayer(i+1,j-1) + Bayer(i+1,j+1)]/4$$

If the pixel in analysis is the Bayer (3,2),

$$RGB(i,j,1) = [Bayer(i-1,j-1) + Bayer(i-1,j+1) +$$
$$\qquad\qquad + Bayer(i+1,j-1) + Bayer(i+1,j+1)]/4$$
$$RGB(i,j,2) = [Bayer(i-1,j) + Bayer(i,j-1) +$$
$$\qquad\qquad + Bayer(i+1,j) + Bayer(i,j+1)]/4 \qquad (7)$$
$$RGB(i,j,3) = Bayer(i,j)$$

If the pixel in analysis is the Bayer (3,3),

$$RGB(i,j,1) = \frac{Bayer(i-1,j) + Bayer(i+1,j)}{2}$$
$$RGB(i,j,2) = Bayer(i,j) \qquad (8)$$
$$RGB(i,j,3) = \frac{Bayer(i,j-1) + Bayer(i,j+1)}{2}$$

Median Based chromatic interpolation algorithm is based on the bilinear interpolation, adding to the model calculation a factor derived from the calculation of the median of the neighbors in a 3 × 3 submatrix (Ramanath *et al.* 2002). This demosaicing algorithm preserves edges well. The **Red** (i, j) matrix is the result of applying the bilinear interpolation algorithm to the RAW data in order to obtain the red component. The **Blue** (i, j) represents the blue component and **Green** (i, j), the green component.

The values of the missing pixels are calculated as follows.

$$RGB(i,j,1) = [(Red(i,j) - Green(i,j))] + Bayer(i,j)$$
$$RGB(i,j,2) = Bayer(i,j) \qquad (9)$$
$$RGB(i,j,3) = [(Blue(i,j) - Green(i,j))] + Bayer(i,j)$$

If the pixel in analysis is the Bayer (2,2),

$$RGB(i,j,1) = Bayer(i,j)$$
$$RGB(i,j,2) = [(Green(i,j) - Red(i,j))] + Bayer(i,j) \qquad (10)$$
$$RGB(i,j,3) = [(Blue(i,j) - Red(i,j))] + Bayer(i,j)$$

If the pixel in analysis is the Bayer (2,3),

$$RGB(i,j,1) = [(Red(i,j) - Blue(i,j))] + Bayer(i,j)$$
$$RGB(i,j,2) = [(Green(i,j) - Blue(i,j))] + Bayer(i,j) \qquad (11)$$
$$RGB(i,j,3) = Bayer(i,j)$$

If the pixel in analysis is the Bayer (3,2),

Smooth Hue Transition interpolation algorithm is similar to the Bilinear method; however, for the calculation of the red and blue channels, it contemplates the values of luminance channel (green) (Dargahi and Daneshpande 2007). To calculate the pixel value, first, it must determine the green value as the interpolation Green Bilinear (i, j).

If the pixel in analysis is the Bayer (2,2),

$$RGB(i,j,1) = \left(\frac{Green(i,j)}{2}\right) \times$$
$$\times \left(\frac{Bayer(i,j-1)}{Green(i,j-1)} + \frac{Bayer(i,j+1)}{Green(i,j+1)}\right)$$

$$RGB(i,j,2) = Bayer(i,j) \qquad (12)$$

$$RGB(i,j,3) = \left(\frac{Green(i,j)}{2}\right) \times$$
$$\times \left(\frac{Bayer(i-1,j)}{Green(i-1,j)} + \frac{Bayer(i+1,j)}{Green(i+1,j)}\right)$$

If the pixel in analysis is the Bayer (2,3),

$$RGB(i,j,1) = Bayer(i,j)$$

$$RGB(i,j,2) = \left(\frac{Green(i,j)}{4}\right) \times \left(\frac{Bayer(i-1,j)}{Green(i-1,j)} + \right.$$
$$+ \frac{Bayer(i,j-1)}{Green(i,j-1)} + \frac{Bayer(i+1,j)}{Green(i+1,j)} +$$
$$\left. + \frac{Bayer(i,j+1)}{Green(i,j+1)}\right) \qquad (13)$$

$$RGB(i,j,3) = \left(\frac{Green(i,j)}{4}\right) \times \left(\frac{Bayer(i-1,j)}{Green(i-1,j)} + \right.$$
$$+ \frac{Bayer(i-1,j+1)}{Green(i-1,j+1)} + \frac{Bayer(i+1,j-1)}{Green(i+1,j-1)} +$$
$$\left. + \frac{Bayer(i+1,j+1)}{Green(i+1,j+1)}\right)$$

If the pixel in analysis is the Bayer (3,2),

$$RGB(i,j,1) = \left(\frac{Green(i,j)}{4}\right) \times \left(\frac{Bayer(i-1,j-1)}{Green(i-1,j-1)} + \right.$$
$$+ \frac{Bayer(i-1,j+1)}{Green(i-1,j+1)} + \frac{Bayer(i+1,j-1)}{Green(i+1,j-1)} +$$
$$\left. + \frac{Bayer(i+1,j+1)}{Green(i+1,j+1)}\right)$$

$$RGB(i,j,2) = \left(\frac{Green(i,j)}{4}\right) \times \left(\frac{Bayer(i-1,j)}{Green(i-1,j)} + \right. \qquad (14)$$
$$+ \frac{Bayer(i,j-1)}{Green(i,j-1)} + \frac{Bayer(i+1,j)}{Green(i+1,j)} +$$
$$\left. + \frac{Bayer(i,j+1)}{Green(i,j+1)}\right)$$

$$RGB(i,j,3) = Bayer(i,j)$$

If the pixel in analysis is the Bayer (3,3),

$$RGB(i,j,1) = \left(\frac{Green(i,j)}{2}\right) \times \left(\frac{Bayer(i-1,j)}{Green(i-1,j)} + \right.$$
$$\left. + \frac{Bayer(i+1,j)}{Green(i+1,j)}\right)$$

$$RGB(i,j,2) = Bayer(i,j) \qquad (15)$$

$$RGB(i,j,3) = \left(\frac{Green(i,j)}{2}\right) \times \left(\frac{Bayer(i,j-1)}{Green(i,j-1)} + \right.$$
$$\left. + \frac{Bayer(i,j+1)}{Green(i,j+1)}\right)$$

Adaptive chromatic interpolation algorithm allows making decisions and simultaneously calculate the missing colors. Some adaptive algorithms are Adaptive Color Plane and Edge-Direct, which are explained next.

Adaptive Color Plane chromatic interpolation algorithm uses classifications (α and β) to estimate the address to which the pixels belong in order to maintain uniformity of the edges (Ramanath *et al.* 2002). The calculation of the components in the green channel must be performed in the following way, with α as the vertical direction classifier and β as the horizontal one.

$$\alpha = \left| -Bayer(i,j-2) + (2 \times Bayer(i,j) - Bayer(i,j+2)) \right| +$$
$$+ \left| Bayer(i,j-1) - Bayer(i,j+1) \right|$$

$$\beta = \left| -Bayer(i-2,j) + (2 \times Bayer(i,j) - Bayer(i+2,j)) \right| +$$
$$+ \left| Bayer(i-1,j) - Bayer(i+1,j) \right|$$

if $\alpha < \beta$

$$RGB(i,j,2) = \left(\frac{Bayer(i,j-1) + Bayer(i,j+1)}{2}\right) + \qquad (16)$$
$$+ \left(\frac{-Bayer(i,j-2) + (2 \times Bayer(i,j) - Bayer(i,j+2))}{4}\right)$$

if $\alpha = \beta$

$$RGB(i,j,2) = \left\{ \begin{array}{l} \{[\ Bayer(i,j-1) + Bayer(i,j+1) + \\ \quad + Bayer(i-1,j) + Bayer(i+1,j)\ ]/4\} + \\ + \{[\ -Bayer(i,j-2) - Bayer(i,j+2) + \\ \quad + (4Bayer(i,j) - Bayer(i-2,j) - \\ \quad - Bayer(i+2,j))\ ]/8\} \end{array} \right.$$

If the pixel in analysis is the Bayer (2,2),

$$RGB(i,j,1) = \left(\frac{Bayer(i,j-1) + Bayer(i,j+1)}{2} \right) +$$
$$+ \left(\frac{- Bayer(i,j-1) + (2 \times Bayer(i,j) - Bayer(i,j+1))}{2} \right)$$

$$RGB(i,j,3) = \left(\frac{Bayer(i-1,j) + Bayer(i+1,j)}{2} \right) +$$
$$+ \left(\frac{- Bayer(i-1,j) + (2 \times Bayer(i,j) - Bayer(i+1,j))}{2} \right) \quad (17)$$

If the pixel in analysis is the Bayer (3,2),

$$\alpha = \left| - Green(i-1,j+1) + (2 \times Green(i,j) - \right.$$
$$\left. - Green(i+1,j-1) \right| + \left| Bayer(i-1,j+1) - \right.$$
$$\left. - Bayer(i+1,j-1) \right|$$

$$\beta = \left| - Green(i-1,j-1) + (2 \times Green(i,j) - \right.$$
$$\left. - Green(i+1,j+1) \right| + \left| Bayer(i-1,j-1) - \right.$$
$$\left. - Bayer(i+1,j+1) \right| \quad (18)$$

if $\alpha < \beta$

$$RGB(i,j,1) = [Bayer(i-1,j+1) + Bayer(i+1,j-1)]/2 +$$
$$+ [- Green(i-1,j-1) + (2 \times Green(i,j) -$$
$$- Green(i+1,j+1))]/2$$

if $\alpha = \beta$

$$RGB(i,j,1) = \begin{cases} [Bayer(i-1,j-1) + Bayer(i-1,j+1) + \\ + Bayer(i+1,j-1) + Bayer(i+1,j+1)]/4 + \\ + [- Green(i-1,j-1) - Green(i-1,j+1) + \\ + (4Bayer(i,j) - Green(i+1,j-1) - \\ - Bayer(i+1,j+1))]/4 \end{cases}$$

If the pixel in analysis is the Bayer (2,3),

$$\alpha = \left| - Green(i-1,j+1) + (2 \times Green(i,j) - \right.$$
$$\left. - Green(i+1,j-1) \right| + \left| Bayer(i-1,j+1) - \right.$$
$$\left. - Bayer(i+1,j-1) \right|$$

$$\beta = \left| - Green(i-1,j-1) + (2 \times Green(i,j) - \right.$$
$$\left. - Green(i+1,j+1) \right| + \left| Bayer(i-1,j-1) - \right. \quad (19)$$
$$\left. - Bayer(i+1,j+1) \right|$$

if $\alpha < \beta$

$$RGB(i,j,3) = \begin{cases} [Bayer(i-1,j+1) + \\ + Bayer(i+1,j-1) \qquad + \\ + [- Green(i-1,j+1) + \\ + (2 \times Green(i,j) - \\ - Green(i+1,j-1))]/2 \end{cases}$$

if $\alpha > \beta$

$$RGB(i,j,3) = \begin{cases} [Bayer(i-1,j-1) + \\ + Bayer(i+1,j+1)]/2 + \\ + [- Green(i-1,j-1) + \\ + (2 \times Green(i,j) - \\ - Green(i+1,j+1))]/2 \end{cases}$$

if $\alpha = \beta$ $\quad (19)$

$$RGB(i,j,3) = \begin{cases} [Bayer(i-1,j-1) + \\ + Bayer(i-1,j+1) + \\ + Bayer(i+1,j-1) + \\ + Bayer(i+1,j+1) \\ \\ [- Green(i-1,j-1) - \\ - Green(i-1,j+1) + \\ + (4 \times Bayer(i,j) - \\ - Green(i+1,j-1) - \\ - Bayer(i+1,j+1))]/4 \end{cases}$$

In the Edge-Direct chromatic interpolation, the green channel is computed first, in the direction of the pixel analyzed by some classifiers (α, β), where α represents the vertical direction and β, the horizontal one; then it continues with the red and blue channels. An example of the calculation is shown next.

If the pixel in analysis is the Bayer (4,3) or Bayer (3,4),

$$\alpha = \left| \frac{(Bayer(i,j-2) + Bayer(i,j+2))}{2} - Bayer(i,j) \right|$$

$$\beta = \left| \left(\frac{Bayer(i-2,j) + Bayer(i+2,j)}{2} - Bayer(i,j) \right) \right| \quad (20)$$

if $\alpha < \beta$

$$RGB(i,j,2) = \frac{Bayer(i,j-1) + Bayer(i,j+1)}{2}$$

if $\quad \alpha > \beta$

$$RGB(i, j, 2) = \frac{Bayer(i-1, j) + Bayer(i+1, j)}{2}$$

if $\quad \alpha = \beta$ $\qquad (20)$

$$RGB(i, j, 2) = [Bayer(i-1, j) + Bayer(i+1, j) + \\ + Bayer(i, j-1) + Bayer(i, j+1)]/4$$

After calculating the green channel, the red and blue channels are computed as follows.

If the pixel in analysis is the Bayer (2,2),

$$RGB(i, j, 1) = \frac{Bayer(i, j-1) + Bayer(i, j+1)}{2} - \\ - \left(\frac{(Green(i, j-1) + Green(i, j+1) - (2 \times Green(i, j)))}{2} \right)$$

$$RGB(i, j, 3) = \frac{Bayer(i-1, j) + Bayer(i+1, j)}{2} - \qquad (21) \\ - \left(\frac{(Green(i-1, j) + Green(i+1, j) - (2 \times Green(i, j)))}{2} \right)$$

If the pixel in analysis is the Bayer (2,3),

$$RGB(i, j, 1) = Bayer(i, j)$$

$\qquad (22)$

$$RGB(i, j, 3) = \begin{cases} \begin{aligned} & \left[Bayer(i-1, j-1) + Bayer(i+1, j+1) + \\ & + Bayer(i-1, j+1) + Bayer(i+1, j-1) \right]/4 \\ & [Green(i-1, j-1) + Green(i+1, j+1) + \\ & + Green(i-1, j+1)]/4 \\ & + \left(\frac{Green(i+1, j-1) - (4 \times Green(i, j))}{4} \right) \end{aligned} \end{cases}$$

If the pixel in analysis is the Bayer (3,2),

$$RGB(i, j, 1) = \begin{cases} \begin{aligned} & [Bayer(i-1, j-1) + Bayer(i+1, j+1) + \\ & + Bayer(i-1, j+1) + Bayer(i+1, j-1) \\ & [Green(i-1, j-1) + Green(i+1, j+1) + \\ & + Green(i-1, j+1)]/4 + \\ & [Green(i+1, j-1) - (4 \times Green(i, j))]/4 \end{aligned} \end{cases}$$

$\qquad (23)$

$$RGB(i, j, 3) = Bayer(i, j)$$

IMAGE QUALITY EVALUATION

In the evaluation of an algorithm, two kinds of metrics can be set: the first is designed to measure the effectiveness of the algorithm, and the quality of the result is evaluated; the second metric evaluates the efficiency of the algorithm for the computational case the complexity is established based on the amount of mathematical operations required for its implementation (Tuya *et al.* 2007). The research development focuses on the identification of analysis methods of the quality of images without reference.

There are several evaluation methods for the chromatic interpolation algorithms used to measure the quality of the interpolated image (RGB). The most common evaluation methods are those which have a reference image, for example, Color Mean Squared Error (CMSE), Color Peak Signal-to-Noise Ratio (CPSNR), CIELAB ΔE, among others. However, there are methods to measure the quality of the image when there is no reference, for example, Blur Metric, False Color Measure and Achromatic Zipper, which evaluate the quantity of artifacts or defects contained in the calculated RGB image.

The false color artifact usually appears on the edges of the images. It occurs when the result of the interpolation calculates color pixels that do not truly belong to the image (Wenmiao and Tan 2003). Figure 5 shows a false color image.

Figure 5. False color.

The blur artifact is caused by a loss of information of high frequencies in the calculated RGB image (Crete *et al.* 2007; INTECO 2009), making it look blurry and without very well defined edges, as shown in Fig. 6.

Zipper effect occurs commonly in the image edges making them look zipper-shaped. Figure 7 shows an image with zipper effect.

The blur effect affects the edges definition in an image. Since the defined edges in an image are the elements that add

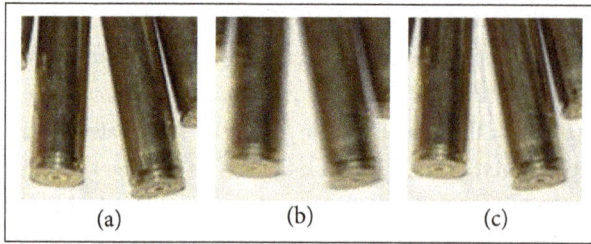

Figure 6. Blur artifact: (a) Original image; (b) Image with high Blur artifact; (c) Image with low Blur artifact.

Figure 7. Zipper effect — image with bilinear interpolation.

high frequency components, the blur artifact identification is based on the comparison of the image against itself, but after filtering the high frequencies; the similarity degree of the images is directly proportional to the presence of the blur artifact. The blur index is calculated in the luminance component of the image, that is, in the green channel (Crete *et al.* 2007). The detailed procedure to calculate the blur index is presented in Fig. 8, which shows that the green channel is taken from the RGB image and passes through a Low-Pass Filter — the result is a "Blur G" image. The vertical and horizontal components refer to the rows and columns of the RGB image matrix with a size of $m \times n$. The applied filter is:

$$F = \frac{[1\ 1\ 1\ 1\ 1\ 1\ 1\ 1\ 1]}{9}$$

$$F' = \text{Transpose}(F) = (F')_{m,n}$$

(24)

where F', for the horizontal components, is the transpose of the filtered vertical components of matrix F.

Then the variations between neighboring pixels are calculated for the green channel of the original image and the "Blur G". This is achieved calculating the absolute difference between rows ($D_{F_{ver}}$ and $D_{B_{ver}}$) and columns ($D_{F_{hor}}$ and $D_{B_{hor}}$) for the original and "Blur G" images.

$$D_{F_{ver(i,j)}} = |F(i,j) - F(i-1,j)|\ ;$$
$$for\ i = 1\ to\ m-1, j = 0\ to\ n-1$$

$$D_{B_{ver(i,j)}} = |B(i,j) - B(i-1,j)|\ ;$$
$$for\ i = 1\ to\ m-1, j = 0\ to\ n-1$$

$$D_{F_{hor(i,j)}} = |F(i,j) - F(i,j-1)|\ ;$$
$$for\ j = 1\ to\ n-1, i = 0\ to\ m-1$$

(25)

$$D_{B_{hor(i,j)}} = |B(i,j) - B(i,j-1)|\ ;$$
$$for\ j = 1\ to\ n-1, i = 0\ to\ m-1$$

where: m is the width and n the height of the image.

This calculation considers the two directions, vertical and horizontal, in order to get the VP_RGB and VP_BlurRGB matrixes, as shown in Fig. 8; the magnitude of the variation between these two matrixes is the blur metric (Crete *et al.* 2007).

The False Color Measure evaluates the amount of pixels with a different color from the neighboring pixels; this algorithm is applied on multiple versions of the same image, calculated from the same data of the Bayer array using different chromatic interpolation algorithms.

For the calculation of False Color Measure, first one must find the edges that are common in the different versions of the image, then this particular version to be evaluated is divided into 5×5 submatrixes which are moved one position, as shown in Fig. 9. The median of the corresponding submatrix is calculated in the analysis of each pixel. Finally, the matrix with

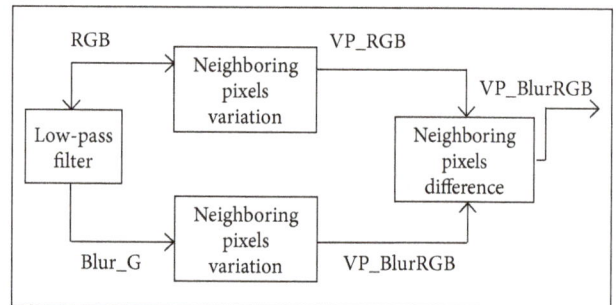

Figure 8. Blur Metric process. (Adapted from Crete *et al.* 2007).

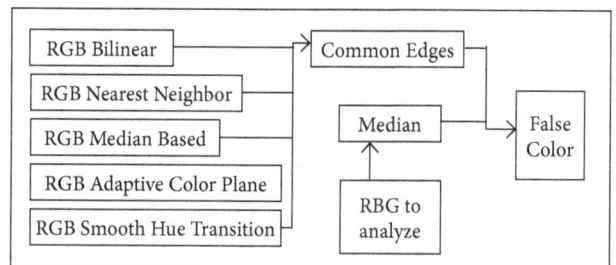

Figure 9. Calculation process for False Color.

the calculated median is analyzed from the information of the identified edges by applying Eq. 24 (Maschal Jr. *et al.* 2010).

$$FC = \frac{\sum_{i,j \in CE}\left[(G_{i,j} - C_{i,j}) - M_{i,j}\right]^2}{nCE} \tag{26}$$

where: *FC* is the false color; *CE* are the common edges; *G* is the green channel; *C* is the analyzed channel (green, red or blue); *M* is the channel median to be analyzed; *nCE* is the number of common edges.

For the calculation of Chromatic and Achromatic Zipper metric, the RGB input image is converted to a grayscale image using Eq. 27:

$$Gray(i,j) = (Red(i,j) \times 0.299) +$$
$$+ (Green(i,j) \times 0.587) + \tag{27}$$
$$+ (Blue(i,j) \times 0.114)$$

Then the edges are identified using the following gradient filter:

$$\begin{cases} GV = \begin{bmatrix} -1 & 1 \end{bmatrix} \\ GH = \begin{bmatrix} -1 \\ 1 \end{bmatrix} \end{cases} \tag{28}$$

where: *GV* is the vertical gradient; *GH* is the horizontal gradient.

The horizontal and vertical masks have to be calculated in the following way:

$$SMapx(x, y) = \begin{cases} 2 & Gx(x, y) < 0 \\ 1 & Gx(x, y) > 9 \\ 0 & Other \end{cases} \tag{29}$$

where: *Gx* is the vertical or horizontal gradient.

Thus, the image information that presents artifacts zipper (*zipRGBx*) is removed using Eq. 30. *SMapy* is calculated in the same way.

$$zipRGBx = (SMapx \times RGB) \tag{30}$$

ZipRGBx and *zipRGBy* are converted to CIELAB color space; finally, the Chromatic Zipper (DC) and Achromatic Zipper (DL) indexes are calculated by applying Eq. 31:

$$DL(x, y) = \left(\Delta L(x, y)^2\right)^{1/2}$$

$$DC(x, y) = \left(\left(\frac{\Delta C(x, y)}{Sc}\right)^2 + \left(\frac{\Delta H(x, y)}{Sh}\right)^2\right)^{1/2}$$

where:

$$\Delta L(x, y) = L(x, y) - L(x, y - 1) \tag{31}$$
$$\Delta C(x, y) = \left(a(x, y)^2 + b(x, y)^2\right)^{1/2} -$$
$$- \left(a(x, y - 1)^2 + b(x, y)^2\right)^{1/2}$$
$$\Delta H(x, y) = \left(\Delta E_{76}(x, y)^2\right) -$$
$$- \left(\Delta L(x, y)^2 - \Delta C(x, y)^2\right)^{1/2}$$

where: *Sc* and *Sh* are the constants of CIE-94 color difference formula — *Sc* = 1 + 0.045 *C**; *Sh* = 1 + 0.015 *C**; *C** is the geometrical mean chroma; *L*, *a* and *b* are CIELAB values; ΔE_{76} is the Euclidean distance between *L*, *a* and *b* values.

CHROMATIC INTERPOLATION ALGORITHMS

For the tests execution, first, the selection of the images database was made, then the tool software was executed using the captured Bayer array and, as a result, the images in the RGB color model were acquired. The RGB images were evaluated by the following methods: Blur Metric, Chromatic and Achromatic Zipper and False Color Measure. Finally, the obtained results were analyzed.

The criteria to select the Bayer array database were:

- The captured images should present similar characteristics in their frequency components.
- Data should be captured with the multispectral imaging sensor OV2640, which is similar to the payload sensor of the Libertad 2 satellite mission.
- The database should contemplate different types of lighting.
- The database should contemplate different color types.

For the tests, 39 images — three groups with 13 images each — were chosen, and the tests scenarios are shown in Table 1. The first group of images was taken with natural lighting; examples of such images are shown in Fig. 10 a,b; the second was captured in a controlled environment with artificial lighting; examples of such images are shown in Fig. 10 c,d; and, finally, the third group was captured in the same controlled environment but with low-frequency images and continuous color regions; examples of such images are shown in Fig. 10 e,f.

Table 1. Scenarios for group of images.

	Group 1	Group 2	Group 3
Edges	Bigger quantity in high frequencies	Bigger quantity in high frequencies	Bigger quantity in low frequencies
Lighting	Natural	Artificial	Artificial
Color	Different shades	Different shades	Continuous color regions

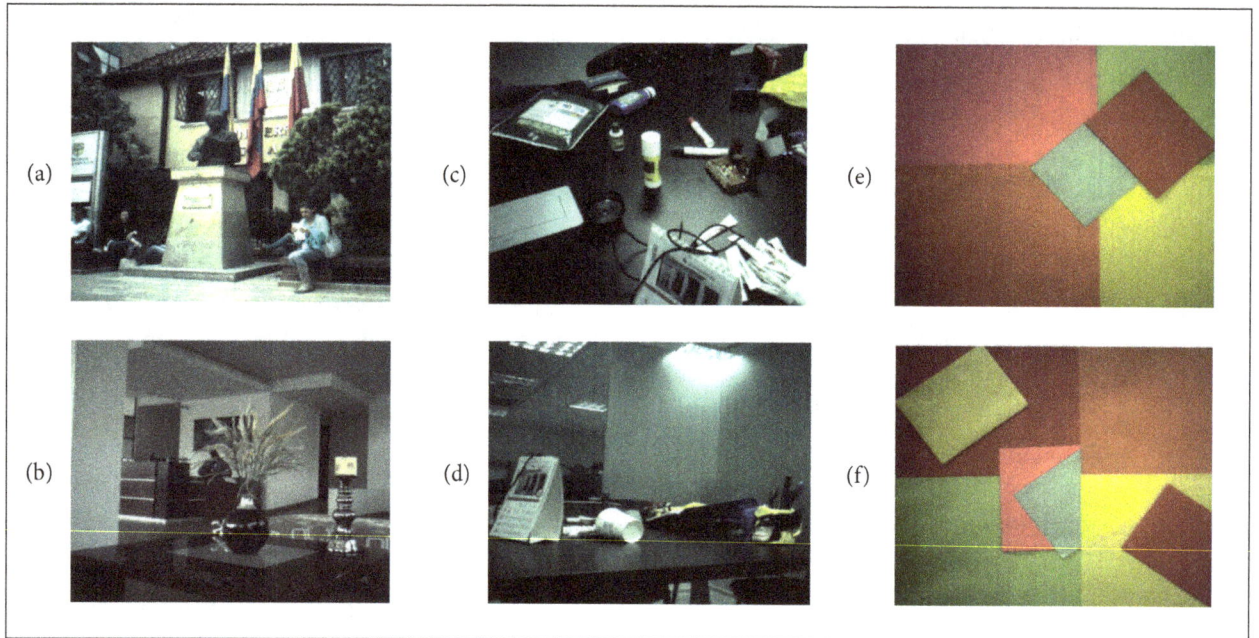

Figure 10. Located image. (a) and (b) at first group; (c) and (d) at second group; (e) and (f) at third group.

The first and second groups present similar characteristics in their frequency components (similar in the amount of edges). These image groups were chosen in order to evaluate the zipper artifacts and false color because they are evaluated at the edges of the image. The third group of images was chosen especially to evaluate the false color artifact.

Then the algorithms of the quality metrics (False Color - Green, False Color - Blue, False Color - Red, Blur Metric, DL and DC) were applied in each one of the images, and the obtained results are presented next.

As shown in Fig. 11 and Table 2, the Nearest Neighbor chromatic interpolation algorithm presents more false color in the red channel, with an average of 30,482; the lowest false color belongs to the Smooth Hue Transition algorithm, with an average of 23,415. The highest blur corresponds to Bilinear, Smooth Hue Transition and Median Based algorithms, with an average of 0.1582; less blur occurs in the Nearest Neighbor algorithm, with an average of 0.1127. In the blue channel, the highest false color is reached by the Nearest Neighbor algorithm, with an average of 12,555, and a lower false color occurs in the Smooth

Hue Transition algorithm, with an average of 9,716. The Adaptive Color Plane has the highest DL, with an average of 40,5593. The Bilinear and Median Based algorithms present the lowest DL, with 15,6431. For DC, the highest value is 1.4029 from the Adaptive Color Plane, while the Bilinear and Median Based reach the lowest value, 0.6799. The false color metric for the green channel gets the highest value with the Bilinear, Smooth Hue Transition and Median Based algorithms, with an average of 22,269, while the Nearest Neighbor presents the lowest value, 22,018.

The algorithm that presents higher false color in the red channel is the Nearest Neighbor, with an index of 30,485. This algorithm presents false color 17.32% higher than the Bilinear, 23.18% higher than the Smooth Hue Transition, 18.60% higher than the Adaptive Color Plane and 17.32% higher than the Median Based. The algorithms that present higher blur are Bilinear, Smooth Hue Transition and Median Based, with an index of 0.1582; these algorithms present blur 11.12% higher than the Adaptive Color Plane and 28.76% higher than the Nearest Neighbor. The algorithm with more achromatic zipper artifact in the images was the Adaptive Color Plane, with a

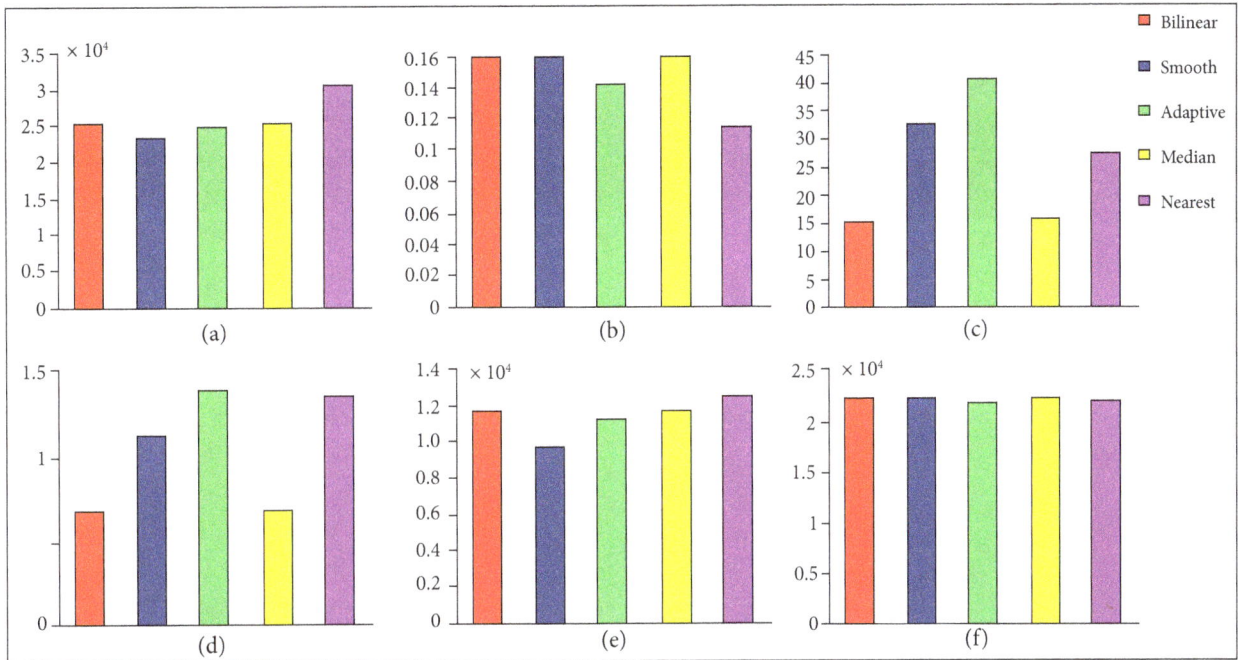

Figure 11. Metric average of 39 images. (a) False Color - Red channel; (b) Blur; (c) Zipper DL; (d) Zipper DC; (e) False Color - Blue channel and (f) False Color - Green channel

Table 2. Metric average of 39 images.

Average	Bilinear	Smooth Hue Transition	Adaptive Color Plane	Median Based	Nearest Neighbor
False Color - Red	25,201	23,415	24,812	25,201	30,482
False Color - Blue	11,735	9,716	11,282	11,735	12,555
False Color - Green	22,269	22,269	21,911	22,269	22,018
Blur Metric	0.1582	0.1582	0.1406	0.1582	0.1127
DL	15.6431	32.6303	40.5593	15.6431	27.3644
DC	0.6799	1.1227	1.4029	0.6799	1.3593

40.5593 index; this algorithm has an achromatic zipper 61.43% higher than the Bilinear and the Median Based, 19.54% higher than the Smooth Hue Transition, 61.43% higher than the Median Based and 32.52% higher than the Nearest Neighbor.

In the test, the algorithm with more chromatic zipper artifact was the Adaptive Color Plane, with a rate of 1.4029; this algorithm has a chromatic zipper 51.53% higher than Bilinear and Median Based, 19.97% higher than the Smooth Hue Transition and 3.10% higher than the Nearest Neighbor. In the test, the algorithm with more false color artifact in the blue channel of the image was the Nearest Neighbor, with an index of 12,555; this algorithm has false color 6.53% higher than Bilinear and Median Based, 22.61% higher than the Smooth Hue Transition and 10.13% higher than the Adaptive Color Plane. Algorithms that have higher false color in the

green channel are Bilinear, Smooth Hue Transition and Median Based, with an index of 22,269; these algorithms have false color 1.60% higher than the Adaptive Color Plane and 1.12% higher than the Nearest Neighbor.

RESULT ANALYSIS

The results of the comparison between chromatic interpolation algorithms performed with the 39 images are presented, applying the Blur Metric, False Color Measure, Chromatic and Achromatic Zipper evaluation methods. The sum of the artifacts is presented in Table 2.

Adaptive Color Plane chromatic interpolation algorithm presents the best performance in the quality analysis for the three evaluation metrics, with a total value of 200,281.3084, as shown in Table 3. The Smooth Hue Transition algorithm has the highest index artifacts values, according to the evaluation metrics, with a total value of 281,347.7491, as shown in Table 3. The Bilinear and Median Based algorithms present equal index artifacts values according to the evaluation metrics, with a total value of 204,970.444, as shown in Table 3. The Smooth Hue Transition chromatic interpolation algorithm presents the highest artifacts values index, 25.21%. The Adaptive Color Plane presents the lowest artifacts values index with respect to the totality of the presented artifacts in the 39 images, 17.94%.

Table 3. Total artifacts in the chromatic interpolation algorithms.

Chromatic interpolation algorithms	Total artifacts	Percentage (%)
Bilinear	204,970.444	18.36
Smooth Hue Transition	281,347.7491	25.21
Adaptive Color Plane	200,281.3084	17.94
Median Based	204,970.444	18.36
Nearest Neighbor	224,325.5096	20.10

CONCLUSIONS

Given the acquisition and download process of the images, the efficiency analysis of the algorithms is irrelevant because the worst case, 8.2593 s of processing, is insignificant for the acquisition and download time of the image. Therefore, the assessment of the algorithms must focus on the quality instead the efficiency, since the artifacts can disturb the researches based on satellite images with badly interpolation procedure.

The chromatic interpolation algorithm to be used must be chosen according to the type of research that will be carried out; for example, research based on region segmentation from satellite images could be affected if the interpolation algorithm does not present low blur artifacts index. On the other hand, in a research based on low-resolution images, it is mandatory to choose an interpolation algorithm with low zipper artifact, due to the low resolution; each pixel means more information, and the artifact can disturb the image interpretation. However, for researches in which the color is essential, as crop recognition, the algorithm with low false color artifact should precede over the algorithm with low zipper artifact.

REFERENCES

Adobe Creative Team (2007) Adobe Illustrator CS3. Berkeley: Peachpit Press.

Cambridge in Colour (2013) Digital camera sensors. A learning community for photographers; [accessed 2014 Nov 23]. http://www.cambridgeincolour.com/tutorials/camera-sensors.htm

Crete F, Dolmiere T, Ladret P, Nicolas M (2007) The blur effect: perception and estimation with a new no-reference perceptual blur metric. Proceedings of SPIE - The International Society for Optical Engineering 12. doi: 10.1117/12.702790

CubeSat Organization (2014) CubeSat; [accessed 2014 Feb 13]. http://www.cubesat.org/

Dargahi N, Daneshpande V (2007) New methods in Bayer demosaicing algorithms, Proceedings Psychology, 221, Applied Vision & Image Systems Engineering, Stanford University, UK.

Elizondo JJE, Maestre LEP (2005) Fundamentos de procesamiento de imágenes. Mexicali: Universidad Autónoma de Baja California.

Greenland S, Clark C (2010) Cubesat platforms as an on-orbit technology validation and verification vehicle. In Proceedings of the Pestana Conference Centre; Funchal, Portugal.

Hurkman AV (2010) Color correction handbook: professional techniques for video and cinema. Berkeley: Peachpit Press.

Maschal Jr. RA, Young SS, Reynolds J, Krapels K, Fanning J, Corbin T (2010) Review of bayer pattern color filter array (CFA) demosaicing with new quality assessment algorithms (No. ARL-TR-5061). Army Research Lab Adelphi Md Sensors and Electron Devices Directorate.

Instituto Nacional de Tecnologías de la Comunicación (2009) Ingeniería del software: metodologías y ciclos de vida; [accessed 2014 Oct 11]. http://www.inteco.es/calidad_TIC/

Li X, Gunturk B, Zhang L (2008) Image demosaicing: a systematic survey. In: International Society for Optics and Photonics. Electronic Imaging 2008 (p. 68221J-68221J).. Boulder: National Institute of Standards and Technology.

Lillesand T, Kiefer RW, Chipman J (2014) Remote sensing and image interpretation. Hoboken: John Wiley & Sons.

Losson O, Macaire L, Yang Y (2009) Comparison of color demosaicing methods. Adv Imag Electron Phys 162:173-265. doi: 10.1016/S1076-5670(10)62005-8

Malvar HS, He L, Cutler R (2004) High-quality linear interpolation for demosaicing of Bayer-patterned color images. Proceedings of the IEEE International Conference on Acoustics, Speech, and Signal Processing; Montreal, Canada.

Martínez J, Díaz A (2005) Percepción remota "Fundamentos de teledetección espacial". Comisión Nacional del Agua Conagua; [accessed 2014 Nov 23]. http://siga.conagua.gob.mx/

Muñoz AV (2012) Principios de color y holopintura. Alicante: Club Universitario.

Pérez C, Aguilera DG, Muñoz AL (2003) Estudio de viabilidad del uso de imágenes comprimidas e procesos de clasificación. Teledetección y Desarrollo Regional. Proceedings of the X Congreso de Teledetección; Cáceres, Spain.

Ramanath R, Snyder WE, Bilbro GL, Sander III WA (2002) Demosaicking methods for Bayer color arrays. J Electron Imaging 11(3):306-315.

Sandau R (2008) Status and trends of small satellite missions for Earth observation. Acta Astronaut 66(1-2):1-12. doi: 10.1016/j.actaastro.2009.06.008.

Selva D, Krejci D (2012) A survey and assessment of the capabilities of Cubesats for Earth observation. Acta Astronaut 74:50-68. doi: 10.1016/j.actaastro.2011.12.014

Tuya J, Román IR, Cosín JD, editors (2007) Técnicas cuantitativas para la gestión en la ingeniería del software. Oleiros: Netbiblo.

Wenmiao L, Tan YP (2003) Color filter array demosaicking: new method and performance measures. IEEE Trans Image Process 12(10):1194-1210. doi: 10.1109/TIP.2003.816004

Wheeler P (2009) High definition cinematography. New York: Taylor & Francis.

Woellert K, Ehrenfreund P, Ricco AJ, Hertzfeld H (2011) Cubesats: Cost-effective science and technology platforms for emerging and developing nations. Adv Space Res 47(4):663-684. doi: 10.1016/j.asr.2010.10.009

2

A New Algorithm for Shock Sensor Calculation at Supersonic Speeds on a 3D-Unstructured Grid

Vahid Tahmasbi[1], Seyed Mohammad Hossein Karimian[1]

ABSTRACT: In this paper, 3-Dsupersonic flow around two types of wings is solved using a new algorithm for shock sensor calculation. A dual-time-stepping implicit method with 2nd-order accuracy is used for time integration of the equations. In each real time step, the non-linear system of equations is solved by iterating in pseudo-time, using a multi-step integration method. A cell-center finite volume scheme is applied to discretize the solution domain. Governing equations are discretized using 2nd-order central scheme of Jameson. Undesirable oscillations are prevented using artificial dissipation terms containing 2nd and 4th-order derivative terms. The second-order derivative term is proportional to shock sensor, which is a function of pressure gradient in general and is devised to capture shock waves correctly. Appropriate calculation of shock sensor is very important especially for the solution of 3-D supersonic flow on unstructured grids. In this study, a simple efficient algorithm is proposed for shock sensor calculation to stabilize solution in supersonic 3-D flows on unstructured grids. The new algorithm, implemented at an in-house code, is evaluated by comparison of its results with wind tunnel test data and upwind-type differencing scheme of Roe for a tailplane model tested at Royal Aircraft Establishment. The results show that supersonic flow with shock waves has been accurately captured.

KEYWORDS: Shock sensor, Supersonic speed, RAE tailplane.

INTRODUCTION

Accurate solution of fluid flow around supersonic vehicles is one of the aerospace engineering concerns. Among the key factors for this purpose is the correct capture of high-gradient phenomena such as shock waves. In order to prevent non-physical oscillations around the shock waves, some numerical methods add artificial dissipation terms to the system of equations. The most well-known among these is the method of Jameson.

The idea of using non-linear viscosity for the numerical simulation of shock waves was introduced by von Neumann and Richtmyer (1950). Jameson *et al.* (1981) implemented a 3rd-order artificial viscosity and a shock sensor to solve Euler equations in a domain discretized by finite volume method. Jameson and Mavriplis (1986) presented an artificial viscosity that was the combination of 2nd- and 4th-order derivatives. A 2-D structured triangular mesh was used in this paper. At each face of a finite volume, 2 directions were defined. Then, shock sensor was considered proportional to the maximum value of pressure gradients in these 2 directions. This method, however, was not effective in a general 3-D unstructured grid.

Jameson and Baker (1987) developed artificial dissipation terms for 2- and 3-D unstructured grids. In this paper, shock sensor at a node was defined proportional to the pressure gradient at that node. Then, shock sensor on each face of a finite volume was considered to be the maximum one of the 2 neighbor nodes of that face. Results showed that this method was successful for the solution of 2-D flows on unstructured grids. Jameson (1995a,b) further developed previous works and analyzed the effect

1.Amirkabir University of Technology – Aerospace Engineering Department – Tehran – Iran.

Author for correspondence: Vahid Tahmasbi | Amirkabir University of Technology – Aerospace Engineering Department | 424 Hafez Ave | P.O. Box: 15875-4413 – Tehran – Iran | E-mail: vtahmasbi@gmail.com

of artificial viscosity on accuracy and convergence of solution. Xu *et al.* (1995) essentially applied Jameson method to calculate artificial viscosity in a gas kinetic finite volume method for the solution of Euler equation on structured grids.

Most of the mentioned references and many others (Siclari 1989; Siclari and Del Guidice 1990; Jameson 1986; Volpe *et al.* 1987) in the literature are related to the calculation of artificial dissipation terms on structured grids. Flow solvers using these algorithms capture shock waves correctly in high-speed flow, eliminate oscillations around them, and converge smoothly. However, since these methods use specific directions in space to calculate shock sensor, they cannot be used on unstructured grids. It should be mentioned that artificial dissipation terms have been also introduced in some finite element methods to solve Euler equations in supersonic flows. However, these forms of artificial dissipation terms cannot be used in the context of finite volume cell-centered methods (Barter 2008; Burbeau 2010).

Other than the mentioned literature that discusses the implementation of shock sensor in artificial viscosity, there are other methods that have been introduced for shock capturing in supersonic flows. A quick review of these methods is given next.

Jiang and Shu (1996) presented a new weighted essentially non-oscillatory (WENO) scheme for Cartesian grids. The smoothness indicators of this scheme can be interpreted as a shock sensor (Visbal and Gaitonde 2005). Ducros *et al.* (1999) developed a new shock sensor, which was more accurate in distinguishing shocks from other flow features, like vortices, and more suitable for unsteady flows. Sjögreen and Yee (2004) presented a new numerical dissipation based on the artificial compression method (ACM) of Harten (1978) and a new shock sensor based on wavelet analysis and multi-resolution method of Harten (1995). Cook and Cabot (2004, 2005) presented a new artificial viscosity with spectral-like viscosity, which could damp oscillations near discontinuities without using the shock sensor. Oliveira *et al.* (2010) introduced a new 2-step shock sensor, which was able to detect shocks based on the truncation error ratio on the coarse and fine grids and the local flow gradients at left- and right-hand side finite volumes. Cook *et al.* (2013) extended the Hyperviscosity Method (Cook and Cabot 2005) to unstructured arbitrary Lagrangian-Eulerian (ALE) grids, and Shen *et al.* (2013) presented a new shock sensor, which, unlike many other shock sensors, is parameter-free and needs no user specified constants.

Against central scheme methods, upwind-type differencing schemes, such as flux-vector splitting of van Leer (1982) and the approximate Riemann solver of Roe (1981), utilize concepts from the method of characteristics in order to determine the direction of spatial differencing to discrete hyperbolic equations. These schemes have the advantage of being naturally dissipative in contrast to central schemes in which separate dissipation terms are added to overcome oscillations arising in regions of strong gradients. Instead, central schemes are less time-consuming in comparison with upwind schemes.

The purpose of this paper, however, was to develop a new method for shock sensor calculation in artificial viscosity of supersonic flow solvers on unstructured grids. Therefore, other methods, including those in the References section, are not intended here.

In the present study, for discretization of the governing equations, finite volume cell-centered based method is applied. Researches carried out by Stolcis and Johnston (1990) indicated that the isotropic behavior of the shock sensor does not increase the effect of 2nd-order artificial dissipation near the discontinuities, but increases the level of the 4th-order artificial dissipation in the zero-gradient areas. This problem has been resolved by definition of non-isotropic shock sensor as a function of special form of pressure gradient on each face of finite volume (Stolcis and Johnston 1990). This shock sensor has made significant improvement in capturing shock waves. Nevertheless, in the present study, we found that the use of this type of shock sensor leads to divergence of solution in supersonic flows on 3D-unstructured grids. Therefore, a new non-isotropic shock sensor and its calculation algorithm are proposed in this paper. In fact, the non-isotropic shock sensor introduced by Stolcis and Johnston (1990) is modified by means of a simple and efficient method to be used in 3-D supersonic flow. Grid information is saved in a face-based data structure which is suitable for the calculation of shock sensor in the algorithm applied in this paper. Comparison of the results obtained in the present algorithm with the experimental data and upwind-type differencing scheme of Roe (1981) illustrates the high accuracy of this algorithm to solve 3-D supersonic flows embedding shock waves.

EULER EQUATIONS

Euler equations are the limit of the Navier-Stokes equations for vanishing viscosity effects. These equations can be written in the Cartesian coordinate system as:

$$\frac{\partial Q}{\partial t}+\frac{\partial F}{\partial x}+\frac{\partial G}{\partial y}+\frac{\partial H}{\partial z}=0 \tag{1}$$

where: Q is the vector of conserved quantities; F, G and H represent the convective fluxes in the following form:

$$Q=\begin{bmatrix} \rho \\ \rho u \\ \rho v \\ \rho w \\ \rho E \end{bmatrix}, \; F=\begin{bmatrix} \rho u \\ \rho u^2+p \\ \rho uv \\ \rho uw \\ (\rho E+p)u \end{bmatrix}, \; G=\begin{bmatrix} \rho v \\ \rho uv \\ \rho v^2+p \\ \rho vw \\ (\rho E+p)v \end{bmatrix},$$

$$H=\begin{bmatrix} \rho w \\ \rho uw \\ \rho vw \\ \rho w^2+p \\ (\rho E+p)w \end{bmatrix} \tag{2}$$

where: ρ, u, v, w, p and E denote density, velocity components, pressure and internal energy, respectively.

Euler equations are augmented by the equation of state, which, for a perfect gas, is given by:

$$p=(\gamma-1)\left[\rho E - \frac{\rho(u^2+v^2+w^2)}{2}\right] \tag{3}$$

where: γ is the ratio of specific heat.

Having integrated Eq. 1 over control volume Ω with boundary $\partial\Omega$, the following equation is obtained:

$$\frac{\partial}{\partial t}\int_{\Omega} Q \, dV + \oint_{\partial\Omega} (F \, dS_x+G \, dS_y+H \, dS_z)=0 \tag{4}$$

where: V is the cell volume; dS_x, dS_y and dS_z denote components of the outward vector normal to the surface.

The implementation of Eq. 4 to each finite volume (or cell) in the solution domain results in a set of ordinary differential equations (ODE) in the following form:

$$\frac{d}{dt}(Q_iV_i)+R_i(Q) - D_i(Q)=0 \tag{5}$$

where: $R_i(Q)$ and $D_i(Q)$ denote sum of the convective fluxes and artificial viscosity terms on the surfaces of ith finite volume, respectively. Note that the value of Q used for the calculation of fluxes on the finite volume surface is averaged between its values at the center of finite volume i and the neighbor finite volume of that surface.

Therefore, on the jth surface of ith finite volume, Q would be:

$$Q_j=\frac{Q_i+Q_k}{2} \tag{6}$$

where k indicates neighbor finite volume of the jth surface of ith finite volume.

IMPLICIT TIME MARCHING SCHEME

A dual-time-stepping scheme is employed to solve the non-linear Eq. 5, which is written in the following implicit form:

$$\frac{d}{dt}(Q_iV_i)+R_i(Q^{n+1}) - D_i(Q^{n+1})=0 \tag{7}$$

The 2nd-order backward difference scheme is applied to discretize the transient term. This results in the following equation for each finite volume:

$$\frac{3}{2\Delta t}\left(Q_i^{n+1} \; V_i^{n+1}\right) - \frac{2}{\Delta t}\left(Q_i^n \; V_i^n\right)+\frac{1}{2\Delta t}\left(Q_i^{n-1} \; V_i^{n-1}\right)+$$
$$+R_i(Q^{n+1}) - D_i(Q^{n+1})=0 \tag{8}$$

where: $i = 1, N_c$, being N_c the total number of finite volumes in the solution domain.

The above system of non-linear coupled equations is solved by iterating it in a pseudo-time, called τ, through the following equation:

$$\frac{dQ_i}{d\tau_i}+R_i^*(Q^n)=0 \tag{9}$$

where $R_i^*(Q^n)$ is the unsteady residual given by:

$$R_i^*(Q^n) = \frac{3}{2\Delta t}\left(Q_i^{n+1} \; V_i^{n+1}\right) - \frac{2}{\Delta t}\left(Q_i^n \; V_i^n\right)+$$
$$+\frac{1}{2\Delta t}\left(Q_i^{n-1} \; V_i^{n-1}\right)+R_i(Q^{n+1}) - D_i(Q^{n+1}) \tag{10}$$

Having known Q^n, 4-stage Runge-Kutta scheme is used to solve Eq. 9 for Q_i^{n+1}, which is first initialized by $Q_i^{n+1}= Q_i^n$, and then the following iteration procedure in pseudo-time, indexed by m, is continued until the convergence of Q_i^{n+1} is reached.

$$Q_i^{(0)} = (Q_i^{n+1})^m$$

$$Q_i^{(1)} = Q_i^{(0)} - \frac{1}{4} \frac{\Delta \tau_i}{V_i} R_i^*(Q^{(0)})$$

$$Q_i^{(2)} = Q_i^{(0)} - \frac{1}{3} \frac{\Delta \tau_i}{V_i} R_i^*(Q^{(1)})$$

$$Q_i^{(3)} = Q_i^{(0)} - \frac{1}{2} \frac{\Delta \tau_i}{V_i} R_i^*(Q^{(2)}) \qquad (11)$$

$$Q_i^{(4)} = Q_i^{(0)} - \frac{1}{3} \frac{\Delta \tau_i}{V_i} R_i^*(Q^{(3)})$$

$$(Q_i^{n+1})^{m+1} = Q_i^{(4)}$$

In order to minimize computational time, numerical dissipation term D_i is calculated only at the first and third stages of Eq. 11. The real-time step is not restricted in the above scheme, but the pseudo-time step is restricted by the following stability criterion at each finite volume:

$$\Delta \tau = \min \left[\frac{CFL_i\ V_i}{\sum_{j=1}^{Nf_i} \lambda_j}\ ,\ \frac{2\Delta t}{3} \right] \qquad (12)$$

where: CFL_I is the Courant number for pseudo-time marching; j denotes the face of the ith finite volume; Nf_i is the number of finite volume faces, and

$$\lambda = u\ dS_x + v\ dS_y + w\ dS_z \qquad (13)$$

BOUNDARY CONDITIONS

At the far field, non-reflecting boundary conditions were applied based on the characteristic analysis. On the solid wall boundary, the slip condition needs to be imposed. Pressure value at the solid wall is obtained by extrapolating from their values of adjacent cells.

ARTIFICIAL DISSIPATION

In this paper, artificial dissipation term is a blend of 2nd- and 4th-order differences of flow variables Q with coefficients that depend on the local pressure gradient to prevent the appearance of high-frequency modes corresponding to odd and even point oscillations and oscillations near the shock waves (Jameson and Mavriplis 1986). For the ith finite volume, artificial dissipation term is defined as:

$$D_i(Q) = \sum_{j=1}^{Nf_i} d_{ij}^{(2)} + \sum_{j=1}^{Nf_i} d_{ij}^{(4)}$$

$$d_{ij}^{(2)} = \frac{\varepsilon_{ij}^2}{2} \lambda_{ij}\ (Q_i - Q_k) \qquad (14)$$

$$d_{ij}^{(4)} = \frac{\varepsilon_{ij}^4}{2} \lambda_{ij}\ (\nabla^2 Q_i - \nabla^2 Q_k)$$

where: subscripts ij denote j^{th} surface of ith finite volume; k denotes the neighbor finite volume across the jth surface of ith finite volume; $\varepsilon_{ij}^{(2)}$ and $\varepsilon_{ij}^{(4)}$ are adaptive coefficients defined as:

$$\varepsilon_{ij}^{(2)} = k^{(2)}\ \upsilon_{ij}$$

$$\varepsilon_{ij}^{(4)} = \max \left[0\ ,\ (k^{(4)} - \varepsilon_{ij}^{(2)}) \right] \qquad (15)$$

where: $k^{(2)}$ and $k^{(4)}$ are constants in the range of $(0.5 < k^{(2)} < 1)$ and $(1/256 < k^{(4)} < 1/32)$, respectively; υ_{ij} is the shock sensor which will be defined later.

The Laplacian operator ∇^2 and the scaling factor λ_{ij} are defined as:

$$\nabla^2 Q_i = \sum_{j=1}^{Nf_i} (Q_k - Q_i) \qquad (16)$$

$$\lambda_{ij} = |u_{ij}\ dS_x + v_{ij}\ dS_y + w_{ij}\ dS_z| +$$
$$+ c_{ij} \sqrt{(dS_x)^2 + (dS_y)^2 + (dS_z}$$

where: c_{ij} is the local speed of sound; λ_{ij} is the scaling factor calculated from the following equation:

$$\lambda_{ij} = \left(\frac{V_i}{\Delta \tau_i} + \frac{V_k}{\Delta \tau_k} \right) \qquad (17)$$

The isotropic shock sensor υ_{ij} is a pressure-sensitive sensor usually defined as:

$$\upsilon_{ij} = \max(\upsilon_i\ ,\ \upsilon_k) \qquad (18)$$

where:

$$v_i = \frac{\left| \sum_{j=1}^{Nf_i} (P_k - P_i)_j \right|}{\left| \sum_{j=1}^{Nf_i} (P_k + P_i)_j \right|} \qquad (19)$$

The term v_k can be calculated in a similar manner. A non-isotropic shock sensor was defined for each face as (Stolcis and Johnston 1990):

$$v_{ij} = \left| \frac{P_i - P_k}{P_i + P_k} \right| \qquad (20)$$

Flow solvers implementing this shock sensor have been successful in shock wave capturing. On unstructured grids, however, they do not perform well. To resolve this shortcoming, a new algorithm is proposed to calculate shock sensor in the present paper. This algorithm includes the following 3 steps:

- Step 1: shock sensor v_{ij} is calculated on all of the finite volume surfaces within the domain using Eq. 20. The shock sensor calculated in this step is denoted by $N = 0$.
- Step 2: shock sensor v_{ij} on each finite volume surface is substituted by the maximum value of v_{ij} with $N = 0$ on this face and on the other faces of 2 finite volumes neighbor to this face. This is done for every finite volume surface within the domain. The shock sensor calculated in this step is denoted by $N = 1$.
- Step 3: if step 2 is repeated for v_{ij} with $N = 1$, then a set of new v_{ij} will be obtained, which we denote by $N = 2$. Shock sensors with higher N can be obtained in a similar manner.

For the sake of clarity, these 3 steps are introduced on a 2-D unstructured grid shown in Fig. 1. Assume that step 1 is done and we have v_{ij} with $N = 0$ on all of the finite volume surfaces in the solution domain. Now v_{ij} with $N = 1$ would be:

$$v_{ij} = \max\left(v_{ij} , v_{i2} , v_{i3} , v_{k2} , v_{k3} \right) \qquad (21)$$

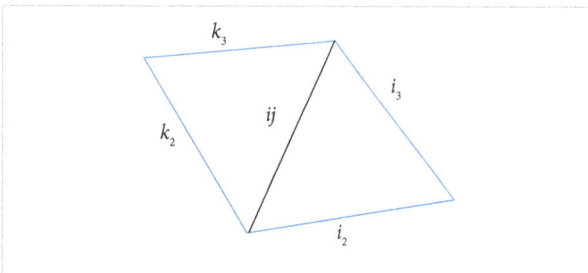

Figure 1. Edges involved in calculation of shock sensor related to edge *ij*.

Now we have v_{ij} with $N = 1$ on all finite volume surfaces within the domain. As mentioned before, if Eq. 21 is implemented for these new v_{ij}, one can obtain v_{ij} with $N = 2$ on all finite volume surfaces. Note that the method offered by Stolcis and Johnston (1990) is in fact the first step of the present algorithm. In this paper, the following test cases are solved using our in-house code in which the present algorithm is implemented. In addition, the minimum value of N leading to the convergence of solution, denoted by N_{min}, is investigated in each case.

NUMERICAL RESULTS

In this section, supersonic flow is simulated around Royal Aircraft Establishment (RAE) tailplane (Mabey *et al.* 1984) and a rectangular wing. The RAE tailplane has panel aspect ratio of 1.2, taper ratio of 0.27, leading edge sweep angle of 50.2° and approximately an airfoil of NACA 64A010.2. For the tailplane wing, lengths are non-dimensionalized with the chord of root. The rectangular wing has panel aspect ratio of 2 and airfoil section of NACA 64A010.2. For this configuration, lengths are non-dimensionalized with its airfoil chord. Initial conditions of the stationary fluid are $\rho = 1$ kg·m^{-3}, $P = 1$ bar and $T = 300$ K.

A tetrahedral mesh is generated around the wings without any additional refinement near the shock. Both wings are located in the middle of the solution domain. The outer boundary of the computational domain, shown in Fig. 2a, is a cube with dimensions of $40 \times 40 \times 40$ root chords on which regular triangular surface mesh is generated. A regular triangular mesh is generated on the wing surface as well. For the RAE tailplane, this grid is shown in Fig. 2b. Grids generated around RAE tailplane and rectangular wing have total finite volumes of 483,245 and 117,030, respectively.

In this paper, grid properties are stored using face-based data structure. In both test cases, the same values of Courant number

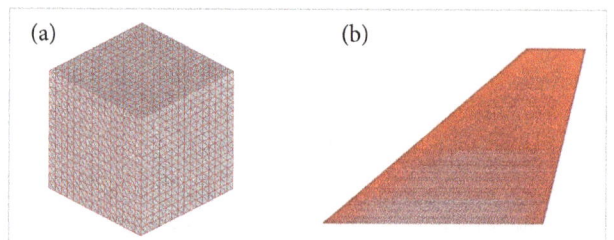

Figure 2. Surface grids on the boundaries of computational domain for RAE tailplane.

in pseudo-time (CFL_I), Courant number in real time (CFL), convergence criteria to stop iterations in pseudo-time (ER_I), and convergence criteria to stop iterations in real time (ER) are used. These values are given in Table 1.

Table 1. Details of solution parameters.

CFL_I	CFL	ER_I	ER
1.0	1.E+6	1.E-3	1.E-7

Supersonic flow around RAE tailplane is simulated under two different conditions: at free stream Mach number (M) of 1.2 and 0° angle of attack (AOA) as the first case and at free stream M of 1.71 and 0.14° AOA as the second case to validate the present algorithm. For the first test case, after 4,045 iterations, solution error defined as $\varepsilon_r = \sum_{i=1}^{Nc} (\rho_i - \bar{\rho}_i)/\rho_i$ was reduced to less than ER, defined in Table 1 to obtain the converged solution. Note that N_c is the total number of cells and $\bar{\rho}_i$ is the density at the previous iteration. The minimum number of pressure-gradient repetition required for the calculation of shock sensor, N_{min}, was 3 and 5 for the first and second test cases, respectively. For the first test case, pressure-coefficient distribution along the chord at semi-span sections of η equal to 14, 42, and 65% from the root is shown in Figs. 3. Having considered that the present solution is inviscid and viscous effects are not included, the agreement observed between the present results and the experimental data (Mabey *et al.* 1984) is very good. At position $\eta = 65\%$ near the tip of the wing, however, some deviation from the experiment is seen near the trailing edge. This may be related to 3-D apex and tip effects that are not predicted very well by the present method at M = 1.2.

For the second test case, the solution was converged after 3,116 iterations. Similarly, pressure results of the second test case are shown in Figs. 4. Again at each semi-span section pressure-coefficient distribution along the chord is compared with results of the experiment (Mabey *et al.* 1984). The present results are in excellent agreement with experimental data. Only small differences are observed between these results in both cases, which is not surprising, since we are solving inviscid flow here. Therefore, one can conclude that the present algorithm works quite well on unstructured grids.

As seen, N_{min} increases with Mach number. To investigate this variation, the tailplane case was simulated for higher Mach number of 2.0. The solution was converged after 13,720 iterations, and the results of this case were obtained with N_{min} = 7. As seen, N_{min} has been increased again with Mach

number. Note that pressures are non-dimensionalized by free-stream pressure.

Following these results, it would be interesting to see if N_{min} increases with Mach number in other configurations, such as rectangular wing. Therefore, supersonic flow at Mach numbers of 1.2, 2.0, and 2.4 over a rectangular wing at 0° AOA is solved using the present algorithm. In these cases, solutions were converged after 3,875, 2,623, and 2,024 iterations, respectively. As a sample of results, the distribution of pressure coefficient along the chord in the mid-section of this wing at M = 2.0, compared with upwind scheme of Roe (1981) for identical grids, is illustrated in Fig. 5. As shown in this figure, a good agreement has been achieved between

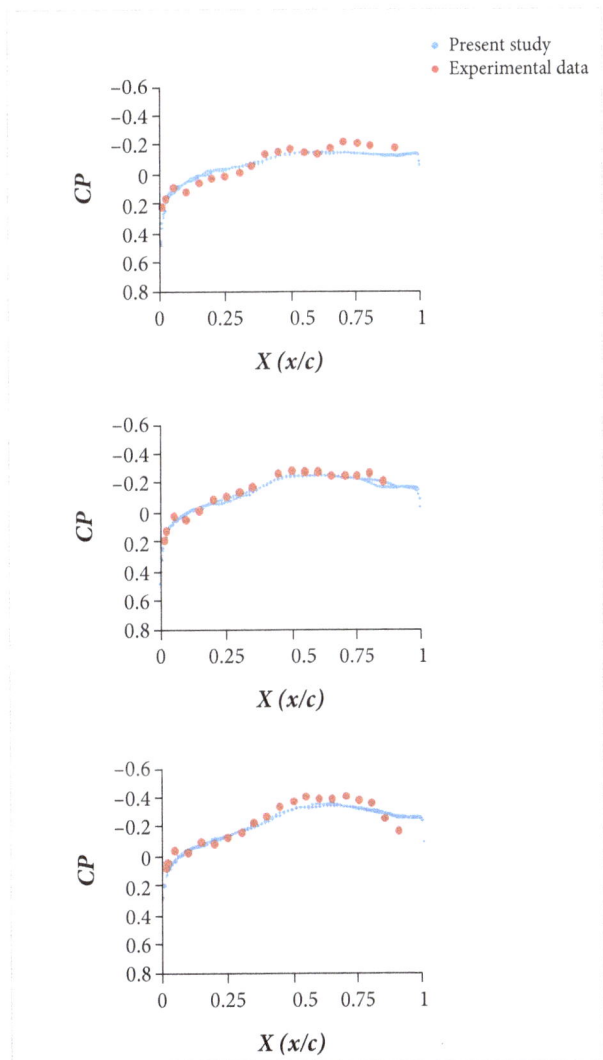

Figure 3. RAE tailplane at M = 1.2 and AOA = 0°. Comparison of pressure coefficient with experimental data (a) η = 14%, (b) η = 42%, (c) η = 65%.

the upwind scheme of Roe (1981) and the present scheme. N_{min}, which led to the convergence of solution for Mach numbers of 1.2, 2.0, and 2.4, is given in Table 2. One more time, its

values have increased with Mach number. The increase in N_{min} should be due to the addition of non-linearities and flow gradients with Mach number in flow regions, such as the shock waves.

Table 2. Variation of N_{min} with Mach for the rectangular wing.

Mach Number	1.2	2	2.4
N_{min}	0	6	10

The effect of N on the solution convergence is investigated as well. For this purpose, convergence rate of solution for rectangular wing at M = 2.0 and AOA = 0° is plotted *versus* the iteration in one real-time step (Fig. 6). Because of the implicit treatment, the solution is obtained in one large real-time step, and the error is plotted *versus* pseudo-time iteration in this figure. According to Table 1, the minimum error to stop iterations in pseudo-time is 10^{-3}. As seen in Fig. 6, the solution convergence is achieved for N not less than 6.0. It is quite interesting that, regardless of the N value, the minimum error is achieved after about 2,600 pseudo-iterations. One should know that the cost of computation in each iteration increases with N. In addition, the results show that the error reduction becomes more monotonic at higher N values.

Obviously, a 3-D oblique shock wave will be formed in the front of RAE tailplane in M = 2.0 and AOA = 0°. It is not an easy task to depict this shock wave, since it is both oblique and 3-D. The strongest segment of this shock wave can be seen in a plane parallel to the free stream flow. As shown in Figs. 7 and 8, non-dimensional pressure and Mach contours on this plane obtained from the present method are compared with the results of upwind scheme of Roe (1981). Note that both results are obtained on the same grid. Contours of both methods confirm formation of oblique shock wave in the

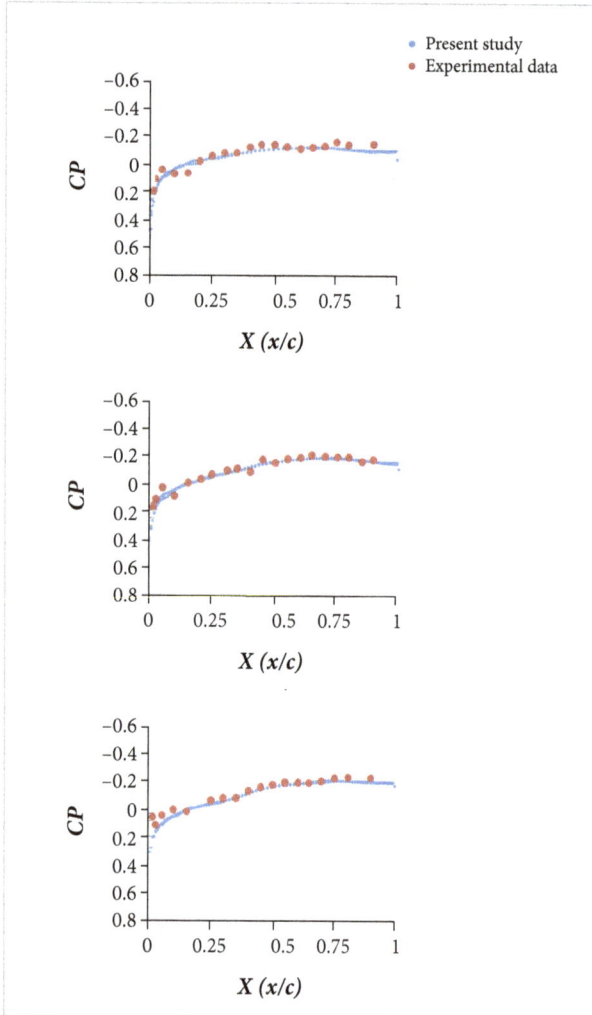

Figure 4. RAE tailplane at M = 1.71 and AOA = 0.14°. Comparison of pressure coefficient with experimental data (a) η = 14%, (b) η = 42%, (c) η = 65%.

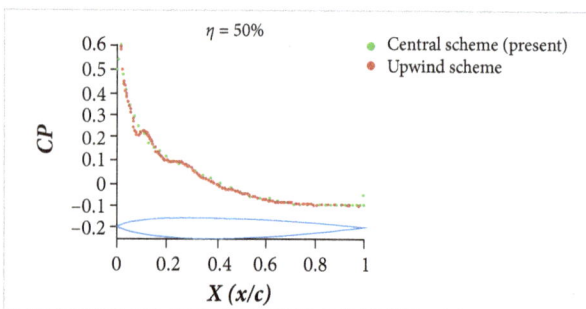

Figure 5. Pressure coefficient along the chord in the middle of rectangular wing at M = 2.0 and AOA = 0°.

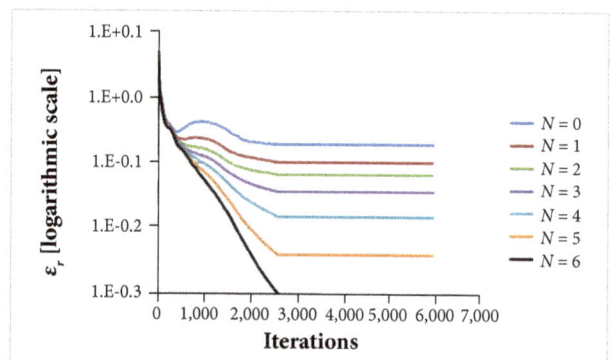

Figure 6. Effect of N on the rate of convergence with iteration at the first real-time step for rectangular wing at M = 2.0 and AOA = 0°.

leading edge region. Since flow behind the oblique shock slips in the spanwise direction along the leading edge, pressure and Mach do not reach their values at stagnation point. The results presented in these figures show that the oblique shock wave is captured thinner and in a location closer to the leading edge in the present method. This observation is quantitatively

demonstrated in Figs. 9, in which non-dimensional pressure and Mach number variations are plotted along a line parallel to the free stream from the leading edge ($L/C_r = 0$) towards the upstream at different semi span positions. L is the distance from leading towards the upstream and C_r is the wing chord at its root. As seen in these figures, the location where gradients of pressure and Mach start is closer to the leading edge in the present method. The difference between the results of the present method and those of the Roe (1981) method is due to the order of methods in space. The present method is 2nd-order in space, but Roe (1981) method is 1st-order in space.

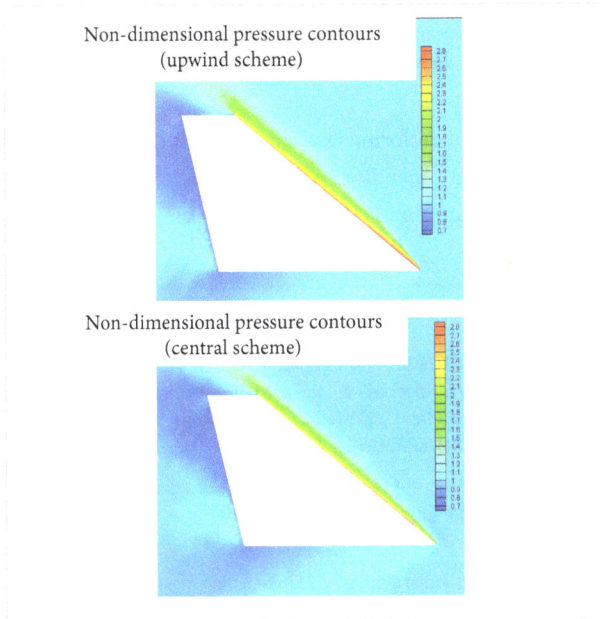

Figure 7. Non-dimensional pressure contours on a plane parallel to the free stream flow for the RAE tailplane at M = 2.0 and AOA = 0°.

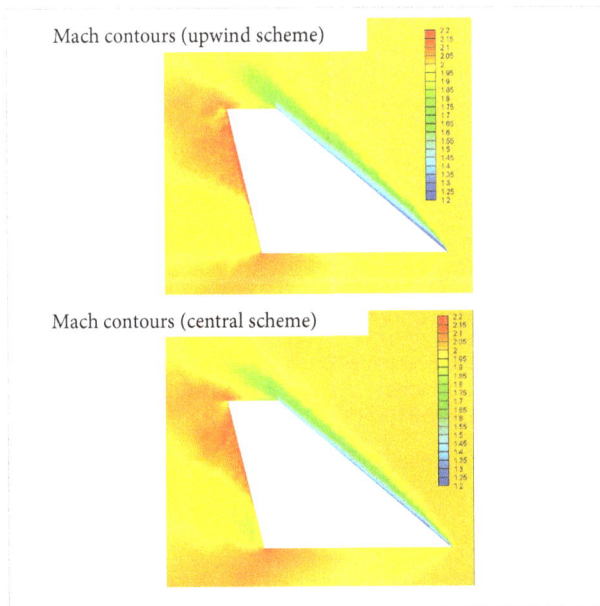

Figure 8. Mach contours on a plane parallel to the free stream flow for the RAE tailplane at M = 2.0 and AOA = 0°.

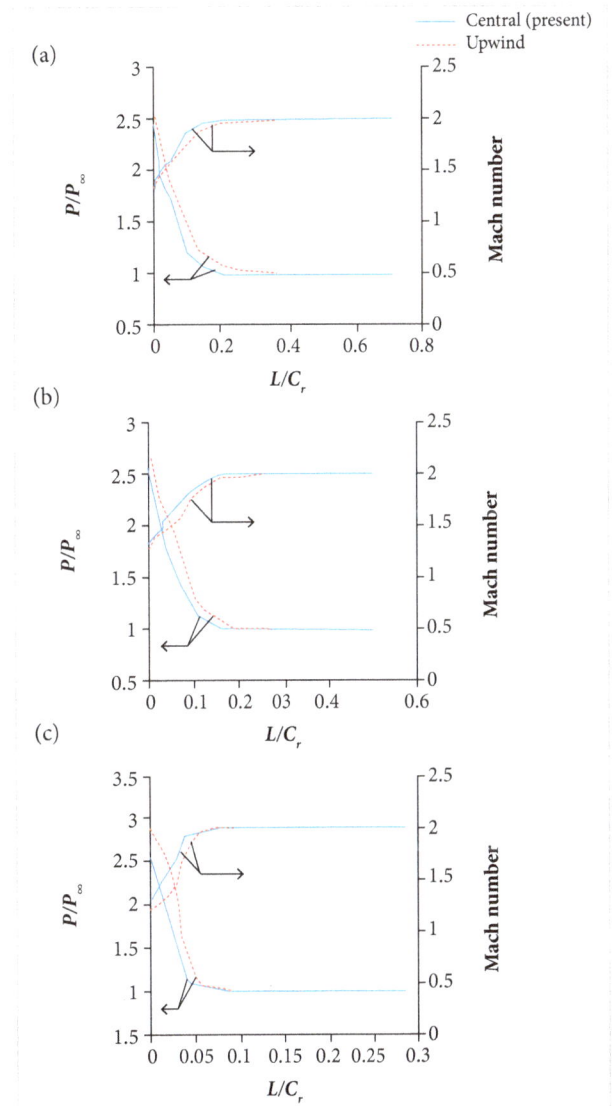

Figure 9. RAE tailplane at M = 2.0 and AOA = 0°. Non-dimensional pressure and Mach number variations along a line parallel to the free stream from the leading edge ($L/C_r = 0$) towards the upstream. (a) η = 90%, (b) η = 50%, η = 10%.

CONCLUSION

In this paper, 3-D supersonic flow around 2 types of wings is solved on unstructured grids using a new algorithm for shock sensor calculation. This shock sensor, which is based on repetitious calculation of pressure-gradient, is very effective in capturing shock waves without oscillations, leading to a stable and converged solution. The present algorithm is a modified version of the method introduced earlier in the literature. Face-based data structure is used here, since it is compatible with the data required for the shock sensor calculation. To validate the present algorithm, its results are compared with the experimental data and the results of Roe (1981) upwind

scheme. A comparison shows that the present algorithm predicts shock waves and the supersonic flow fields accurately. Analysis of the results indicates that the minimum number of pressure-gradient repetition required for the calculation of shock sensor in the present algorithm increases with Mach number at least for the 2 types of wings tested here.

AUTHOR'S CONTRIBUTION

Tahmasbi V performed the simulation and Karimian SMH wrote the main paper. Both authors discussed the results and implications and commented on the manuscript at all stages.

REFERENCES

Barter GE (2008) Shock capturing with PDE-based artificial viscosity for an adaptive, higher-order discontinuous Galerkin finite element method (PhD thesis). Cambridge: Massachusetts Institute of Technology.

Burbeau A (2010) A node-centered artificial viscosity method for two-dimensional Lagrangian hydrodynamics calculations on a staggered grid. Comm Computat Phys 8(4):877-900. doi: 10.4208/cicp.030709.161209a

Cook AW, Cabot WH (2004) A high-wavenumber viscosity for high-resolution numerical methods. J Comput Phys 195(2):594-601. doi: 10.1016/j.jcp.2003.10.012

Cook AW, Cabot WH (2005) Hyperviscosity for shock-turbulence interactions. J Comput Phys 203(2):379-385. doi: 10.1016/j.jcp.2004.09.011

Cook AW, Ulitsky M, Miller D (2013) Hyperviscosity for unstructured ALE meshes. Int J Comput Fluid Dynam 27(1):32-50. doi: 10.1080/10618562.2012.756477

Ducros F, Ferrand V, Nicoud F, Weber C, Darracq D, Gacherieu C, Poinsot T (1999) Large-eddy simulation of the shock/turbulence interaction. J Comput Phys 152(2):517-549. doi: 10.1006/jcph.1999.6238

Harten A (1978) The artificial compression method for computation of shocks and contact discontinuities. III. Self-adjusting hybrid schemes. Math Comput 32(142):363-389. doi: 10.2307/2006149

Harten A (1995) Multiresolution algorithms for the numerical solution of hyperbolic conservation laws. Comm Pure Appl Math 48(12):1305-1342. doi: 10.1002/cpa.3160481201

Jameson A (1986) A vertex based multigrid algorithm for three-dimensional compressible flow calculations. Proceedings of the ASME Symposium on Numerical Methods for Compressible Flow; Anaheim, USA.

Jameson A (1995a) Analysis and design of numerical schemes for gas dynamics. 1: artificial diffusion, upwind biasing, limiters and their effect on accuracy and multigrid convergence. Int J Comput Fluid Dynam 4(3-4):171-218. doi:10.1080/10618569508904524

Jameson A (1995b) Analysis and design of numerical schemes for gas dynamics. 2: artificial diffusion and discrete shock structure. Int J Comput Fluid Dynam 5(1-2):1-38. doi: 10.1080/10618569508940734

Jameson A, Baker T (1987) Improvements to the aircraft Euler method. Proceedings of the 25th AIAA Aerospace Sciences Meeting; Reno, USA.

Jameson A, Mavriplis D (1986) Finite volume solution of the two-dimensional Euler equations on a regular triangular mesh. AIAA J 24(4):611-618. doi: 10.2514/3.9315

Jameson A, Schmidt W, Turkel E (1981) Numerical solutions of the Euler equations by finite volume methods using Runge-Kutta time-stepping schemes. Proceedings of the 14th Fluid and Plasma Dynamics Conference; Palo Alto, USA.

Jiang G, Shu C (1996) Efficient implementation of weighted ENO schemes. J Comput Phys 1(126):202-228. doi:10.1006/jcph.1996.0130

Mabey DG, Welsh BL, Cripps BE (1984) Measurements of steady and oscillatory pressures on a low aspect ratio model at subsonic and supersonic speeds. Bedford: British Royal Aerospace Establishment. Report No.: TR-84095.

Oliveira M, Lu P, Liu X, Liu C (2010) A new shock/discontinuity detector. Int J Comput Math 87(13):3063-3078. doi: 10.1080/00207160902919284

Roe PL (1981) Approximate Riemann solvers, parameter vectors, and difference schemes. J Comput Phys 43(2):357-372. doi: 10.1016/0021-9991(81)90128-5

Shen Y, Cui K, Yang G, Zha G (2013) Parameter-free shock detector and high order hybrid algorithm for shock/complex flowfield interaction. Proceedings of the 51st AIAA Aerospace Sciences Meeting; Texas, USA.

Siclari MJ (1989) Three-dimensional hybrid finite volume solutions to the Euler equations for supersonic-hypersonic aircraft. Proceedings of the 25th AIAA Aerospace Sciences Meeting; Reno, USA.

Siclari MJ, Del Guidice P (1990) Hybrid finite volume approach to Euler solutions for supersonic flows. AIAA J 28(1):66-74. doi: 10.2514/3.10354

Sjögreen B, Yee HC (2004) Multiresolution wavelet based adaptive numerical dissipation control for high order methods. J Sci Comput 20(2):211-255. doi:10.1023/B:JOMP.0000008721.30071.e4

Stolcis L, Johnston L (1990) Solution of the Euler equations on unstructured grids for two dimensional compressible flows. Aeronaut J 94(930):181-195.

Van Leer B (1982) Flux vector splitting for the Euler Equations. In: Krause E, editor. Eighth International Conference on Numerical Methods in Fluid Dynamics: proceedings of the Conference, Rheinisch-Westfälische Technische Hochschule, Aachen, Germany, June 28-July 2, 1982. Berlin, New York: Springer-Verlag. p. 507-512. [Lecture Notes in Physics, 170].

Visbal MR, Gaitonde DV (2005) Shock capturing using compact-differencing-based methods. Proceedings of the 43rd AIAA Aerospace Sciences Meeting; Reno, USA.

Volpe G, Siclari MJ, Jameson A (1987) A new multigrid Euler method for fighter-type configurations. Proceedings of the 8th AIAA Computational Fluid Dynamics Conference; Honolulu, USA.

Von Neumann J, Richtmyer RD (1950) A method for the numerical calculation of hydrodynamic shocks. J Appl Phys 21(3):232-237. doi: 10.1063/1.1699639

Xu K, Martinelli L, Jameson A (1995) Gas-kinetic finite volume methods, flux-vector splitting and artificial diffusion. J Comput Phys 120(1):48-65. doi: 10.1006/jcph.1995.1148

3

The ITASAT CubeSat Development and Design

Valdemir Carrara[1,2], Rafael Barbosa Januzi[3], Daniel Hideaki Makita[4], Luis Felipe de Paula Santos[5], Lidia Shibuya Sato[5]

ABSTRACT: Because they are inexpensive platforms for satellites, CubeSats have become a low-cost way for universities and even developing countries to have access to space technology. This paper presents the ITASAT design, particularly the Attitude Determination and Control Subsystem, the Onboard Software, and the Assembly, Integration and Testing program. The ITASAT is a 6U CubeSat nano-satellite in development at the Instituto Tecnológico de Aeronáutica, in São José dos Campos, Brazil. The platform and its subsystems will be provided by industry while the payloads are being designed and developed by the principal investigators. The ITASAT Attitude Determination and Control Subsystem will rely on a 3-axis magnetometer, 6 analog cosine sun sensors, 3-axis MEMS gyroscopes, 3 magnetic torque coils, and 3 reaction wheels. The Attitude Determination and Control Subsystem operating modes, control laws, and embedded software are under the responsibility of the Instituto Tecnológico de Aeronáutica. A Kalman filter shall be employed to estimate the quaternion attitude and gyroscope biases from sensor measurements. The Attitude Determination and Control Subsystem operating modes are the nominal mode, with geocentric pointing attitude control and the stabilization mode, in which only the satellite angular velocity is controlled. The nominal mode will be split into 2 sub-modes: reaction wheel control plus magnetic wheel de-saturation and 3-axis magnetic attitude control. Simulation results have shown that the attitude can be controlled with 1-degree accuracy in nominal mode with the reaction wheels, but these errors grow as much as 20 degrees or higher with the 3-axis magnetic control.

KEYWORDS: Satellite attitude control, CubeSat, Attitude determination, Kalman filter.

INTRODUCTION

CubeSat platforms use Commercial Off-The-Shelf (COTS) components that cost tenths of the expensive radiation-resistant and space qualified components. One of the greatest advantages of these nano-satellites is the possibility of using them as technological development platforms, in which new equipment, new ideas, and new concepts can be tested, qualified and optimized before going through the high reliability commercial systems. This technology is ready to develop and to perform in orbit testing of sensors, actuators, and control logic at low cost for the Attitude Determination Control System (ADCS) of CubeSats. With few adaptations, control logics can be integrated to large satellites, with the advantage of already having been qualified in flight. Brazil has placed 3 CubeSats platforms in orbit, all of them with passive attitude stabilization, without active control. Four CubeSats are under development, and these should have 3-axis autonomous attitude stabilization and control.

As a recent technology, there are still few publications concerning CubeSats attitude stabilization and control. In recent years, however, several space missions based on nano-satellites have been launched, with some of them with onboard active attitude control. Vega *et al.* (2009), for instance, showed a magnetic actuation system for precession control of a 3U spin stabilized CubeSat, with magnetometer and coarse sun sensors for attitude determination. Li *et al.* (2013) describe the design of an ADCS for CubeSat, with a non-linear fuzzy controller (Adaptive Fuzzy Sliding Magnetic Control) based on reaction wheels, which showed superior performance for a conventional Proportional, Integral

1.Instituto Nacional de Pesquisas Espaciais – Engenharia e Tecnologias Espaciais – Divisão de Mecânica Espacial e Controle – São José dos Campos/SP – Brazil. **2.**Departamento de Ciência e Tecnologia Aeroespacial – Instituto Tecnológico de Aeronáutica – Divisão de Engenharia Aeronáutica – São José dos Campos/SP – Brazil. **3.**Universidade Federal de São Paulo - Instituto de Ciência e Tecnologia - Departamento de Ciência da Computação – São José dos Campos/SP – Brazil. **4.**Universidade Federal de São Paulo - Instituto de Ciência e Tecnologia - Departamento de Engenharia da Computação – São José dos Campos/SP – Brazil. **5.**Departamento de Ciência e Tecnologia Aeroespacial – Instituto Tecnológico de Aeronáutica – Divisão de Engenharia Eletrônica e Computação – São José dos Campos/SP – Brazil.

Author for correspondence: Valdemir Carrara | Departamento de Ciência e Tecnologia Aeroespacial – Instituto Tecnológico de Aeronáutica – Divisão de Engenharia Aeronáutica | Praça Marechal Eduardo Gomes, 50 – Vila das Acácias | CEP: 12.230-901 – São José dos Campos/SP – Brazil | Email: val.carrara@gmail.com

and Derivative (PID) controller in simulation. Oluwatosin *et al.* (2013) demonstrate the stability of an attitude control for CubeSat using 4 reaction wheels with quaternion feedback error. Babcock and Bretl (2011) suggested a Kalman filter with state supplied by the truncated quaternion, as originally formulated by Lefferts *et al.* (1982). They applied the filter for the attitude estimation process of a CubeSat, with vectors provided by a magnetometer and coarse sun sensors based on current generated in the solar panels. A similar procedure was employed by Vinther *et al.* (2011) to estimate the attitude of a CubeSat with an Unscented Kalman Filter whose state was defined by the attitude quaternion and angular rate. To get rid of the singularity in the covariance matrix, that authors utilized the delta-quaternion, as formulated by Lefferts *et al.* (1982). In a thorough job, Quadrino (2014) presented the ADCS design and testing plan for the MicroMas CubeSat. This satellite employs 6 coarse sun sensors, 4 infrared horizon sensors, an Inertial Measurement Unit (IMU) with a 3-axis magnetometer, an accelerometer, and a 3-axis gyroscope. Attitude determination is performed by a TRIAD algorithm (Black 1964), together with an algorithm that selects the best combination of 2 vectors from the 3 sensors. A Kalman filter based on the study of Lefferts *et al.* (1982), with state given by the reduced quaternion (obtained by TRIAD), estimates and filters the attitude quaternion and the gyroscope bias. The MicroMas attitude control is performed by a set of three reaction wheels and three magnetic torquers.

Attitude control by purely magnetic means faces 2 major problems: the generated torque is always perpendicular to the direction of Earth's magnetic field, and the magnitude of the torque varies with the orbital position of the satellite. Both prevent the linear control theory to be applied in proving the attitude stability. Due to this reason, several studies have been proposed to demonstrate the stability of purely magnetic controllers with state feedback (Byrnest and Isidori 1991; Lovera and Astolfi 2004), with or without attitude determination (Bushenkov *et al.* 2002). Silani and Lovera (2005) adopt an H_∞ periodic controller to demonstrate the stability of the magnetic control, and Psiaki (2001) describes a purely magnetic control that uses a periodically asymptotic linear quadric regulator.

The ITASAT shall be placed in a sun-synchronous low altitude circular orbit, around 600 km, with 97° inclination. Launch campaign is scheduled to last from March till May 2017. The main payload is a transponder that receives and transmits environmental data collected by autonomously operated platforms on ground. The ITASAT satellite is based on a CubeSat

architecture with 6U shape, arranged in a matrix of 3 × 2 cubic units, with mass around 4.5 kg. The ITASAT bus was acquired from industry, including the Power, Telemetry, Tracking and Command (TT&C), OnBoard Computer (OBC) subsystems, and ADCS components. The main objective of this paper is to present the current status of the ITASAT design, particularly on the Assembly, Integration and Testing (AIT) program and the ADCS, which is the most complex subsystem on the satellite. Although the ADCS components were bought from CubeSat industry, the control logic, attitude determination, and operating modes shall be developed by an in-house program. Attitude determination is based on 2 vector measurements, provided by a 3-axis magnetometer and coarse sun sensors (CSS). Angular rate will be achieved by a set of MEMS gyroscopes. The magnetometer and gyroscope signals are digitally acquired by a microprocessor and transferred to the OBC by an I²C interface, which also reads the CSS and computes the control logic. The OBC board comprises also a second magnetometer. This paper describes the ADCS design of ITASAT, together with details of the TT&C software architecture and the AIT plan. Some results from attitude simulation of the ITASAT in each operating mode will be shown, together with the efforts that were carried out to accomplish the requirements for the mission. Currently the flight model is already ready to be launched, although some minor changes in the flight software are still to be made. Next section will present the ADCS subsystem, the ITASAT operating modes, and the simulation results. In sequence the Onboard Software design and the AIT program design will be presented.

ATTITUDE DETERMINATION AND CONTROL SUBSYSTEM

The ITASAT attitude was specified to be 3-axis stabilized with the yaw axis (*x*-axis) pointing towards the zenith, as shown in Fig. 1. The *z*-axis shall remain aligned to the orbit angular

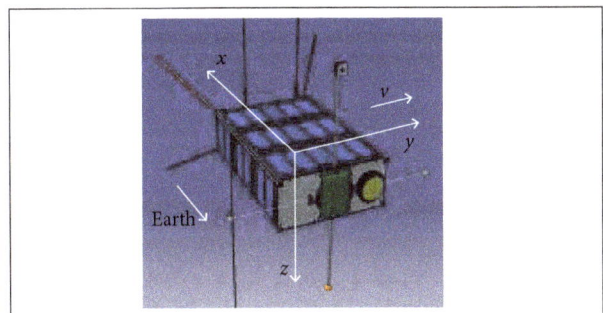

Figure 1. ITASAT CubeSat body-fixed axes: *x* is yaw, *y* is roll, and *z* is pitch.

momentum, perpendicular to the orbit plane. The *y*-axis (roll) completes the body-fixed frame, aligned to the satellite velocity vector. Although there is no requirement to align both the roll and pitch axes with the orbit velocity vector and angular momentum, attitude stability is achieved only when the axis of the largest moment of inertia, that is, the *z*-axis remains aligned to the normal of the orbit plane. The body-fixed frame yaw-roll-pitch is parallel to the orbit frame *x-y-z* when the satellite pointing error is null.

Attitude determination relies on a set of 5 coarse analog sun sensors (CSS) on each satellite side, except the face pointing to Earth, where there is no sensor, together with a 3-axis magnetometer (MAG) and a 3-axis MEMS gyroscope (GYR). Attitude control is achieved by means of a set of 3 magnetic torque coils (MT), 2 of which uses a magnetic core material and one is an air core coil, and 3 reaction wheels (RW). All sensors and actuators have their axes aligned to the satellite axes.

There are no restrictions concerning the attitude pointing error and determination accuracy. However, since one of the payloads is a camera, it was established that a pointing error below 15 degrees is acceptable. Requirement on the attitude stability is more restringing, actually, and shall be better than $0.06°/s$ in order to guarantee a sharp picture. This is not an easy task, since MEMS gyroscope outputs barely reach this magnitude.

To fulfill the control requirements, four attitude control modes were envisaged. Just after orbit injection and starting the onboard computer, the satellite automatically enters in Safe Mode (SM), in which no actions for attitude control are taken. This mode allows ground commands and sends onboard telemetry; however, sensors and actuators are switched off. By ground command the SM can be switched to Attitude Stabilization mode (ASM) or alternatively to the Attitude Determination and Estimation mode (ADE), as shown in Fig. 2. In fact, the ADE is not a real operating mode, since the computations for attitude determination and estimation are also done in all modes, except the SM. It is considered a true mode due to the

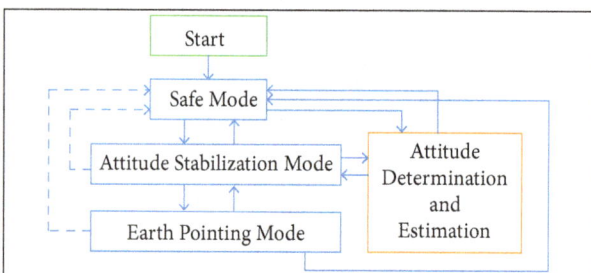

Figure 2. Attitude control modes for ITASAT.

necessity of attitude knowledge even when the attitude is not being stabilized or controlled. Moreover, by checking the attitude determination error from telemetry (by means of the filter residue, for instance), the Earth Pointing Mode (EPM) can be activated without taking risk of instability in the attitude control. In Fig. 2, the filled lines represent ground commands while the dotted lines mean autonomous reconfiguration onboard commands.

In Attitude Determination and Estimation the procedures to perform both attitude determination and estimation, as explained in next section, will be carried out. The attitude knowledge will be further used in Earth Pointing Mode to properly point the satellite to Earth. However, in the ADE mode only, all the actuators are switched off and no actions on attitude control are done. Two conditions are needed to allow EPM command: the attitude shall be known with high confidence and the satellite angular rate shall be small. The first condition is acquired by the ADE mode and last condition is achieved by the ASM mode. In the ASM the satellite angular rate shall be damped till a maximum specified value of $0.6°/s \approx 0.01$ rad/s in any axis. This low rate assures the necessary condition for entrance in the three-axis attitude control mode (EPM), since a large angular momentum of the satellite could saturate the reaction wheels. Two different algorithms shall be implemented to ASM mode: the B-dot algorithm (Bushenkov *et al.* 2002), and a Rate Reduction mode (RR), similar to the B-dot but based on gyro measurements instead of the Magnetometer time derivative.

The EPM aims to point the negative *x*-axis to Earth and to keep the angular rate below the specified value. Again, 2 types of control laws were implemented in the onboard computer. The first one employs a Reaction Wheel Controller (RWC) with a conventional PID with gains adjusted to drive the reaction wheels whenever an attitude pointing error is detected. The angular rate is also adjusted to track the orbital rate, around $0.06°s$, so as to keep the right attitude when the satellite travels its orbit. Reaction wheel speed will be damped by means of magnetic torques, provided by the MT, in order to maintain the RW rate in a safe range. The second control law is a 3-axis Magnetic Torque Controller (MTC), which also uses a control based PID exclusively on the magnetic torque coils to drive the satellite attitude. Due to the small torque provided by this type of actuator, it is expected that the attitude error will be larger than the previous controller with the RW. Moreover, the inability of the MT to generate torques in 3 axes can jeopardize this type of control, and makes gain adjusting a real hard job. These 2 types of controller are needed due to the energy restrictions for operating the RW during the whole orbit. From the EPM the satellite can be commanded to enter in any of the remaining attitude operating modes by ground command or in the SM by the autonomous reconfiguration command, which employs specific algorithms to detect attitude anomalies. These anomalies are then analyzed by the on-ground specialists and actions are taken in order to overcome the problem, if any. Figure 3 illustrates the entrance condition and stability range of each operating mode and control algorithm, as function of the satellite's angular rate. As seem in this figure, either the B-dot or the RR can manage to drive the angular rate to a condition that enables the EPM operation with the reaction wheels (RWC). Although the MTC has a large operating range, in terms of angular rate, it lacks controllability in low velocities. The next sections will present respectively the ADE, the ASM, and the EPM algorithms for ITASAT.

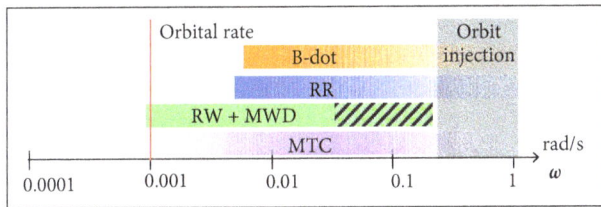

Figure 3. Entrance condition for each operating mode and control algorithm as function of the angular rate.

ADE MODE

The ADE algorithm employs 2 methods do compute the satellite attitude: the TRIAD (Black 1964; Markley 1998; Markley and Crassidis 2014), or QUEST (Shuster and Oh 1981), alternatively, and a Kalman filter with reduced order covariance matrix (Lefferts et al. 1982), or with incremental quaternion (Garcia et al. 2016). Alternatively the Kalman filter can estimate not only the attitude, but also the disturbance torques (Söken and Hajiyev 2014). Either a low pass filter (LPF) on the gyroscope measurements or a numeric derivative of the attitude can be employed to compute the satellite angular rate, and since the attitude can be estimated by several methods, the number of combinations to compute the attitude and rate is high.

Attitude determination is performed with 2 vectors attitude determination algorithms, based on the magnetometer and solar sensor readings, which needs also knowing these vectors in a reference frame, normally the inertial or the orbital frame. For ITASAT these reference vectors are computed by the onboard computer based on numeric models of the geomagnetic field and the Sun's direction in inertial coordinates. However, since

the geomagnetic field rotates together with the Earth, the model computes the field strength in Earth centered coordinates as function of the geocentric latitude, longitude and altitude (or the radius from Earth's center) of the true satellite position in its orbit. Therefore the satellite orbit shall also be onboard propagated to feed the geomagnetic field with proper data. The SGP4 (Hoots and Roehrich 1980) analytical orbit propagator will be used to compute the satellite inertial position, whereas the geomagnetic field will be calculated by the International Geomagnetic Reference Field (IGRF) model (International Association of Geomagnetism and Aeronomy 2015). The SGP4 takes as input the Two Line Elements (TLE) provided by NORAD (Kelso 2015), with 2 days delay, normally. Figure 4 shows a simplified block diagram of the ADE mode for attitude estimation with Kalman filter, which estimates the inertial attitude quaternion, q, and the gyroscope biases, b. The angular rate ω is computed by a low pass filter applied to the gyro measures compensated from the biases. The remaining attitude determination methods follow similar procedures. Care must be taken when the satellite is in the Earth's shadow, due to the unavailability of CSS data. The time and date necessary for computations are provided by an onboard real time clock (RTC). The complete filter model and equations can be found in Carrara (2016) as well as the control laws for the operating modes, except for the Attitude Stabilization Mode, whose control law will be presented in next section. It is assumed that a prior on ground calibration of the magnetometer was carried out, so the expected in orbit magnetometer bias is negligible (Amorim and Martins Filho 2016).

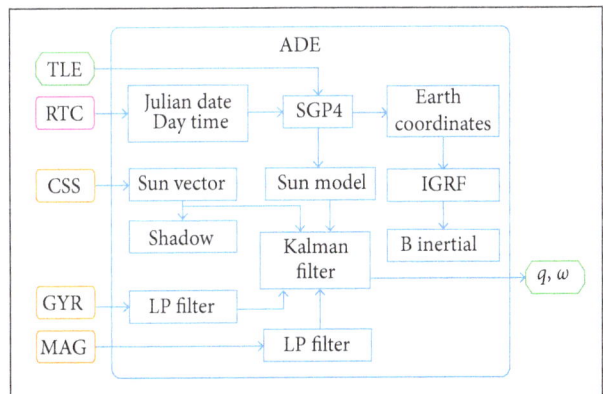

Figure 4. Simplified block diagram of the Attitude Determination and Estimation mode.

ATTITUDE STABILIZATION MODE

The purpose of the ASM is to reduce the initial angular rate of the satellite, around 10 rpm, to a specified value of

0.04 rpm, which is 4 times larger than the controlled rate in EPM. As mentioned before, 2 methods will be implemented; each can be selected by ground command. The B-dot algorithm employs the time derivative of the Earth's magnetic field as measured by the magnetometer to generate a torque by the magnetic coils that tries to break the satellite. Since the magnetic torque is given by the cross product between the coil magnetic moment and the Earth's field strength, as

$$\mathbf{g}_m = \mathbf{m}_{mag} \times \mathbf{B} \qquad (1)$$

so the magnetic moment vector can be obtained by

$$\mathbf{m}_{mag} = -k_{bdot} \frac{\dot{\mathbf{B}}}{|\mathbf{B}||\dot{\mathbf{B}}|} \qquad (2)$$

and remembering that by the time derivative of a vector in a rotating frame is such that

$$\dot{\mathbf{B}}\Big|_i = \dot{\mathbf{B}}\Big|_b + \boldsymbol{\omega} \times \mathbf{B} \qquad (3)$$

where the subscripts i and b mean the derivative taken in the inertial and body fixed frames, respectively, it results, after substituting Eqs. 3 and 2 in Eq. 1,

$$\mathbf{g}_m = \frac{k_{bdot}}{|\mathbf{B}||\dot{\mathbf{B}}|} (\boldsymbol{\omega} \times \mathbf{B}) \times \mathbf{B} \qquad (4)$$

considering, of course, that $\dot{\mathbf{B}}\big|_i \ll \dot{\mathbf{B}}\big|_b$ and therefore $\dot{\mathbf{B}}\big|_i$ can be neglected. The magnetic torque is opposed to the component of the angular rate perpendicular to the magnetic field $\mathbf{B}$, and can decrease this component. Even considering that the component of $\boldsymbol{\omega}$ in the direction of $\mathbf{B}$ can not be reduced, it will always be possible to generate torque in three-axis, although not at same time, since $\mathbf{B}$ changes slowly as the satellite travels its orbit.

The Rate Reduction algorithm is similar to the B-dot, except that it takes directly the satellite's angular rate as measured by the gyroscope, instead of the time derivative of the geomagnetic field, to compute the magnetic moment to be applied to the MT:

$$\mathbf{m}_{mag} = \frac{k_{rr}}{|\boldsymbol{\omega}||\mathbf{B}|} \boldsymbol{\omega} \times \mathbf{B} \cdot \qquad (5)$$

Both modes were simulated in MATLAB with the PROPAT Toolbox (Carrara 2015), and gave similar results. The B-dot simulation is shown in Fig. 5. It can be seen that attitude stability

is achieved in less than 6 h after orbit injection. It was adopted in simulation a sun-synchronous orbit at 630 km altitude, and the sensors were modeled with Gaussian noise and bias compatible with the information provided by the supplier. A constant residual magnetic moment of 0.001 Am² magnitude on the satellite was considered as a single perturbation on attitude. A maximum dipole of 0.04 Am² (20% of the maximum available) in each magnetic torque coil was adopted to avoid fast control action, which could destabilize the attitude. Considering the expected satellite inertia, the kinetic rotational energy corresponding to the maximum angular rate of 0.01 rad/s is around 5 μJ. The simulated kinetic energy after orbit injection is shown in Fig. 5. The stability condition is achieved after 7 h.

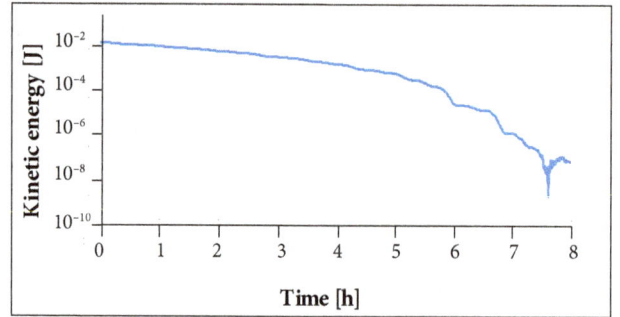

Figure 5. Simulated rotational kinetic energy during attitude stabilization with B-dot algorithm.

EARTH POINTING MODE

The EPM aims to point the satellite axis to Earth and to keep the angular rate closed to the specified value. Due to limitations on the energy available for the ADCS, the reaction wheels can not be operated the whole orbit. So it is necessary to switch between the 2 control laws of the EPM whenever the satellite needs fine Earth pointing to acquire pictures, for instance. This can be done only when the satellite is in contact with mission center, located in São José dos Campos, Brazil. For the remaining of the orbit 3-axis magnetic attitude stabilization and control will be employed. The control laws and the simulated results are shown in sequence.

EPM with Reaction Wheel Control

When in fine pointing mode, the wheel's speeds need to be damped by a magnetic control law. The block diagram of the RW control is shown in Fig. 6. The RW controller, C_{RW}, computes the attitude error in Euler angles from the attitude quaternion $\mathbf{q}$ with respect to the reference attitude, besides the satellite angular velocity with respect to the orbital frame, and applies them to a conventional PD controller, which, in turn, commands the torque to the reaction wheels. The magnetic controller takes the wheel speed and applies a proportional gain to drive the TR and to keep the speed inside the specified range.

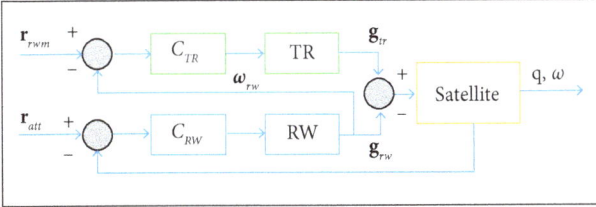

Figure 6. EPM control with reaction wheels and wheel speed damping with magnetic control.

The RW model comprehends a dead band of 50 rpm around the 0 speed, according to supplier data, besides a torque and speed hardware limits. The dead band proved to be significant in attitude stabilization and pointing accuracy. A simulation of the EPM was carried out using the Kalman filter to estimate the attitude and a low pass filter on the simulated gyroscope. The gyro model considers a drift, or bias, a bias instability, a Gaussian noise, sensitivity scale factor error and crossover coupling sensitivity. A low pass filter is applied to the simulated gyro measurements prior to Kalman filtering. Magnetometer and CSS models are based on Gaussian noise, scale factor errors and cross-axis coupling only. On Earth shadow the attitude is still estimated using MAG but with reduced accuracy. The proportional magnetic control to unload the RW momentum is composed of 2 individual controllers: to adjust the speed of the pitch wheel, and to control the speed in the roll-yaw plane. So it is possible not only to damp the wheel's speed, but also to set a non-null angular rate on pitch axis or on the roll-yaw plane.

Simulation results of the EPM mode are shown in Figs. 7 and 8. Figure 7 presents steady state of the attitude pointing

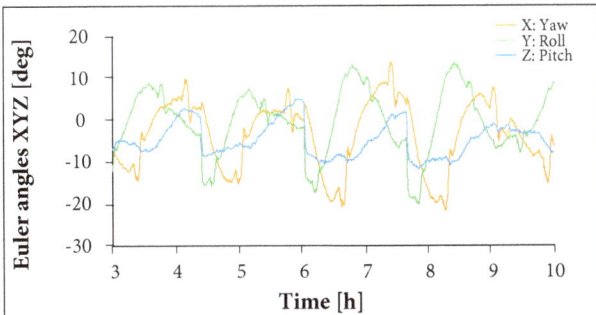

Figure 7. Simulated attitude error in EPM mode with reaction wheel and magnetic control for wheel momentum unloading.

error in Euler angles of a yaw-roll-pitch sequence lasting 7 h (4 orbits, approximately). Although the error exceeds the specified pointing requirement, it is still inside the allowed range most of the time. The reaction wheel speed is shown in Fig. 8. An angular momentum corresponding to a 200 rpm ($\approx$21 rad/s) was commanded in the magnetic controller for both pitch axis and roll-yaw plane.

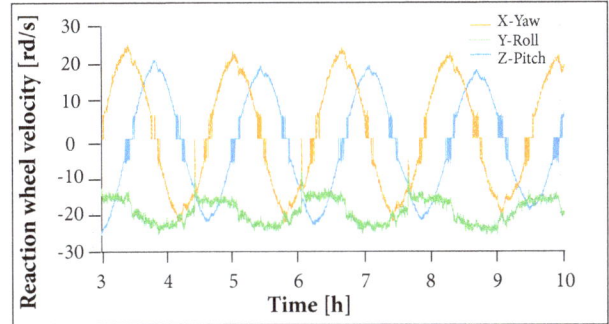

Figure 8. Simulated reaction wheel speed, with magnetic wheel speed control set on −21 rad/s for pitch and 21 rad/s in roll-yaw plane.

EPM with Magnetic Control

Due to energy saving, the EPM with magnetic control, which will be switched by ground command, is the main attitude controller running on the satellite. It employs a PID controller based on the Conventional Cross Product Law (CCPL) (Camillo and Markley 1980) to obtain the magnetic moment to be applied on the MT. The magnetic moment will be commanded by PWM (Pulse Width Modulation) acting on the coils. Since this control is unable to generate torque around the Earth's magnetic field, the controller to be applied has to be very weak, just enough to compensate the disturbance torques. So, the slow variation of the Earth's field with respect to the orbit frame can be used to allow torques in all axes. The derivative gain was chosen relatively large, to stop the satellite attitude motion with respect to the orbital frame. The integral torque, on its turn, has to be very small and acts to compensate the external disturbances. The proportional gain remains in the middle.

The EPM magnetic control is highly influenced by the noise coming from the sensors, particularly on the gyroscope. Due to this reason, similar results were achieved when either the TRIAD attitude determination or the Kalman filter provided the attitude reference for control logic, even considering that the latter provides attitude information in the whole orbit, while the CSS measurements necessary to the TRIAD algorithm is available only on the sunlit part of the orbit. During shadow the magnetic three-axis control takes no action, except by the integral control, which remains active with the last value computed before the satellite enters in Earth shadow. Additionally, satellite angular rate was obtained with a low pass filter applied to a numeric derivative of the attitude, resulting in a less noisy control. To avoid unnecessary control action that happens when the control torque vector is almost aligned to the Earth's magnetic field, a dead band on the control was introduced, which turns off the coils whenever the angle between these vectors is less than 53°. The gains were adjusted to each axis: the higher the moment of inertia the higher the gain. Attitude error was calculated again by the Euler angles (yaw-roll-pitch) with respect to the orbital frame. A maximum duty-cycle of just 10% was adopted in the PWM of the magnetic torque coils, to avoid sudden changes in control action. However, the 0.001 Am² residual magnetic dipole on the satellite proved to be strong enough to destabilize the attitude when the MT is the only actuator. That means that a magnetic cleanliness on the satellite during integration shall be necessary in order to reduce the dipole to at least 10% of that value. The simulation results assume that the maximum residual dipole is, therefore, 0.0001 Am². Figure 9 shows the attitude error in EPM with 3-axis magnetic control, in terms of the Euler angle, from an Euler-axis and angle attitude representation. From an unknown initial attitude, the satellite reaches the steady state after 15 h, approximately. The final error is still large, around 30° maximum, but it is acceptable for this type of control. There are several reasons that explain this large final error, but the inability of the gyroscope to sense such small angular rates, of 10^{-4} rad/s magnitude, is the major one.

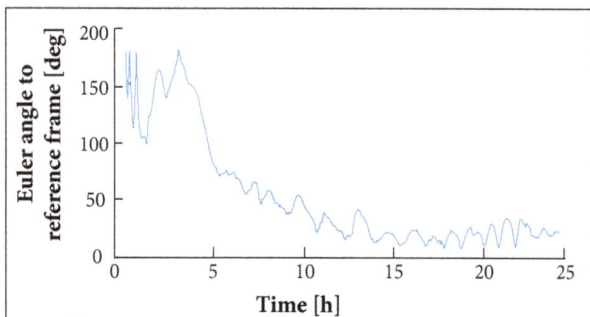

Figure 9. Euler angle, from Euler axis and angle attitude representation, of the simulated attitude in EPM with 3-axis magnetic control.

ONBOARD SOFTWARE

The architecture of the ITASAT onboard software was designed following the modularity applied to the CubeSat structure and hardware; in this context, modularity means the capability to isolate software modules (minor parts of the software) of the remaining system. This modularity is a key point for ITASAT, since one of the major objectives of this project is to develop a multi-mission platform. The modularity allows exchanging parts of the system with small impact in the whole system, thus reducing some eventual rework and helping in the execution of the tests, because once the module is validated, it integrates with the onboard software and only the interfaces need to be validated, usually during integration tests. Other advantage of the modular system is the ability to identify independent parts and to develop them in a parallel way, thus allowing an optimized development of the whole system. However, this kind of development imposes new challenges, because each module of the system shall be clearly defined, detailing its features, vulnerabilities and interfaces, which is not easily done. In the case of ITASAT, Unified Modeling Language (UML) (Object Management Group 2016a) and Systems Modeling Language (SysML) (Object Management Group 2016b) were used to design all the system characteristics: the first one was used strictly for the software and the second one for general purpose of the system.

SOFTWARE COMPONENTS

The ITASAT onboard software was structured in components, modules and units, with the components being sets of modules and the modules as sets of units. The modules inside a component contain specific features of that component. Generally, the software components have many modules with similar features. For instance, in the ITASAT there is a software component called Internal Communication; this component has many software modules with features related to communication between parts of the system, either the communication among different equipment or components of the same software on the onboard computer. Figure 10 illustrates the UML modeling of the Internal Communication component. All the software components of ITASAT were represented using component diagrams. Two components are shown in Fig. 10: Internal Communication and General Function Management, and also 3 interfaces are shown: getIcData, setIcOpMode, and getEquipmentData.

Each interface is available in the Internal Communication and can be used by the General Function Manager. Therefore each component can be individually modified, while keeping the interfaces unchanged, without affecting the functionalities of other components.

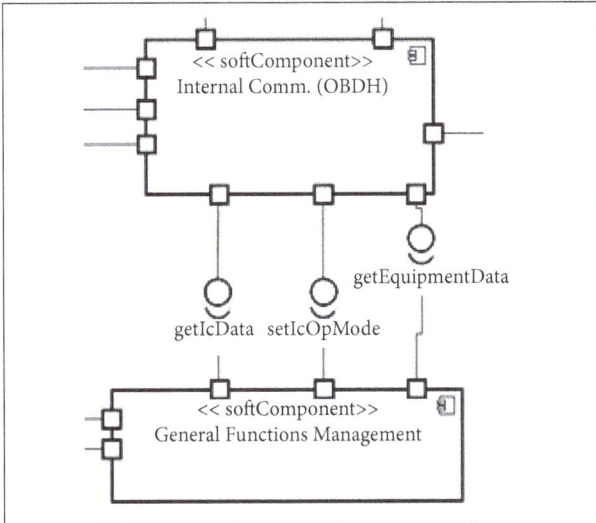

Figure 10. Example of modeling components using UML.

SOFTWARE EXECUTION

Another characteristic of the software components are the tasks controlled by the real-time operating system FreeRTOS (Real Time Engineers 2016). These tasks control the software operation, and mostly, can be configured independently, which brings modularity not only on design level, but also on execution level. Figure 11 illustrates the software layers for the ITASAT. It shows that lower-level tasks such as hardware configuration and drivers are hidden from the software components by the real-time operating system, which is responsible for managing the execution of those components. Each software component has tasks to control its own operation, and therefore the components can be developed and tested apart each other, by activating or deactivating the tasks of interest.

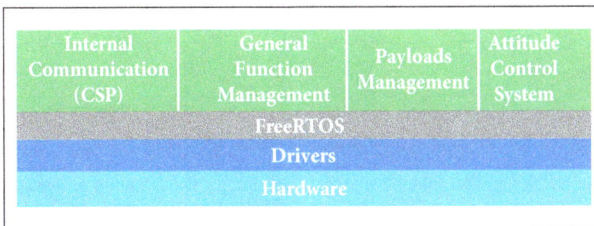

Figure 11. Software layers.

ASSEMBLY, INTEGRATION AND TESTING

The ITASAT-AIT program aims to establish system integration procedures and to apply standard methods for validation and verification testing. With these 2 objectives in mind, the program seeks to develop a modular platform that minimizes the physical interaction between components. The modular goal impose restrictions to the satellite resulting a system composed of several self-contained units that can be easily removed or replaced without significant changes in the platform. This concept allows cost reduction by 10 to 15% and enables the platform to be employed in different tasks with different payloads, besides schedule reduction of AIT activities, mission customization, design heritage, and technology development and consolidation (Kingston 2005).

The ITASAT has a structure volume of six CubeSat units, or 6U, organized in a matrix layout of 2 × 3 units. The internal subsystems were arranged in the units according to (Fig. 12):

- 1U for Power and Energy Subsystem.
- 1U for Onboard Computer Subsystem.
- 2U for Attitude Determination and Control Subsystem.
- 2U for Payloads.

There are several models that can be used for development and verification process (European Cooperation for Space Standardization ECSS 2012). The philosophy of model defines the

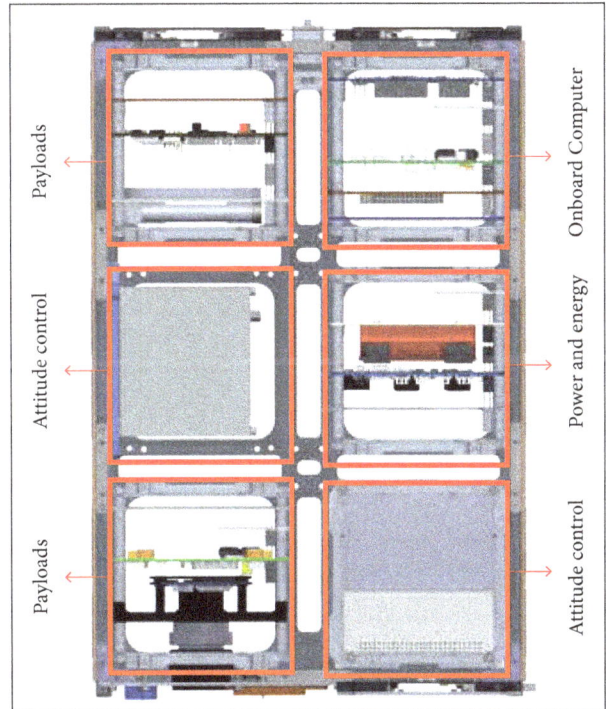

Figure 12. ITASAT internal subsystem distribution.

optimal number and characteristics of physical models needed to achieve confidence in product verification and a smaller impact on costs and risks. ITASAT design will rely in 2 models: the engineering model for functional testing and staff training, and the proto-flight model for acceptance testing and system qualification. However, according to the CubeSat philosophy, the components used for both models shall be the same, which simplifies the qualification and acceptance procedures. In fact, the proto-flight model is needed just because the components of the satellite have been stressed during qualification and acceptance tests to validate the platform, and therefore can not be used for flight or proto-flight models.

The ITASAT AIT campaign foresees random and sinusoidal vibration testing, shock analysis and thermo-vacuum cycles, together with mass properties and magnetic balancing. All these tests are required to increase product quality and to prevent failures that usually can occur in CubeSats. It is estimated that the ITASAT AIT campaign lasts 48 days.

CONCLUSIONS

The ITASAT project is being carried out by university students with funds provided by the Brazilian Space Agency (AEB). The solution adopted for the platform, a 6U CubeSat, can accommodate several payloads with different attitude pointing and stability requirements. Subsystems were purchased directly from the industry, while the payloads are being developed, tested and qualified by leading researchers. The solution adopted for the ADCS, with attitude determination based on two vectors combined with a filter estimation process, which uses the information provided by the gyros, proved to be adequate to meet the modest pointing requirements, with attitude determination accuracy around 10 degrees. The attitude control is performed by magnetic coils that provides stability during most of the orbit, and reaction wheels which allow high pointing accuracy but for limited time due to availability of onboard power. Simulations of the attitude control of ITASAT were quite realistic, with the inclusion of different effects that contribute to assure realism, based on the characteristics of the sensors and actuators. A Kalman filter that estimates the attitude quaternion with the gyro bias allowed a smooth control action either with the reaction wheels or with the magnetic coils. An important point to emphasize is that the ITASAT shall be the first 3-axis attitude stabilized satellite with reaction wheels whose control logic and software developments, allied to testing and qualification procedures, were totally carried out in Brazil.

AUTHOR'S CONTRIBUTION

Carrara V carried out the attitude simulation and control laws; Carrara V, Makita DH, and Januzi RB conducted the onboard attitude control software design and programming; Januzi RB, Sato LS, and Makita DH designed the onboard software for data handling; Sato LS and Santos LFP conceived the AIT and satellite layout; Carrara V, Januzi RB, and Santos LFP wrote the main text, with close colaboration of Sato LS.

REFERENCES

Amorim J, Martins Filho LS (2016) Experimental magnetometer calibration for nanosatellites' navigation system. J Aerosp Technol Manag 8(1):103-112. doi: 10.5028/jatm.v8i1.586

Babcock EP, Bretl T (2011) CubeSat attitude determination via Kalman filtering of magnetometer and solar cell data. Proceedings of the 25th Annual AIAA/USU Conference on Small Satellites; Logan, USA.

Black HD (1964) A passive system for determining the attitude of a satellite. AIAA J 2(7):1350-1351. doi: 10.2514/3.2555

Bushenkov VA, Ovchinnikov MY, Smirnov GV (2002) Attitude stabilization of a satellite by magnetic coils. Acta Astronaut 50(12):721-728. doi: 10.1016/S0094-5765(02)00011-5

Byrnest CI, Isidori A (1991) On the attitude stabilization of rigid spacecraft. Automatica 27(1):87-95. doi: 10.1016/0005-1098(91)90008-P

Camillo PJ, Markley FL (1980) Orbit-averaged behavior of magnetic control laws for momentum unloading. J Guid Control Dynam 3(6):563-568.

Carrara V (2015) An open source satellite attitude and orbit simulator toolbox for Matlab. Proceedings of the 17th International Symposium on Dynamic Problems of Mechanics; Natal, Brazil.

Carrara V (2016) Attitude determination and control of ITASAT CubeSat. Proceedings of the 26th AAS/AIAA Space Flight Mechanics Meeting; Napa, USA.

European Cooperation for Space Standardization (2012) Space Engineering: verification guidelines; Noordwijk, The Netherlands.

Garcia RV, Kuga HK, Zanardi MCFPS (2016) Unscented Kalman filter for determination of spacecraft attitude using different attitude parameterizations and real data. J Aerosp Technol Manag 8(1):82-90. doi: 10.5028/jatm.v8i1.509

Hoots FR, Roehrich RL (1980) Models for propagation of NORAD Element Sets. Spacetrack Report No. 3. Colorado Springs: Aerospace Defense Center.

International Association of Geomagnetism and Aeronomy (2015) International Geomagnetic Reference Field; [accessed 2017 Jan 19]. http://www.ngdc.noaa.gov/IAGA/vmod/igrf.html

Kelso TS (2015) Celestrack NORAD two-line element sets: current data; [accessed 2017 Jan 27]. http://www.celestrak.com/NORAD/elements/

Kingston J (2005) Modular architecture and product platform concepts applied to multipurpose small spacecraft. Proceedings of the 19th Annual AIAA/USU Conference on Small Sallite; Logan, USA.

Lefferts EJ, Markley FL, Shuster MD (1982) Kalman filtering for spacecraft attitude estimation. J Guid Contr Dynam 5(5):417-429. doi: 10.2514/3.56190

Li J, Post M, Wright T, Lee R (2013) Design of attitude control systems for CubeSat-class nanosatellite. Journal of Control Science and Engineering 2013(2013):Article ID 657182. doi: 10.1155/2013/657182

Lovera M, Astolfi A (2004) Spacecraft attitude control using magnetic actuators. Automatica 40(8):1405-1414. doi: 10.1016/j.automatica.2004.02.022

Markley FL (1998) Attitude determination using two vector measurements. NASA Goddard Space Flight Center; Report. 99-001; [accessed 2017 Jan 27]. https://ntrs.nasa.gov/archive/nasa/casi.ntrs.nasa.gov/19990052720.pdf

Markley FL, Crassidis JL (2014) Fundamentals of spacecraft attitude determination and control. New York: Springer.

Object Management Group (2016a) Systems Modeling Language (UML) resource page; [accessed 2017 Jan 27]. http://www.omg.org/spec/UML/

Object Management Group (2016b) OMG Systems Modeling Language; [accessed 2017 Jan 27]. http://www.omgsysml.org/

Oluwatosin AM, Hamam Y, Djouani K (2013) Attitude control of a CubeSat in a circular orbit using reaction wheels. Proceedings of the AFRICON, 2013; Pointe-aux-Piments, Mauritius.

Psiaki ML (2001) Magnetic torquer attitude control via asymptotic periodic linear quadratic regulation. J Guid Control Dynam 24(2):386-394. doi: 10.2514/2.4723

Quadrino MK (2014) Testing the attitude determination and control of a CubeSat with hardware-in-the-loop [Master's thesis]. Cambridge: Massachusetts Institute of Technology.

Real Time Engineers (2016) FreeRTOS™; [accessed 2017 Jan 19]; http://www.freertos.org/

Shuster MD, Oh SD (1981) Three-axis attitude determination from vector observations. J Guid Control Dynam 4(1):70-77. doi: 10.2514/3.19717

Silani E, Lovera M (2005) Magnetic spacecraft attitude control: a survey and some new results. Contr Eng Pract 13(3):357-371. doi: 10.1016/j.conengprac.2003.12.017

Söken HE, Hajiyev C (2014) Estimation of pico-satellite attitude dynamics and external torques via Unscented Kalman Filter. J Aerosp Technol Manag 6(2):149-157. doi: 10.5028/jatm.v6i2.352

Vega K, Auslander D, Pankow D (2009) Design and modeling of an active attitude control system for CubeSat class satellites. Proceedings of the AIAA Modeling and Simulation Technologies Conference; Chicago, USA.

Vinther K, Fuglsang Jensen K, Larsen JA, Wisniewski R (2011) Inexpensive CubeSat attitude estimation using quaternions and Unscented Kalman Filtering. Automatic Control in Aerospace 4(1).

4

Climate Forecasts at the Centro de Lançamento de Alcântara Using the Climate Model RegCM4

Cleber Souza Corrêa[1], Gerson Luiz Camillo[1], Vinicius Milanez Couto[1], Gilberto Fisch[1], Felipe do Nascimento Correa[1], Fabricio Harter[2]

ABSTRACT: This study uses climate modeling (RegCM4 Climate Model) to provide a wind forecast average behavior at low levels, close to the surface. The model was used to generate an estimate of the average vertical wind profile lasting 5 months, from August to December 2015, attempting to observe intra-seasonal variations, with the presence of persistence in the wind field. The results of climate modeling of the wind profile near the surface have the great potential for great operational significance during launch campaigns at the Centro de Lançamento de Alcântara. Three average results were generated for the month of November 2015, while operating in São Lourenço. A dynamical downscaling nested with global models with RCP4.5 and RCP8.5, using 3 different global conditions initialization datasets. It used subsets with the models from the Met Office Hadley Centre (HadGEM2-ES), the Centre National de Recherches Météorologiques (CNRM-CM5), and the Commonwealth Scientific and Industrial Research Organization (CSIRO-Mk3.6), making a downscaling with the RegCM4 Climate Model for the Centro de Lançamento de Alcântara region. The results are preliminary but show great potential. Since the RegCM4 Climate Model can show variations of high and low intensity, its temporal frequency in the average vertical wind profile and the duration of temporal variation are of the order of 3 – 5 days. The RegCM4 Climate Model had better results with the one from the Met Office Hadley Centre, HadGEM2-ES, when it was qualitatively compared with observational data (ERA Interim Model) of the campaign period and reanalysis data.

KEYWORDS: Climate prediction, Centro de Lançamento de Alcântara, RegCM4 Climate Model.

INTRODUCTION

This preliminary study investigates the use of climate model RegCM4 in the generation of future estimates, average vertical wind profiles near the surface regions in the Centro de Lançamento de Alcântara (CLA) region. An estimate of the vertical profile of the average wind behavior in advance of a month or more, revealing features such as frequency and intensity of high wind that represents a very useful intra-seasonal weather scale which has direct strategic influence in aerospace activities.

The use of climate modeling allows temporal extrapolation and climate forecasting of meteorological variables that are of operational interest in the planning of aerospace activities. The importance for the climatological knowledge of Brazilian tropical region, where the CLA, Maranhão State, is located, is due to the influence of meteorological phenomena in the assembly steps and launching of rockets and/or space vehicles, and includes such conditions as strong winds, precipitation and atmospheric electrical discharges (Marques and Fisch 2005).

The use of climatic modeling at the moment has a certain degree of development, as there are studies with other models such as the Brazilian the Regional Atmospheric Modeling (BRAMS), ETA model and Climate extension - Wheather Research and Forcasting (CWRF) for climatic purposes (Freire et al. 2015; Chou et al. 2011; Lo et al. 2008; Nicolini et al. 2004).

The climatic modeling using the RegCM4 climate model is already well developed; at present, there are significant studies in South America, studies on the variability of the convection

1.Departamento de Ciência e Tecnologia Aeroespacial – Instituto de Aeronáutica e Espaço – Divisão de Ciências Atmosféricas – São José dos Campos/SP – Brazil.
2.Universidade Federal de Pelotas – Faculdade de Meteorologia – Pelotas/RS – Brazil.

Author for correspondence: Cleber Souza Corrêa | Departamento de Ciência e Tecnologia Aeroespacial – Instituto de Aeronáutica e Espaço – Divisão de Ciências Atmosféricas | Praça Marechal Eduardo Gomes, 50 – Vila das Acácias | CEP: 12.228-901 – São José dos Campos/SP – Brazil | Email: clebercsc@iae.cta.br

process in Amazonia and northeastern Brazil (De Souza *et al.* 2016) and climatic change on the South America on the changes in rainfall variability (Da Rocha *et al.* 2012; Reboita *et al.* 2010); CMIP5 of valuation models that best express the cyclonic vortices of activity at high levels (VCANS) in northeastern Brazil (Pinheiro *et al.* 2014) and other regions as in Europe (Alexandri et al. 2015; Giorgi *et al.* 2012; Artale *et al.* 2010) and in Africa (Malavelle *et al.* 2011; Konaré *et al.* 2008).

Research has the strategy of using the RegCM4 climate model in this study aims to investigate and to provide fine scale regional information in the Brazilian Northeast, prospecting statistics and intensity of climatological information for the use of average values that characterize the period, that it is a dry season for the Brazilian northeast. The application of the RegCM4 model would assist in aerospace planning activities performed by the Departamento de Ciência e Tecnologia Aeroespacial (DCTA). The validation of this application of the RegCM model would enable a better understanding of the prevailing structures of atmospheric circulation at low levels and measuring the ability of RegCM4 model to represent the main features of the vertical wind profile for the CLA region.

METHODOLOGY
PHYSICAL AND DYNAMIC ASPECTS: ATMOSPHERIC CIRCULATION OVER AND NEIGHBORING THE REGION OF STUDY

The dynamic structure of the atmospheric circulation in Brazil's tropical region is directly linked to atmospheric processes and on the oceanic processes, specifically those that occur in Pacific and Atlantic Oceans. This structure can be described in two important mechanisms of planetary level: Hadley and Walker circulations (Walker and Bliss 1932; Moura and Shukla 1981; Johanson and Fu 2009).

According to Molion and Bernardo (2000), the tropical convection is essentially controlled — enhanced or inhibited — by the general circulation of the atmosphere, through the global scale phenomena, resulting from the complex interaction between the planet's surface and the atmosphere. Physical characteristics that particularly affect tropical convection are the distribution of continents and oceans, unequal power balance in the atmosphere, the different topographies and the various types of vegetation.

Ferreira and Melo (2005) showed that the atmospheric circulation over the tropical region is strongly modulated and modified by thermodynamic patterns in the basins of the

Pacific and Tropical Atlantic. Thus, in years in which there are anomalies, positive or negative, of the sea surface temperature (SST) in the basins of these oceans, the cell of Hadley, which operates in southern direction (ascending limb of the tropics and branches descendants in latitudes subtropical), and the cell of Walker, that works in the zonal direction (ascending branch in the western Pacific and descending branch in the eastern Pacific), are disturbed, causing strong anomalies in atmospheric circulation over the tropics. As a result, these cells are displaced from their climatological positions.

Although episodes of the *El Niño*-Southern Oscillation (ENSO), the Walker Cell, weaken, in a situation with circulation changes, result in the generation of an anomalous Hadley Cell circulation in the Pacific Ocean east, with upward movement within the tropics and downward movements in the upper troposphere, toward the poles (middle and high latitudes), with a return on such movements to the tropics in the lower troposphere. During these anomalies when atmospheric circulation year are set with *El Niño* (warm phase of ENSO), which represent the abnormal warming of the surface and sub-surface Equatorial Pacific Ocean. In South America, northern and northeastern Brazil, due to these changes in the Pacific Walker Cell, in *El Niño* years it is observed precipitation patterns with values below the climatological normal (Souza *et al.* 2009). Years with strong *El Niño* are extremely dry, as occurred in 2010 in the Brazilian Amazon.

In this definition, it is considered not only the presence of warm water in the Pacific the *Corriente El Niño*, but also changes in the atmosphere near to the ocean surface, with the weakening of the trade winds (which blow from east to west) in the region equatorial.

With the warming of the ocean and the weakening of the winds, atmospheric circulation changes begin to be observed in the low and high levels, causing changes in the moisture transport patterns and therefore are variations in the distribution of rainfall in tropical regions, middle and high latitudes. In some regions of the world, it is also observed increase or drop in temperature. Note that the surface winds in the equatorial region, in some cases, will even change direction, or are from west to east. There is a shift in the region with more cloud formation and the Walker Cell is split over the Pacific. In the Equatorial Pacific Ocean, warm waters can be observed in almost all its extension. The decrease in temperature per unit depth along the west coast of South America, mainly due to weakening of the trade winds, results in this sector getting warmer. In the Western Pacific Equatorial, temperature values of the surface of the sea are found to be bigger than average. Furthermore, the warmer water will be impounded more westerly than normaling greater evaporation, extending the upward movements, which in turn generate rain clouds, characterizing the Cell of Walker. The region with the larger amount of rainfall is the northeastern Indian Ocean, west of the Pacific Ocean through the Indonesia and the region with downward movements of the cell of Walker, occurs in the central and eastern Equatorial Pacific. Importantly, such downward movements of the cell of Walker in the Eastern Equatorial Pacific become more intense than usual, which inhibits the formation of rain clouds.

La Niña years, in South America and, more specifically, the northern region of Brazil, due to changes in the Walker Cell, present rainfall patterns with values above the climatological normal. Years with strong *La Niña* are extremely wet years in the northern region, as was observed during late 2011 and early 2012 in the Brazilian Amazon.

Wang (2002a,b) showed differences in the intensity and location of Walker and Hadley cells over the Pacific Ocean Tropical associated with the ENSO phenomenon in conjunction with the occurrence of the two main modes of thermal variability of sea surface temperature (SST) in the Tropical Atlantic Ocean (south gradient (north-south) and Equatorial (west-east)).

Vimont (2005) provided a plausible physical explanation for the spatial structure of decadal variability associated with ENSO, presenting on average the existence of long intercycle variations in the variability of ENSO, indicating the existence of outstanding features in spatial variations in decades with ENSO.

Its spatial variations are generated through the physical mechanisms that operate during the interannual ENSO cycle. Its findings provide an important framework for interpreting the modeling decadal variability associated with ENSO. In an analysis of the effect of the Hadley cell in South America, one can verify the existence of positive or negative anomalies in SST gradient in the Atlantic basin in the southern direction.

Nobre and Molion (1986) provides an explanation for the behavior of the Inter-Tropical Convergence Zone (ITCZ) associated with variations of SST in the Atlantic.

The position of the ITCZ and the circulation of the North Atlantic subtropical anticyclone influence the trade winds of northeast of the Brazil. The movement of ITCZ has intensified the northeast trade winds. When the North Atlantic waters are colder than normal, the subtropical North Atlantic anticyclone and the northeast trade winds intensify. If in this same period the South Atlantic is warmer

than usual, the South Atlantic subtropical anticyclone and the trade winds weaken southeast.

This pattern favors the shift of the ITCZ to positions further south of the Equator and is conducive to the occurrence of normal years, rainy or very rainy for the northern sector of the Northeast of Brazil.

When the South Atlantic waters are cooler than normal, the South Atlantic subtropical anticyclone and the trade winds from southeast intensify. If in this same period the waters in the North Atlantic are warmer than normal, the subtropical North Atlantic anticyclone and the northeast trade winds are weaken (Nobre and Shukla 1996; De Souza and Nobre 1998; Andreoli and Kayano 2007; Alves *et al.* 2012). This pattern favors the shift of the ITCZ to the northernmost position of the Equator and is contributory to the occurrence of dry or very dry years in the northern part of Northeast Brazil. As a result, monitoring of oceanic and atmospheric patterns during the rainy season is of fundamental importance for the weather and climate forecasts to be generated with the highest possible degree of reliability.

The interaction between the Walker and Hadley cells over the tropical Pacific associated with ENSO and jointly the occurrence of the two main thermal variability modes of the Tropical Atlantic Ocean (the southern and equatorial gradient of SST) create complex behaviors dynamics on the north and the Brazilian northeast. Besides, the existence of a breeze of wind field affects the CLA region. The use of climatic modeling to simulate the dynamics of the predominant circulation of the Alcântara region, it will be important to test the RegCM4 model will have the ability to represent this dynamic atmospheric condition.

STATE-OF-THE-ART MULTIMODEL DATASET: CMIP5

The fifth phase of the Coupled Model Intercomparison Project (CMIP5) has produced a state-of-the-art multimodel dataset designed to improve our knowledge of climate variability and climate change. Researchers worldwide are analyzing the model output and will produce results likely to underlie the forthcoming Fifth Assessment Report by the Intergovernmental Panel on Climate Change (IPCC-AR5). Unprecedented in scale and attracting interest from all major climate modeling groups, CMIP5 includes long-term simulations of twentieth-century climate and projections for the twenty-first century and beyond. Conventional atmosphere-ocean global climate models and Earth system models of intermediate complexity are for the first time being joined by more recently developed Earth system models under an experimental design that allows both types of models to be compared to observations on an equal footing. Besides the long-term experiments, CMIP5 calls for an entirely new suite of near-term simulations focusing on recent decades and the close future to the year 2035. These "decadal predictions" are initialized based on observations and will be used to explore the predictability of climate and to assess the forecast system's predictive skill. The CMIP5 experiment design also allows for the participation of stand-alone atmospheric models and includes a variety of idealized experiments that will improve understanding of the range of model responses found in the more complex and realistic simulations (Taylor *et al.* 2012). There were used in this study the scenarios of the IPCC/CMIP5, Representative Concentration Pathways (RCP) from the current concentration of CO_2, and the two projections of Emissions of Greenhouse Gases (EGG) used were RCP4.5 and RCP8.5. The RCP4.5 presents a trajectory of increasing atmospheric CO_2 concentration to 520 ppm (parts per million) in 2070, increasing more slowly by the end of the 21st century. The RCP8.5 shows a similar growth trajectory till the middle of the century, followed by a rapid and sharp increase, reaching a CO_2 concentration of 950 ppm at the end of the century.

Therefore, in this study, the Model RegCM4 with using the following data from the subset CMIP5, described below. The information concerning of the CMIP5 multi-model dataset can be seen in Table 1.

Dataset from 3 global models to perform simulations were obtained on the following internet site: http://clima-dods.ictp.it/data/Data/RegCM_Data. Each global model is a set of files with simulations, historical data and initialization parameters. The HadGEM2 model, for example, is a 2-stage development from HadGEM1, representing improvements in the physical model and the addition of earth system components and coupling. The HadGEM2 project targeted two key features of physical performance: ENSO and northern continent

land-surface temperature biases. The latter had a particularly high priority in order for the model to be able to adequately model continental vegetation. Through focused working groups, it was identified a number of mechanisms that improved the performance. Some known systematic errors in HadGEM1, such as the Indian monsoon, were not targeted for attention in HadGEM2, which has substantially improved mean SSTs and wind stress and improved tropical SST variability compared to HadGEM1. The northern continental warm bias in HadGEM1 has been significantly reduced. The power spectrum of *El Niño* is made worse, but other aspects of ENSO are improved. Overall, there is a noticeable improvement from HadGEM1 to HadGEM2 when comparing global climate indices, putting HadGEM2 in a leading position compared to the CMIP3 models. The HadGEM2 model has a publication describing the methodology used and the model structure (Collins *et al.* 2008).

The CNRM-CM5 model is described in Voldoire *et al.* 2011. The CNRM-CM5 includes the atmospheric model ARPEGE-Climat (v5.2), the ocean model NEMO (v3.2), the land surface scheme ISBA and the sea ice model GELATO (v5) coupled through the OASIS (v3) system. The main improvements since CMIP3 are the following. The horizontal resolution has been increased both in the atmosphere (from 2.8° to 1.4°) and in the ocean (from 2° to 1°). The dynamical core of the atmospheric component has been revised. A new radiation scheme has been introduced and the treatments of tropospheric and stratospheric aerosols have been improved. Particular attention has been devoted to ensuring mass/water conservation in the atmospheric component. The land surface scheme ISBA has been externalized from the atmospheric model through the SURFEX platform and includes new developments such as a parameterization of sub-grid hydrology, a new freezing scheme, and a new bulk parameterization for ocean surface fluxes. The ocean model is based on the state-of-the-art version of NEMO, which

Table 1. Subset of the CMIP5 multi-model dataset.

Modeling center	Model	Resolution	Institution
CNRM-CERFACS	CNRM-CM5	1.5° × 1.5°	Centre National de Recherches Météorologiques/Centre Européen de Recherche et de Formation Avancée en Calcul Scientifique
CSIRO-QCCCE	CSIRO-Mk3.6	1.8° × 1.8°	Commonwealth Scientific and Industrial Research Organization in collaboration with the Queensland Climate Change Centre of Excellence
MOHC	HadGEM2-ES	1.8° × 1.2°	Met Office Hadley Centre (additional HadGEM2-ES realizations with contribution of the Instituto Nacional de Pesquisas Espaciais (INPE)

has greatly progressed since the OPA8.0 version used in the CMIP3 version of CNRM-CM.

The CSIRO-Mk3.6 model is described by the following references for publication: Rotstayn *et al.* 2012 and Jeffrey *et al.* 2013. This global climate model (GCM) was developed from the earlier Mk3.5 version, which was described in detail by Gordon *et al.* (2002; 2010). It is a coupled atmosphere-ocean model with dynamic sea-ice. It also has a soil-canopy scheme with prescribed vegetation properties. The ocean, sea-ice and soil-canopy models are unchanged between Mk3.5 and Mk3.6. The main differences between Mk3.5 and Mk3.6 are the inclusion of an interactive aerosol treatment and an updated radiation scheme in Mk3.6. Rotstayn *et al.* (2010) gave an overview of Mk3.6, and also assessed the model's simulation of Australian mean climate and natural rainfall variability associated with ENSO, with generally favorable conclusions. The atmospheric component is a spectral model, which utilizes the flux form of the dynamical equations (Gordon 1981). It has eighteen vertical levels and horizontal resolution of approximately $1.875° \times 1.875°$ (spectral T63). The ocean model is based on version 2.2 of the Modular Ocean Model (MOM2.2) (Pacanowski 1996). Every atmospheric grid-box is coupled to 2 oceanic grid-boxes: enhanced north-south resolution in the ocean model was implemented with the aim of improving the representation of tropical variability. The ocean model thus has a resolution of approximately $1.875° \times 0.9375°$ and has 31 vertical levels. The sea-ice model is based on O'Farrell (1998), with revised numerics as described by Gordon *et al.* (2010).

The data reanalysis obtained from the ERA Interim Model was used for comparison of RegCM4 model and is available on the site http://apps.ecmwf.int/datasets/data/interim-full-daily/levtype=sfc/, on The European Centre For Medium-Range Weather Forecasts (ECMWF). The system includes a 4-dimensional variational analysis (4D-Var) with a 12-hour analysis window. It shows global atmospheric and surface parameters from 1 January 1979 to present, at T255 spectral resolution (~ 80 km) on 60 vertical levels, 6-hourly atmospheric fields on model levels, pressure levels, potential temperature and potential vorticity, 3-hourly surface fields and daily vertical integrals and Synoptic monthly averages at 0 UTC; 6 UTC; 12 UTC; 18 UTC. The detailed documentation of the parameters in this dataset is presented in Berrisford *et al.* (2011).

MODEL REGCM4: A REGIONAL CLIMATE MODEL SYSTEM

The Regional Climate Model System (RegCM4), originally developed at the National Center for Atmospheric Research (NCAR), is maintained in the Earth System Physics (ESP) section of the International Center for Theoretical Physics (ICTP). The latest version of model, RegCM4, is now fully supported by the ESP (http://gforge.ictp.it/gf/project/regcm/). This model can be applied to any region of the world, with a grid spacing of up to about 10 km (hydrostatic limit), and, for a wide range of studies, from process studies to paleoclimate and future climate simulation. This RegCM4 version include major upgrades in the structure of the code and its pre- and post-processors, along with the inclusion of some new physics parameterizations and upgrades of some physics schemes (convection, planetary boundary layer (PBL), and cloud microphysics) and uses Common Land Model (CLM) surface schemes (Giorgi *et al.* 2015). The RegCM4 was configured with a single grid with 80×100 points (Fig. 1), with a low resolution of 50 km and 23 vertical levels, with Normal Mercator projection, with the grid center focused on the CLA region.

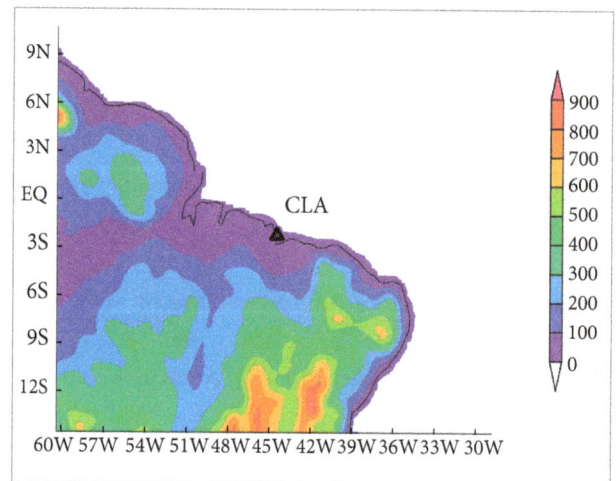

Figure 1. Domain (80 × 100 points) applied to generate the simulations centered on the region of the Centro de Lançamento de Alcântara, with data from the topographic surface, in meters, used by the RegCM4 model.

The namelist of terrain parameters was set to perform an extra smoothing in boundaries and a satellite soil moisture dataset (ESA CCI - SOILMOISTURE.nc) for initialization of soil moisture. The soil moisture data was generated by Europe Space Agency (ESA), which is part of the Climate Change Initiative

(CCI) project of the ESA Programme on Global Monitoring of Essential Climate Variables. The modeling physical parameters were set with the parameters which can be seen in Table 2. The physicsparam namelist controls, the best way to select the physical model is to follow the methodology described in the reference manual RegCM4 (Giorgi *et al.* 2011). Some RegCM4 parameters settings have been made different from the default, these differences are shown in Table 2. The nesting using the FNEST option in coreparam namelist in RegCM4 model, is supported in this configuration, hydrostatic nested into hydrostatic. The RegCM4 was configured to run a single grid with a spatial resolution of 50 km, the CMIP5 models used have a approximate resolution of 150 to 200 km.

Table 2. Parameters used in Physics parameters namelist in RegCM4 model.

Parameter	Value
Lateral boundary conditions scheme	Relaxation, exponential technique — Marbaix *et al.* (2003)
Planetary Boundary Layer (PBL) scheme	Holtslag PBL — Holtslag *et al.* (1990)
Cumulus convection scheme	
Over land	Emanuel (1991)
Over ocean	Emanuel (1991)
Moisture scheme	Explicit moisture Nogherotto/ Tompkins — Elguindi *et al.* (2013).
Ocean flux scheme	Zeng *et al.* (1998)
Zeng ocean model roughness formula to be used	1-> (0.0065*ustar*ustar)/egrav
Scenario	RCP4.5/RCP8.5

THE COMMUNITY LAND MODEL

The Community Land Model (CLM) (Oleson *et al.* 2008) is the land surface model developed by the National Center of Atmospheric Research (NCAR) as part of the Community Climate System Model (CCSM), described in detail in Collins *et al.* (2006). CLM contains 5 possible snow layers with an additional representation of trace snow and 10 unevenly spaced soil layers with explicit solutions of temperature, liquid water and ice water in each layer. To account for land surface complexity within a climate model grid cell, CLM uses a tile or mosaic approach to capture surface heterogeneity. Each CLM grid cell contains up to 4 different land cover types (glacier, wetland, lake, and vegetated), where the vegetated fraction

can be further divided into 17 different plant functional types. Hydrological and energetical balance equations are solved for each land cover type and aggregated back to the gridcell level. Since CLM was developed for the global scale, several input files and processes were modified to make it more appropriate for regional simulations, including the use of high-resolution input data, soil moisture initialization, and an improved treatment of grid cells along coastlines. For the model input data, CLM requires several time-invariant surface input parameters: soil color, soil texture, percent cover of each land surface type, leaf and stem area indices, maximum saturation fraction, and land fraction (Lawrence and Chase 2007). The CLM model has better performance than the Biosphere-Atmosphere Transfer scheme (BATS) model and the CLM model was chosen to use in this simulation. The RegCM4 model was installed on an Intel machine with XEON E5649 CPU, 36 GB of RAM in an environment with Linux Ubuntu 14.04, and the model was compiled with G-Fortran. The code uses MPI2 library (OpenMPI) and it was configured to use 24 threads to permit use the parallelism of underline processor. In the simulation, it was discarded two initial months, July and August, using the period September-November 2015 period for the study. For the post-processing of the result, it was used GrADS software for generation of the average wind vertical profile over the CLA area. By using the scripts the average.sh and dailymean.sh functions of the model postprocessing, the RegCM output data in netCDF were treated and generated the figures on GrADS format.

STATISTICS INFORMATION

It was used to statistical analysis calculate the mean and standard deviation of the time series of the months from September to November 2015, and it was also estimated the Mean Squared Error. Between the median values generated by RegCM4 model and data from anemometers of micrometeorological tower at four levels (70, 43, 28.5, and 11.78 m) at the launch area the CLA:

$$(MSE) = \frac{(X_M - X_O)^2}{N} \qquad (1)$$

where: X_M is the result of RegCM model at level $z = 995$ hPa with median wind for model integration; X_O is the mean value observed in each level, in similar hours; N ($N = 360$) is the maximum number of observations used time series.

The Root Mean Squared Error (RMSE) is an estimate of the magnitude of bias between the model and the observed data for the period analyzed. Equation 2 is used:

$$RMSE = \sqrt{\frac{(X_M - X_O)^2}{N}} \qquad (2)$$

RESULTS

Model RegCM4 was set to the tropical Atlantic coast region northeast of Brazil with good performance for the location of the CLA, which is in an area on the coast and where the prevailing atmospheric circulation is the ocean to the continent, and could represent the dynamics of predominant circulating in the dry the season for the region analyzed.

Figures 2 and three present the images generated by the 3 global models used to initialize the RegCM4 model. The best qualitative result was generated for the HadGEM2 model, an intermediate result was the CNRM-CM and the CSIRO-Mk3.6. Compared to the observational results of the same period with Era Interim Model (part 4A in Fig. 2 and part 4B in Fig. 3). The HadGEM2 model presented a temporal timing of maxima and minima of the average vertical wind profile for the month of November 2015, consistent with the dynamics of the observed circulation (part 3A in Fig. 2 and part 3B in Fig. 3). These periods with maxima and minima in the wind profile have a frequency of every 3 to 5 days. This situation indicates the possibility that the convective processes that occur over the ocean have lower fluctuations frequency than a week.

An interesting aspect that can be observed in the figures generated by the average vertical profile of the wind directions from the RegCM4 model is its change associated with wind

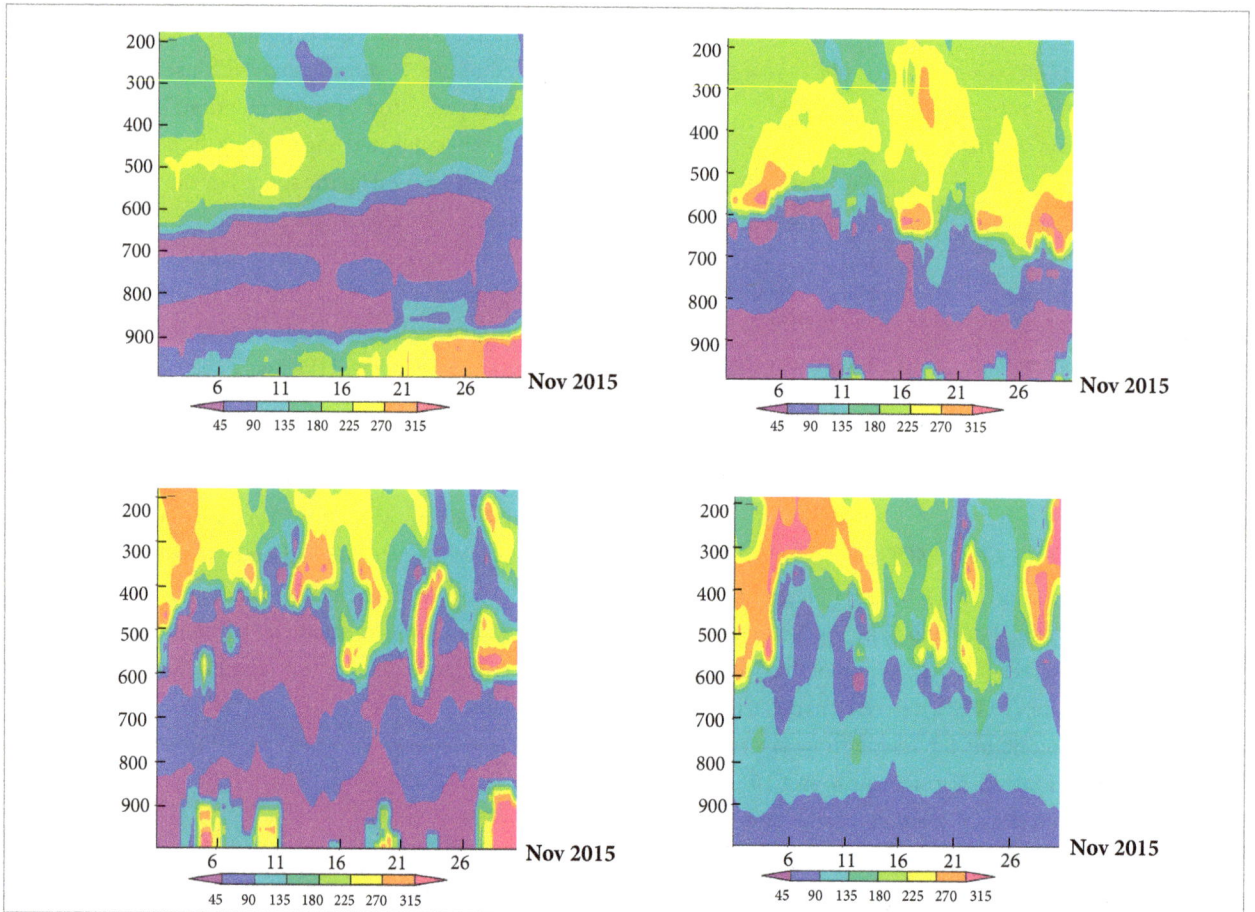

Figure 2. Results of the RegCM4 model with the RCP4.5 scenario, with the average wind direction profile (CNRM-CM5 –1A; CSIRO-Mk3.6.0 – 2A and HadGEM2-ES – 3A models) for the month of November 2015, and the average wind direction profile of ERA Interim Model for the same period (4A).

speed. When the wind is weaker, the predominant direction is from north/northwest, but when it is intense, it comes from northeast, whose direction is climatologically prevalent. This behavior is observed in the lower levels near the surface. The preliminary indications of wind direction median profiles, near the surface, suggest that the RegCM4 model has a certain north bias, meridionally, in future studies that possibility and its estimation will be investigated. The wind direction profile showed more complexity and the RegCM4 model is not represented as well relative to the intensity of the wind.

The vertical Y axis of the images in Figures 2 and 3 are in pressure levels, the RegCM4 model defaults on 23 levels, the level 23 near the surface corresponds to the level of 995 hPa and the level 1 value of 25 hPa in height top of the model. The figures were generated in the geographical position of the CLA and their variation in time, for the month of November 2015.

The simulations generated by RegCM4 with RCP8.5 scenarios originated outcomes associated with wind profile,

with magnitudes higher, well above the climatological reality in 3 different conditions of global initializations. In view of that, the simulations with RCP8.5 scenarios were discarded. The result obtained in RCP8.5 scenario was very sensitive to the initialization conditions, which are more intense and shows the effects of higher concentrations of greenhouse gases.

The results in Table 3 are interesting because the estimate generated by RegCM model in the period from September to November 2015 shows that the average wind velocity during this period was an estimate of the order of magnitude of the value of the average observation.

Table 4 shows the calculation of the RMSE between the mean values generated by the model RegCM4 and the mean values in the different heights of the anemometer tower, is an estimate of the magnitude of bias between them, of the order of 3 to 4 m/s, in the magnitude of wind velocity — a reasonable value in magnitude for the analyzed period.

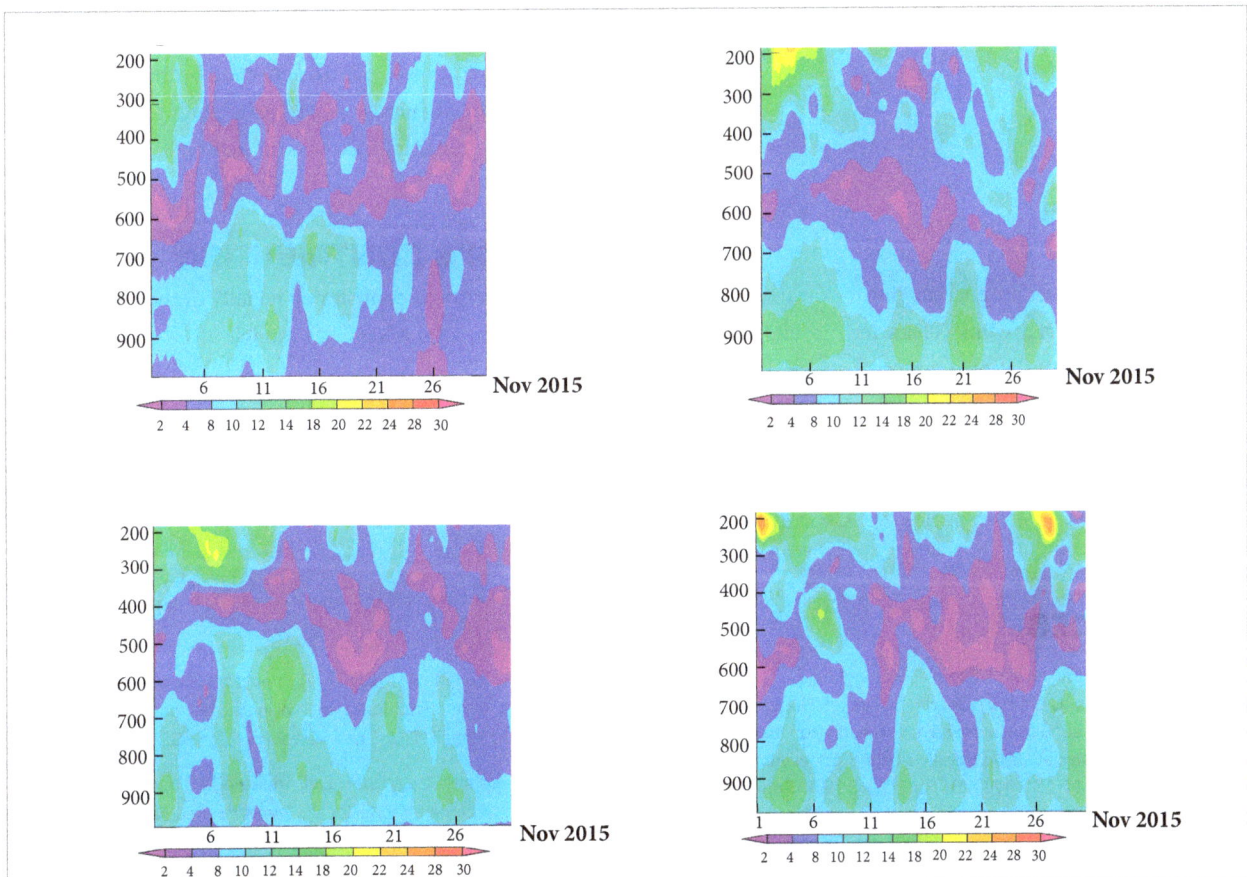

Figure 3. Results of the RegCM4 model with the RCP4.5 scenario, with the average of the vertical wind velocity profile (CNRM-CM5-1B, CSIRO-Mk3.6.0-2B, and HadGEM2-ES-3B models) for the month of November 2015, and the average vertical wind profile of ERA Interim Model for the same period (4B).

Table 3. Statistical information calculated from the average values of RegCM model (RCP4.5 scenario with the HadGEM-ES model) and average data of wind velocity at anemometer tower in Centro de Lançamento de Alcântara, in the period from September to November 2015.

RegCM model level 995 hPa	Wind average velocity (m/s)	Anemometer tower			
		70 (m)	43 (m)	28.5 (m)	16.3 (m)
Mean	7.44	9.00	7.78	7.25	6.50
Standard deviation	2.98	2.18	2.00	1.84	1.64

Table 4. Mean Squared Error and Root Mean Squared Error results in the period from September to November 2015, between the RegCM model (RCP4.5 scenario with the HadGEM-ES model) and the observed data in the anemometer tower in Centro de Lançamento de Alcântara.

Anemometer tower (m)	Mean Squared Error	Root Mean Squared Error
70	17.65	4.20
43	13.80	3.71
28.5	12.68	3.56
16.3	11.78	3.43

DISCUSSION

Preliminary results are encouraging for the use of climatic modeling for prospect dynamics characteristics and persistence in atmospheric circulation features, as well as the wind direction, velocity and frequency, and persistence with maxima and minima, which has a great operational role in the release of activities in the CLA region, provide fine-scale regional information. The RCP4.5 scenarios (IPCC/CMIP5) can have a very interesting application for the use of climatic modeling when applied in simulations that attempt to recognize prevailing characteristics of atmospheric circulation, for time intervals of 3 to 6 months. The months from October to November are ideal for the rocket launch operation in northeastern Brazil. This guide the choice of the months for generate up the study in more detail.

This methodology will use, through global reanalysis initializations, more detailed regional climatic parameters in the region of interest in CLA, allowing to obtain climatological results of past periods and the use of IPCC scenarios studies to quantify possible influences in wind strength, with the changes observed in the increase of the greenhouse effect.

The RegCM4 model can be used with different physical parameterizations and also generate grids with higher spatial resolution and can use the non-hydrostatic configuration of order 10 km or lower. This first test was significant as it showed a great potential for use, but it can be made further advances and improvements to future research. This paper shows a possible application scenarios generated by the IPCC/CMIP5 and productions of bias estimates for the study region to provide fine-scale regional information in the Brazilian Northeast.

AUTHOR'S CONTRIBUTION

All the authors contributed equally in the configuration and modeling of the RegCM4 model.

ACKNOWLEDGEMENTS
The authors thank the support of the Instituto de Aeronáutica e Espaço.

REFERENCES

Alexandri G, Georgoulias AK, Zanis P, Katragkou E, Tsikerdekis A, Kourtidis K, Meleti C (2015) On the ability of RegCM4 regional climate model to simulate surface solar radiation patterns over Europe: an assessment using satellite-based observations. Atmos Chem Phys 15:13195-13216. doi:10.5194/acp-15-13195-2015

Alves JMB, Souzal EB, Costa AA, Martins ESPR, Silva EM (2012) Sobre o sinal de um downscaling dinâmico às oscilações intrassazonais de precipitação no setor norte do Nordeste do Brasil. Rev Bras Meteorol 27(2):219-228. doi: 10.1590/S0102-77862012000200008

Andreoli RV, Kayano MT (2007) A importância relativa do atlântico tropical sul e pacífico leste na variabilidade de precipitação do Nordeste do Brasil. Rev Bras Meteorol 22(1):63-74.

Artale V, Calmanti S, Carillo A, Dell'Aquilla A (2010) An atmosphere-ocean regional climate model for the Mediterranean area: assessment of a present climate simulation. Climate Dynamics 35:721-740.

Berrisford P, Dee DP, Poli P, Brugge R, Fielding K, Fuentes M, Kallberg PW, Kobayashi S, Uppala S, Simmons A (2011) The ERA-Interim archive Version 2.0. ERA Report Series. The European Centre for Medium-Range Weather Forecasts.

Collins WD, Bitz CM, Blackmon ML, Bonan GB, Bretherton CS, Carton JA, Chang P, Doney SC, Hack JJ, Henderson TB, Kiehl JT, Large WG, McKenna DS, Santer BD, Smith RD (2006) The Community Climate System Model version 3 (CCSM3). Journal of Climate 19:2122-2143.

Collins WJ, Bellouin N, Doutriaux-Boucher M, Gedney N, Hinton T, Jones CD, Liddicoat S, Martin G, O'Connor F, Rae J, Senior C, Totterdell I, Woodward S, Reichler T, Kim J (2008) Hadley Centre technical note 74. Met Office Hadley Centre, Exter, UK; [accessed 2016 Nov 26]. http://www.metoffice.gov.uk/media/pdf/8/7/HCTN_74.pdf

Chou SC, Marengo JA, Lyra A, Sueiro G, Pesquero J, Alves LM, Kay G, Betts R, Chagas D, Gomes JL, Bustamante J, Tavares P (2011) Downscaling of South America present climate driven by 4-member hadCM3 runs. Climate Dynamics. doi: 10.1007/s00382-011-1002-8

Da Rocha RP, Cuadra SV, Reboita MS, Kruger LF, Ambrizzi T, Krusche N (2012) Effects of RegCM3 parameterizations on simulated rainy season over South America. Climate Research 52:253-265.

De Souza EB, Nobre P (1998) Uma revisão sobre o padrão de dipolo no Atlântico Tropical. Rev Bras Meteorol 13(3):1-44.

De Souza EB, Carmo AMC, Moraes BC, Nacif A, Ferreira DBS, Rocha EJP, Souza PJOP (2016) Sazonalidade da precipitação sobre a Amazônia Legal Brasileira: clima atual e projeções futuras usando o modelo RegCM4. Revista Brasileira de Climatologia 18. doi: 10.5380/abclima.v18i0.43711

Elguindi N, Bi X, Giorgi F, Nagarajan B, Pal J, Solmon F, Giuliani G (2013) Regional Climate Model RegCM user's manual version 4.4. Trieste: The Abdus Salam International Centre for Theoretical Physics.

Emanuel KA (1991) A scheme for representing cumulus convection in large-scale models. J Atmos Sci 48:2313-2329. doi: 10.1175/1520-0469(1991)048<2313:ASFRCC>2.0.CO;2

Freire JLM, Freitas SR, Coelho CAS (2015) Calibração do modelo regional BRAMS para a previsão de eventos climáticos extremos. Rev Bras Meteorol 30:158-170.

Ferreira AG, Mello NGS (2005) Principais sistemas atmosféricos atuantes sobre a Região Nordeste do Brasil e a influência dos oceanos Pacífico e Atlântico no clima da região. Revista Brasileira de Climatologia 1(1):15-28.

Giorgi F, Copola E, Solmon F (2012) RegCM4: model description and preliminary tests over multiple CORDEX domains. Climate Research 52:7-29.

Giorgi F, Elguindi N, Cozzini S, Giuliani G (2011) Regional Climatic Model RegCM user's manual version 4.2. Trieste: International Centre for Theoretical Physics.

Giorgi F, Elguindi N, Cozzini S, Solmon F (2015) Regional Climatic Model RegCM User's Guide Version 4.4..

Gordon HB (1981) A flux formulation of the spectral atmospheric equations suitable for use in long-term climate modeling. Mon Weather Rev 109:56-64.

Gordon HB, Rotstayn LD, McGregor JL, Dix MR, Kowalczyk EA, O'Farrell SP, Waterman LJ, Hirst AC, Wilson SG, Collier MA, Watterson IG, Elliott TI (2002) The CSIRO-Mk3 Climate System Model. Technical Paper No. 60. CSIRO Atmospheric Research; [accessed 2016 Nov 26]. http://www.cmar.csiro.au/e-print/open/gordon_2002a.pdf

Gordon HB, O'Farrell SP, Collier MA, Dix MR, Rotstayn LD, Kowalczyk EA, Hirst AC, Watterson IG (2010) The CSIRO-Mk3.5 Climate Model. Technical Report No. 21. Centre for Australian Weather and Climate Research; [accessed 2016 Mar 16]. http://www.cawcr.gov.au/technical-reports/CTR_021.pdf

Holtslag AAM, De Bruijn EIF, Pan H-L (1990) A high resolution air mass transformation model for short-range weather forecasting. Mon Wea Rev 118:1561-1575. doi: 10.1175/1520-0493(1990)118<1561:AHRAMT>2.0.CO;2

Jeffrey S, Rotstayn L, Collier M, Dravitzki S, Hamalainen C, Moeseneder C, Wong K, Syktus J (2013) Australia's CMIP5 submission using the CSIRO Mk3.6 model. Australian Meteorological and Oceanographic Journal 63:1-13.

Johanson CM, Fu Q (2009) Hadley cell widening: model simulations versus observations. Journal of Climate 22:2713-2725.

Konaré A, Zakey AS, Solmon F, Giorgi F, Rauscher S, Ibrah S, Bi X (2008) A regional climate modelling study of the effect of desert dust on the West African monsoon, J Geophys Res 113:D12206. doi: 10.1029/2007JD009322

Lawrence P, Chase T (2007) Representing a new MODIS consistent land surface in the Community Land Model (CLM3.0). J Geophys Res 112:g01023.

Lo JCF, Yang ZL, Pielke RA (2008) Assessment of three dynamical climate downscaling methods using the Weather Research and Forecasting (WRF) model. J Geophys Res 113:D09112. doi: 10.1029/2007JD009216

Malavelle F, Pont V, Mallet M, Solmon F, Johnson B, Leon JF, Liosse C (2011) Simulation of aerosol radiative effects over West Africa during DABEX and AMMA SOP-0. J Geophys Res 116:D08205. doi: 10.1029/ 2010JD014829

Marbaix P, Gallee H, Brasseur O, van Ypersele JP (2003) Lateral boundary conditions in regional climate models: a detailed study of the relaxation procedure. Mon Wea Rev 131:461-479.

Marques RFC, Fisch GF (2005) As atividades de Meteorologia Aeroespacial no Centro Técnico Aeroespacial (CTA). Boletim da Sociedade Brasileira de Meteorologia 29(3):21-25.

Molion LCB, Bernardo SO (2000) Dinâmica das chuvas no Nordeste Brasileiro. Proceedings of the 11th Congresso Brasileiro de Meteorologia; Rio de Janeiro, Brazil.

Moura AD, Shukla J (1981) On the dynamics of droughts in northeast Brazil: observations, theory and numerical experiments with a general circulation model. Journal of the Atmospheric Science 38(12):2653-2675.

Nobre CA, Molion LCB (1986) Climanálise Especial. Edição Comemorativa de 10 anos. Cachoeira Paulista: CPTEC/INPE.

Nobre P, Shukla J (1996) Variations of sea surface temperature, wind stress, and rainfall over the tropical Atlantic and South America. J Climate 9(10):2464-2479. doi: 10.1175/1520-0442 (1996)009<2464:VOSSTW>2.0.CO;2

Nicolini M, Salio P, Ulke G, Marengo J, Douglas M, Paegle J, Zipser E (2004) South American low-level jet diurnal cycle and three-dimensional structure. CLIVAR Exchanges. Newsletter of the Climate Variability and

Predictability Programme (Special Issue Featuring SALLJEX). N° 29, Vol. 9, N° 1, 6-8 y 16; [accessed 2016 Nov 28]. http://www.clivar. org/publications/exchanges

O'Farrell SP (1998) Investigation of the dynamic sea-ice component of a coupled atmosphere-sea-ice general circulation model. J Geophys Res 103:15751-15782.

Voldoire A, Sanchez-Gomez E, Salas y Mélia D, Decharme B, Cassou C, Sénési S, Valcke S, Beau I, Alias A, Chevallier M, Déqué M, Deshayes J, Douville H, Fernandez E, Madec G, Maisonnave E, Moine MP, Planton S, Saint-Martin D, Szopa S, Tyteca S, Alkama R, Belamari S, Braun A, Coquart L, Chauvin F (2011) The CNRM-CM5.1 global climate model: description and basic evaluation. Climate Dynamics 40(9):2091-2121. doi: 10.1007/s00382-011-1259-y

5

Visual Experimental and Numerical Investigations Around the VLM-1 Microsatellite Launch Vehicle at Transonic Regime

Henrique Oliveira da Mata[1], João Batista Pessoa Falcão Filho[2], Ana Cristina Avelar[2], Leonardo Motta Maia de Oliveira Carvalho[2], João Luiz F. Azevedo[2]

ABSTRACT: It is performed and presented an experimental and numerical investigation over the flow patterns around the forebody section of a microsatellite launch vehicle in development at Instituto de Aeronáutica e Espaço. The experimental investigation with a VLM-1 model in 1:50 scale is carried out at the Brazilian Pilot Transonic Wind Tunnel, located in the Aerodynamics Division of the mentioned Institute, using the classical *schlieren* flow visualization technique. *Schlieren* images are obtained for nominal Mach number varying from 0.9 to 1.01. Numerical simulation using Stanford's SU2 code is conducted together with the experimental investigation in order to improve the understanding of the complex physical phenomena associated with the experimental results of this particular regime. The combination of the 2 techniques allowed the assessment of some important aspects on the flow field around the vehicle in the conditions considered in this study, such as shock wave/boundary-layer interaction. The numerical simulation is also very important, allowing the quantification of some important parameters and confirming the shock wave formation patterns observed in the simulation when compared with the *schlieren* images. A good agreement regarding the position of the shock wave, when compared with the *schlieren* images, with a maximum error of about 6%, is observed over the VLM model.

KEYWORDS: Sounding rocket, Transonic Wind Tunnel, Experimental results, *Schlieren* images, CFD.

INTRODUCTION

The VLM-1 microsatellite launch vehicle is a project being developed at the Instituto de Aeronáutica e Espaço (IAE) since 2010 in partnership with the German Space Center (Deutsches Zentrum für Luft- und Raumfahrt e. V. — DLR) (Da Mata 2013). The idea was based on an up-to-date market analysis related to the applicability, frequency of launches, and price of microsatellites in the technological development. Among the applications the most important are the support and/or complementation of the present usages of large-scale satellites and the provision of better assistance for short-time revisiting spatial installations, allowing low-cost missions both related to launcher ground platform as well as design and manufacture of the prototype. The US Federal Aviation Administration (FAA) reports an average rate of 3 microsatellites launches per year, but other studies suggest more than 20 annual launches in a near future, creating a high demand that cannot be supplied anymore by old and costly large-scale projects, such as old missile-based large-scale Russian satellites and some expensive new projects like Angara, PSLV, and Falcon 1. A new era of microsatellite projects has emerged in the last 10 years, as confirmed by the ICBM-based vehicles from Russia, such as Dnepr, Cosmos 3M, and Start, as well as the American Pegasus. Therefore, with the development of the VLM-1 vehicle, the IAE has the possibility of putting Brazil into the promising market of microsatellites launching.

1.Departamento de Ciência e Tecnologia Aeroespacial – Centro de Lançamento de Alcântara – Seção de Segurança de Vôo – Alcântara/MA – Brazil. **2.**Departamento de Ciência e Tecnologia Aeroespacial – Instituto de Aeronáutica e Espaço – Divisão de Aerodinâmica – São José dos Campos/SP – Brazil.

Author for correspondence: Ana Cristina Avelar | Departamento de Ciência e Tecnologia Aeroespacial – Instituto de Aeronáutica e Espaço – Divisão de Aerodinâmica | Praça Marechal Eduardo Gomes, 50 – Vila das Acácias | CEP: 12.230-904 – São José dos Campos/SP – Brazil | Email: anacristina.avelar@gmail.com

The first configuration of the VLM was inspired by a successful Israeli rocket named Shavit. Although there are many configurations with different payloads, types of engines, and missions, the baseline vehicle consists of 3 stages and a payload that can vary from 140 to 350 kg. The main dimensional parameters of the adopted model configuration in 1:50 scale are defined in the scheme of Fig. 1. Its geometry is quite simple and aerodynamic, as it is expected a low drag coefficient because of the well-shaped nose and aspect ratio of 12.3 in a cylindrical body.

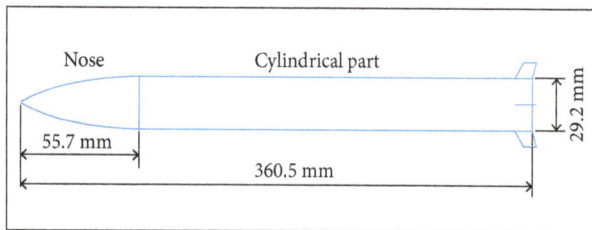

Figure 1. Main dimensions of the model adopted in 1:50 scale.

As the vehicle is developed in partnership with DLR, its evolution has certain particularities, both in organizational part and in research/manufacturing process. Some difficulties during the project must be solved by proving that the correct choice was adopted. One example is the fin conception. In the preliminary conception of the vehicle, no fins were predicted, and the rocket control would be done only by the nozzle. However, during the separation of the first and the second stages, the vehicle loses its control because the 1st-stage thrust is already terminated. In order to solve the problem, the Brazilian technical team proposed a set of fins, and it is up to the IAE staff to prove good stability characteristic of the vehicle, preserving a lower total drag.

In order to assess the aerodynamic behavior of the vehicle and to support other design decisions regarding many aspects of the rocket design, such as dynamic stability and aerodynamic loads, a test campaign was planned to take place in the Brazilian Pilot Transonic Wind Tunnel (TTP), located in the Aerodynamics Division (ALA) of IAE.

The good quality of *schlieren* image collection obtained during the tests, although very helpful for physical phenomenon assessment, was not considered sufficient for the deep understanding of the regime needed by the engineers during their design decisions. In order to improve such understanding, the examination of the problem with another methodology was taken into account. A useful tool that can give information about the whole flow field is based on the numerical techniques applied to the Reynolds-Averaged Navier-Stokes (RANS) equations. The

so-called Computational Fluid Dynamics (CFD) approach, which is nowadays sufficiently matured for rocket design purposes, proved to be the right choice. Stanford's SU2 code (Economon *et al.* 2016) was used to perform the numerical simulations. The RANS equations were solved using the Spalart-Allmaras model (Spalart and Allmaras 1992) to account for the turbulence closure required by the regime of the VLM-1 model (Wilcox 2006). The present paper presents and analyzes the *schlieren* images comparing them with numerical simulation results obtained with the SU2 code. The important insights obtained when using these tools are presented in the following sections.

Fluid flow forms complicated patterns that can only be completely understood with adequate and complementary approaches. The forming flow patterns can become quite unexpected, and it is almost never possible to predict the real flow characteristics relying just on classical aerodynamic theories. So other tools are always needed. One of the most efficient is Flow Visualization, or the direct observation of the flow field. Visualization is an important tool in establishing flow models as a basis for mathematical simplifications. It can be used for the direct solution of engineering problems, as well as to get insights about the concepts of fluid motion (Kline 1943). Although considered many times only as a qualitative method, Flow Visualization is of extraordinary value as it can reveal flow parameter behavior for the entire field. Conventional measuring techniques applied to real problems are generally very limited, because of both the model construction difficulties and the intrusiveness consequence, interfering with the flow pattern. Thus, it is worth noting the importance of experiments in flow visualization because it can assess flow properties for the entire field, which is almost impossible using conventional flow measurement techniques based on discrete points.

Since the advent of sufficiently powerful computers, in the last decades a considerable expansion in flow analysis has been observed as performed by the fluid mechanics academic community. Although CFD solutions can sometimes be considered inaccurate due to its deep numerical nature, it can still reveal some important information about the entire flow field. The best way one can imagine the application of this methodology is to verify its use for specific experimental study cases, improving the capabilities of both techniques: Flow Visualization and CFD. Therefore, the present study compared *schlieren* and CFD results in order to investigate the better physical phenomena in transonic regime for the VLM-1 vehicle model.

THEORETICAL APPROACH

Most of the simple geometry vehicles, such as the VLM-1, present a very typical aerodynamic flow pattern at transonic regime. Figure 2 shows the vehicle at typical transonic flow, when one can distinguish 4 different regions in which the boundary-layer (in dashed blue line) and the supersonic region (in dashed red line) with a shock wave at its end (in solid red line) play important roles. The distinctive regions are: (I) far upstream the vehicle the flow is at undisturbed condition; (II) approaching to the vehicle and along part of its nose the flow feels the presence of the model nose, rising the local pressure level as the boundary-layer starts developing from the nose tip; (III) to overcome the imposed geometry, at some point on the vehicle surface, and before the cone end, the Mach number is sonic because a supersonic region was created, ending with an approximately normal shock wave; (IV) after the shock wave the flow tends to return to the undisturbed condition.

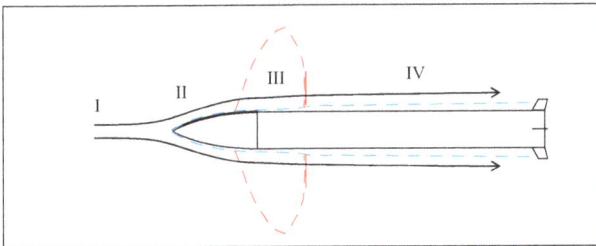

Figure 2. Diagram of the VLM-1 at typical transonic flow, showing the supersonic region and its interaction with the boundary-layer developed over the model surface.

The way the boundary-layer reacts by the impingent shock wave determines the flow locally. The interaction between the shock wave and the boundary-layer has been the topic of much scientific research since the 1950s (Dolling 2001). The diagram of Fig. 2 indicates how the boundary-layer grows significantly at the shock wave impinging point, which is caused by the obvious pressure level increase due to the shock wave. Depending on local characteristics of the flow, that is, whether it is laminar or turbulent, and the shock wave strength, the boundary-layer reacts differently, which may cause multiple shock wave formation, delta formation, as well as the boundary-layer may detach and later on being reattached or not.

Some special cases can be categorized considering weak or strong shock wave impinging over laminar or turbulent boundary-layer. At transonic regime, when the undisturbed flow condition is subsonic, a supersonic region will appear, due to the geometric constraints, and its boundaries will be generally limited by a sonic line and by a normal shock wave at the end, where the supersonic flow returns to subsonic flow conditions. However, the boundary-layer along the vehicle surface has always a subsonic region very close to the wall, and it allows the characteristic parameters of the flow travel upstream. Across the shock wave, the boundary-layer will experience a sudden adverse pressure which will cause its growth locally. The supersonic flow external to the boundary-layer will react as if a geometric change had occurred, giving rise to compression and expansion waves. Depending on the strength of the shock wave, the boundary-layer will react differently, which will change the flow pattern outer the boundary-layer and the shock wave structure.

WEAK SHOCK WAVE IMPINGING ON LAMINAR BOUNDARY-LAYER

Figure 3 shows a diagram of the typical result of a weak shock wave impinging over a laminar boundary-layer. The geometry considered is the VLM-1 front part. In the figure, thin solid lines represent compression waves and the thick solid ones, the shock waves; dashed lines represent expansion waves and the dotted ones, sonic lines. In this case, the laminar boundary-layer is greatly affected by a shock wave and responds by increasing its thickness. Since the shock wave is weak, it is not capable of causing the boundary-layer complete detachment, although a small recirculation region may appear at the base of the stronger shock wave. The combination of weak shock wave with weak boundary-layer provides the appearance of a first shock wave followed by a number of weaker shock waves (Houghton and Carpenter 2003).

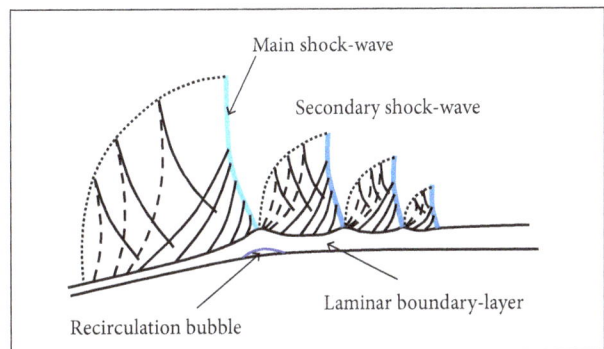

Figure 3. Weak shock wave/laminar boundary-layer interaction. Diagram modified from Houghton and Carpenter (2003).

As the flow travels along the vehicle surface, it reaches sonic condition at some point, and then it enters in a supersonic region where a sonic line and a shock wave are the constraints. Inside this region, outer of the boundary-layer, the flow behaves like a potential flow, where viscous forces are almost negligible. As the region expands geometrically, the flow outside the boundary-layer accelerates until a shock wave abruptly decelerates it to subsonic condition.

Nevertheless, inside the supersonic region, there are 2 facts worth noting. The first one is that, at the beginning, the boundary-layer growth is small and its outer limit defines a convex line for the flow above the boundary-layer. This frontier creates a series of expansion waves, which propagates upward until reaching the sonic line. As the expansion waves reaches the sonic line it returns from it as a compression wave, being propagated downward on the boundary-layer. The expansion and the compression waves are curved lines because of the so many interactions with all other waves.

The second fact is the presence of compression waves at some point downstream, because of the impressive boundary-layer growth, which is due to the adverse pressure rise imposed by the impinging shock. Depending on the strength of the shock wave, a recirculation bubble can occur.

The subsonic flow after the shock wave experiences a new expansion due to the accommodation of the boundary-layer outer limit, and it may accelerate to supersonic regime, as indicated in Fig. 3. Therefore, the same supersonic flow region pattern can be repeated, but now with smaller intensity. The same behavior is repeated until the flow has finally settled down to the undisturbed condition.

STRONG SHOCK WAVE IMPINGING ON LAMINAR BOUNDARY-LAYER

The main difference when compared with the last case is that the strong shock wave perturbs the laminar boundary-layer so much that it generally changes to turbulent, and sometimes it may cause even its complete detachment. Figure 4 shows in a diagram the flow pattern for this case. Because the first shock wave is very strong the other shock waves will disappear, only remaining a weak shock wave. A laminar recirculation bubble is likely to occur in this case, which normally will provoke the changing to turbulent flow regime ahead. It can be observed that the same pattern is found inside the supersonic regions where, whenever the outer flow undergoes, convex geometry expansion waves and concave geometry compression waves appear.

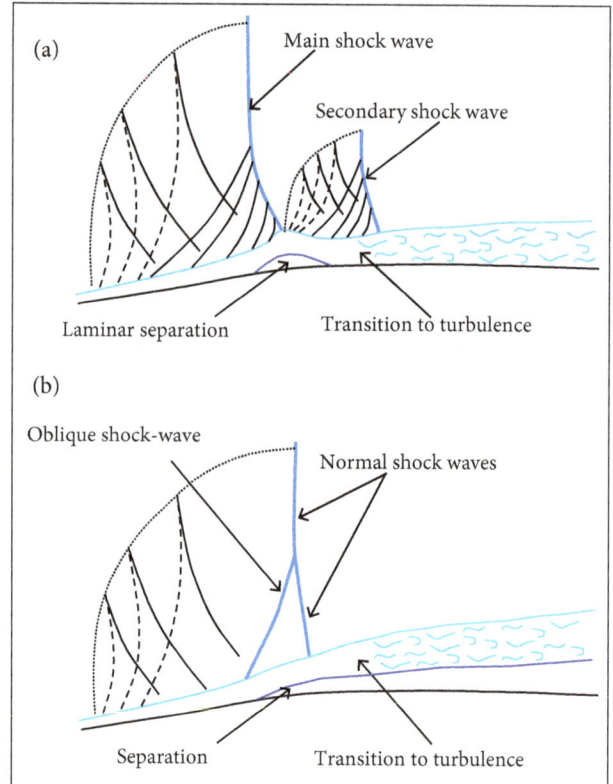

Figure 4. Strong shock wave impinging on laminar boundary-layer, resulting in turbulent boundary-layer: (a) With flow reattachment and (b) Completely separated flow. Diagrams modified from Houghton and Carpenter (2003).

STRONG SHOCK WAVE IMPINGING ON TURBULENT BOUNDARY-LAYER

Based on experiments, it is observed that the necessary force to detach a laminar boundary-layer is considerably lower than for a turbulent boundary-layer (Houghton and Carpenter 2003). For the present study cases, the most likely situation to occur in the presence of turbulent boundary-layer is the one with reattachment.

Some of the physical phenomena described herein were confirmed by the experiments undertaken in TTP, using the *Schlieren* Technique. Nevertheless, numerical simulations were also developed to better understand them by assessing other flow parameters that are not possible to observe just through *schlieren* images.

EXPERIMENTAL AND NUMERICAL APPROACHES
THE PILOT TRANSONIC WIND TUNNEL

The TTP is a scaled-down version (1:8) from an industrial transonic project idealized in the 1980s by the IAE to provide

Brazil with a transonic facility to support the aeronautical development of the country, reaching strategic goals of safety and up-to-date testing capabilities for new Brazilian aerospace projects. The transonic wind tunnel would be driven not only by a main compressor (continuously) but also by an injection system (intermittently), which would help to enlarge the operational tunnel envelope, without penalizing the installed power. Because of this new feature, a pilot transonic facility was also conceived in order to test this challenging idea (Falcão Filho *et al.* 2009). For several reasons, mostly related to budget restrictions, only the pilot facility design was completely built and it became fully operational in 2002.

TTP has also been built to train people and serve as a research tool for tests with profiles and models of simple geometries, like the aerospace vehicles. Figure 5a shows a view of the aerodynamic tunnel installed in a dedicated building, where one can see the tubing that connects the circuit with auxiliary compressors installed in another building to control tunnel pressure. Figure 5b presents the open plenum chamber showing the test section (in red) where the test article is installed and other tunnel components to establish the flow into the test section.

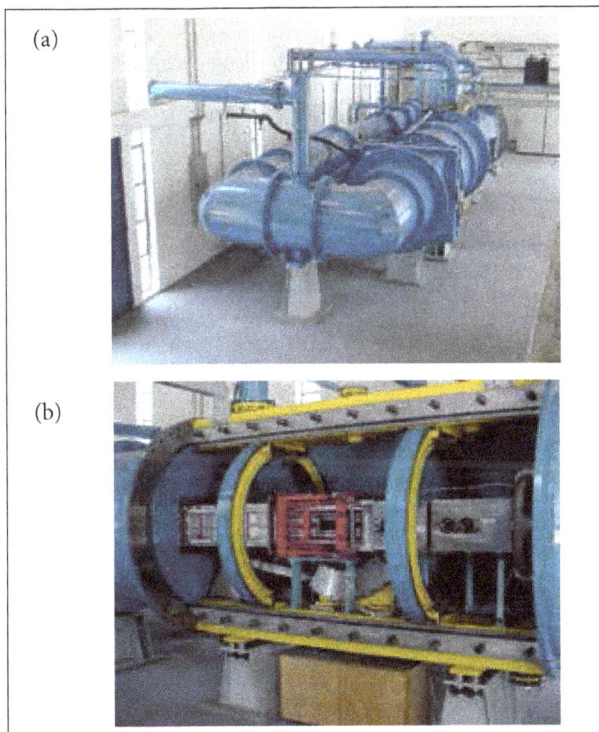

Figure 5. TTP installation: (a) Aerodynamic circuit, with 17 m long, showing tubing for auxiliary control systems; (b) Plenum chamber open showing the test section (in red), first and second throats.

TTP has a conventional closed circuit with test section of 0.30 m wide, 0.25 m high, and 0.81 m long, and it is continuously driven by a main axial compressor with 830 kW of power, attaining flows in the Mach number range from 0.2 to 1.2. An intermittent injection system operates in combined action with the main compressor reaching up to Mach number 1.3 during at least 30 s. Automatic controls of pressure (from 0.5 to 1.2 bar) and temperature guarantee stable operation to settle Mach and Reynolds numbers in the test section, which are the 2 parameters necessary to completely represent the flow for steady-state conditions (Barlow *et al.* 1999). In fact, the Reynolds number is quite difficult to settle in tests with small models, and even for industrial wind tunnels Reynolds numbers are normally of one order lower than those of real flight, requiring adaptations in the test procedure and in the model installation to diminish the so-called Reynolds number effects (Pope and Goin 1978).

Figure 6 shows a diagram of the plenum chamber with devices idealized to perfectly establish the flow into the transonic test section: the first throat which accelerates the flow coming from the stagnation chamber to the test section entrance; the slotted test section in which the article is mounted; the re-entry flap section which can change the angle and vary the mass flow through the slots; the second throat to adjust conditions whenever supersonic tests are performed; and the injector mixing chamber. There are ten injector beaks, which receive compressed air and operate in a choking condition at Mach number 1.9 to transfer momentum to the main stream. In Fig. 6 the curved arrows show the flow direction coming out from the test section and being re-admitted by the flaps by the pressure decrease due to Venturi effect. Normally the flow, represented by its streamlines, it deviates from the model in its proximity. In closed wall test section the streamlines are closer to the model than in real flight flow whilst in open wall test

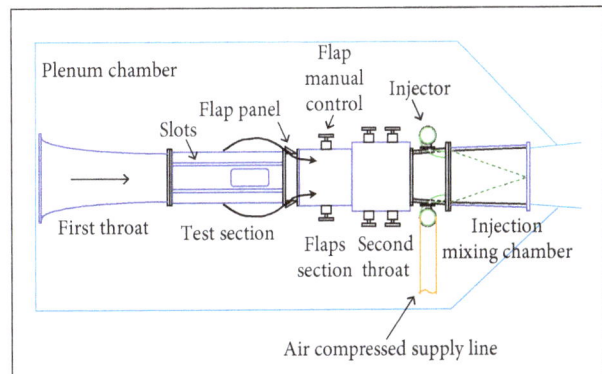

Figure 6. Detail of the plenum chamber inner parts.

section they are more distant from the model then in real flight. In both cases the perfect similarity between flow conditions in wind tunnel and in free flight are prevented. In a transonic semi-open wall test section, even with the flaps close, when the flow upstream the model feels its blockage it responds causing the streamlines to bulge out in an intermediate behavior, in some cases reproducing quite well the real flight condition (Goethert and Nelson 2007).

Figure 7 shows the tunnel operational envelope in terms of the numbers of Mach and Reynolds, for a typical chord of 27.4 mm, which represents 10% of the square-root of the cross sectional test section area (Davis *et al.* 1986). It is clear that the continuous operation of TTP goes up to Mach number 1.23 and to reach Mach number 1.3 it is necessary to use the injection system. The envelope describes all operational limits for the tunnel. It is important to emphasize that, for a fixed Mach number condition, it is possible to vary the Reynolds number, which is accomplished by varying the stagnation pressure of the test section. The Reynolds number variation can be used to give some insight about extrapolating procedures to the real flight condition for some characteristic parameters.

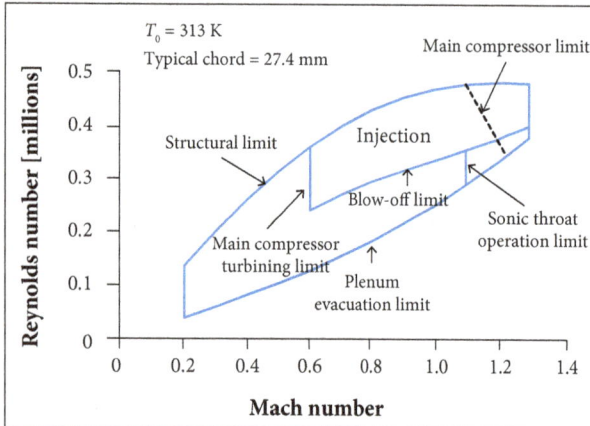

Figure 7. Operational envelope of TTP for stagnation temperature of 313 K and typical chord of 27.4 mm.

Some experiments undertaken in TTP were reported in Da Mata (2013), and they included determination of drag force with study of transition using Carborundum strips with different grit sizes, as stated in Pope and Goin (1978) from Mach number 0.3 to 1.15, and flow visualization using the *Schlieren* Technique to exploit the transonic regime from Mach number from 0.90 to 1.01.

Figure 8 shows the model installed inside the test section of TTP, from which the side wall was removed. It is possible

to see the longitudinal slots of the test section which will have its pressure equalized by the plenum chamber. The 3 basic objectives with the slots are to establish uniform transonic regime flow preventing choking, to diminish the shock and expansion wave reflections from the walls, and to control the mass flow through the walls to improve flow uniformity. The model is instrumented with a 6-component internal balance, and its diameter of 29.2 mm represents 0.9% of blockage area (ratio between the highest model cross sectional area and the test section area). According to Pope and Goin (1978), a blockage ratio below 1% assures negligible test section wall corrections, simplifying enormously the data reduction procedures. The model has four small fins at its base to allow the vehicle stabilization. For this particular test, a special device was used to fix the model in the sting support to move the model 30 mm upper from the central line of the test section, placing it practically at the end of the visualization window, allowing more free space to capture the shock wave formations.

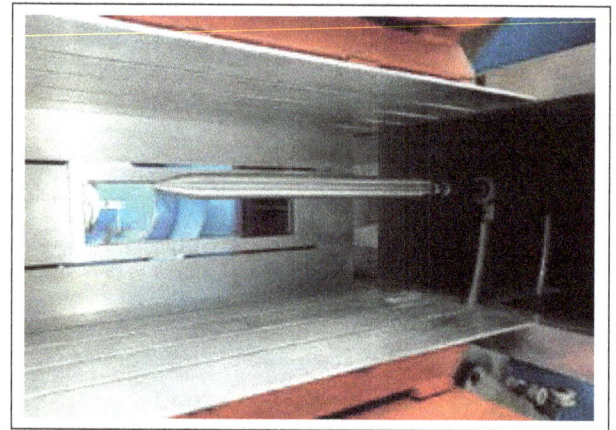

Figure 8. VLM-1 model in 1:50 scale installed in the TTP Test Section.

The TTP test section is relatively small and the scaled model constructed with 29.2 mm of diameter results in very low Reynolds numbers. The *schlieren* images were taken with the model without any transition strip and at 0° angle of attack. Therefore, it is expected that a laminar boundary-layer will extend for a relatively long region on the model surface. Table 1 shows the main parameters related to the test section of TTP for transonic regime, considering the characteristic length of 55.6 mm, which corresponds to the VLM-1 ogive length. Considering that the transition from laminar to turbulent flow in a flat plate will typically occur from 100,000 to 500,000, one can see that the boundary-layer will be laminar approximately up to half-way along the ogive for all experiments. Depending on where the shock wave is located, it is expected local laminar or turbulent boundary-layer characteristics at the foot of the shock.

Table 1. Typical transonic test parameters undertaken in TTP, with corresponding Reynolds numbers related to the ogive length (55.6 mm).

Mach number	Stagnation pressure (kPa)	Stagnation temperature (K)	Re (thousands)
0.7	94	303	658
0.8	94	303	708
0.9	94	303	746
1.0	94	303	773
1.1	94	303	788

SCHLIEREN TECHNIQUE

The experiments were undertaken using the *Schlieren* Technique. The method is based on flow visualization of the light intensity difference as function of the local density gradient in the flow field. For example, high density gradients indicating shock waves around the test object can be visualized by different intensity of light scattered in a recording plane. Figure 9 shows a diagram of the basic physical installation used. The point light source located precisely at the focal point of the first parabolic lens forms a parallel light beam that crosses the test object region. Different density regions refract differently the light beam because of the refrangibility degree variation in the local medium. Therefore, the image on the recording plane will show the high density regions with shadow close to more illuminated areas. The *Schlieren* Technique uses another parabolic lens to converge light rays at the focal point, where a knife edge is approximated, and, as its location is altered, it changes the image contrast, as it blocks some of the light rays which were diverted from the focal point. The combination of the two parabolic lenses produces an image with illumination rate as function of density gradient. Figure 9 shows in dashed lines 2 light rays which experienced symmetric deflections because

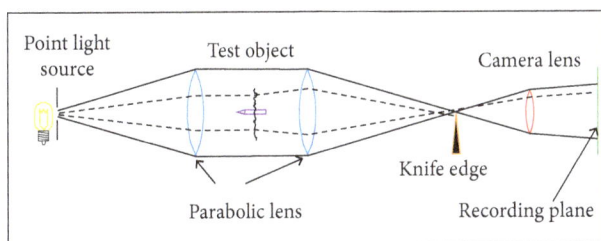

Figure 9. Installation diagram of the *Schlieren* Technique.

of density variation in the flow field. The knife selected one direction instead of the other to obtain an image contrast. Both lens work together to yield an image in which the luminosity is a function of the first derivative of the density in the field (Tropea *et al.* 2007). However, because of the difficulty in reducing the results to obtain density distribution in the field, the images are considered here only to determine the shock wave and expansion locations.

NUMERICAL SIMULATION

It is usual to use numerical simulations when trying to understand complex physical phenomena and experimental results. Since numerical simulations are not, typically, considered alone during important design decisions, some care is usually exercised by the CFD engineers to reduce the grounds for large errors in the results. Hence, typically, mesh independence studies and the enforcement of adequate convergence criteria are emphasized. The discussion about even the substitution of experimental results by numerical simulations is a modern theme, as stated by Kraft (2010), which redirects the debate by pointing both approaches as parts of an integrated solution to reduce the overall cycle time for development of aeronautical systems.

To simulate numerically for steady state condition the experimental configurations, the Navier-Stokes equations (Anderson *et al.* 1984) were solved by a RANS approach with a 2nd-order finite-volume approximation and using an implicit method. The turbulence effects were accounted for by using the Spalart-Allmaras model (Spalart and Allmaras 1992; Spalart 2000). To simulate the flow field around the fore-body region of the VLM-1 model and precisely capture shock waves, the SU2 open source computational code was used (Economon *et al.* 2016).

Although the boundary-layer developed over the model may be locally laminar, resulting in distinct interaction with the shock-wave, using the code for laminar condition is out of question, because the physical field is distinguished turbulent, and a laminar approach does not correspond to the physics. The authors must admit, however, that even the use of the Spalart-Allmaras model with a quadratic constitutive relation (Spalart 2000) may not represent the ideal situation for the present flow condition. Actually, probably no eddy-viscosity model would be really adequate in this case. The ideal situation would be to couple the eddy viscosity model with some transition model, such as, for instance, the Langtry-Menter model (Langtry and

Menter 2009; Halila *et al.* 2016). Unfortunately, however, the effort involved in the coding and validation of such an approach would be beyond the available resources here.

MESH INDEPENDENCE STUDY

Since CFD is highly dependent on numerical methods, errors are always included by the approximate nature of these numerical schemes. Numerical schemes are needed in different parts of the solver to deal with the complex fluid dynamic equations. Unfortunately, some of the errors brought by the numerical algorithm are intrinsic to these numerical schemes and cannot be directly reached by the user without further work in the original derivations. In an effort to minimize the errors included by these approximations, mesh independence studies are needed in order to isolate the influence of the mesh in the final solution.

Before proceeding with the actual simulations of interest, 5 meshes and their influence in the solution were studied. These grids had sizes varying from about 19,000 to approximately 1,000,000 internal hexahedral elements. During the study, 2 major issues were brought to the mesh independence study as practical constrains. The first one was concerned with spatial resolution of the shock-wave and the second was concerned with convergence. Spatial resolution of the shock wave was needed for better comparisons between the post-processing images and *schlieren* images. On the other hand, the grid could not be extremely fine. This refinement excess could add too much numerical stiffness to the solution and, as a consequence, the numerical stiffness would cause trouble when seeking for high quality steady state solutions. Table 2 shows the developed meshes used in the refinement analysis. In all cases, the $y^+ < 1$ for the first boundary-layer calculation point was observed, as prescribed by the turbulence model. The results in terms of shock-wave location and Cp distribution over the model body revealed practical convergence between the fine and the finer meshes.

Table 2. Number of elements in each mesh used during the refinement analysis.

Mesh name	Internal number of mesh elements
Badly refined	19,266
Coarse	32,393
Medium	131,946
Fine	735,513
Finer	1,172,222

In order to select the best mesh for the simulations, pressure coefficients results over the VLM model wall in the longitudinal direction from all meshes were compared with each other, along with a qualitative analysis of the shock wave resolution using Mach number contours. Figure 10 shows an example of the comparison using 3 meshes: (a) the poorest one (poorly refined according to Table 2), (b) the fine, and (c) the finer meshes. Mach number fields using the same range and number

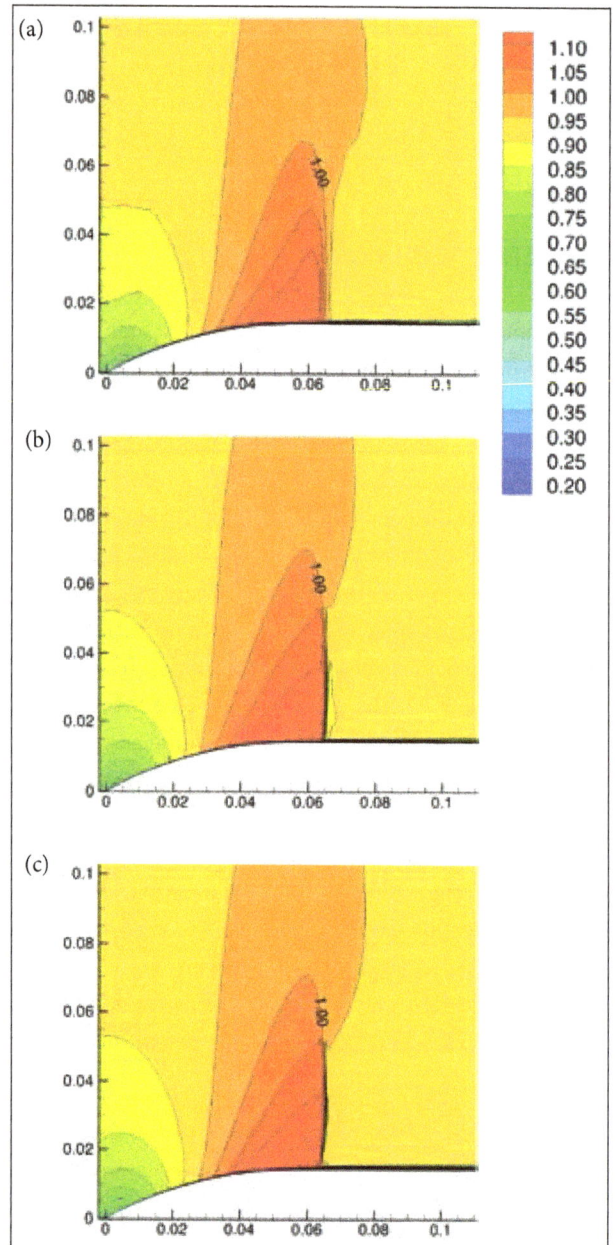

Figure 10. Mach number fields for 3 meshes: (a) Badly refined, (b) Fine, and (c) Finer meshes, according to the specification from Table 2.

of levels are plotted, as they reveal the lack of precision in Fig. 10a, and the good agreement between results from fine (Fig. 10b) and finer (Fig. 10c) meshes.

The finer mesh did not completely converged, only allowing four digits of decrease in the L_∞ norm of the density residue. The "fine" mesh was selected based on the shock wave location and thickness, and also on the convergence characteristics and computing time, presenting good shock resolution together with reasonable computational cost and convergence compared to the other meshes. Figure 11 shows the convergence history for the solution using the fine mesh for nominal free-stream Mach number of 0.9.

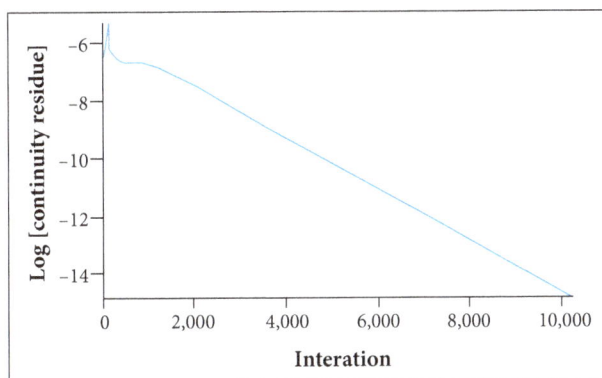

Figure 11. Convergence history for the continuity equation.

CHARACTERISTICS OF THE ADOPTED MESH

During the development of the computational mesh, some simplifications were adopted to make the problem tractable to the computing power available when compared to the full experimental configuration. These were:

- The tunnel walls were not represented as free-stream condition was applied to the flow. With a model blockage area ratio of 1%, practically no wall interference effects for zero angle of attack are expected.
- Since the fore-body region of the model is axisymmetric and the angle of attack for all cases was equal to 0°, the computational field could be restricted to 20° in the azimuthal direction with a simulation plane at each 1°, provided that symmetrical side plane conditions were imposed — previous simulations proved that 10° dihedral angle was enough by comparing it with a 20° solution.
- The model was represented without the fins at its base and the model geometry finishes at the end of the mesh, without the representation of the wind tunnel

sting support or a truncated geometry, considering that the flow phenomena at the model base will not affect the model fore-body region, which is the object of the present research.

Besides the symmetrical lateral plane conditions required by the dihedral created, over the model surface, non-slipping adiabatic flow conditions were imposed, and elsewhere far-field conditions were imposed. Although a 2-D mesh could be used because of the symmetrical characteristics of the problem, a 3-D approach is a better solution because of the inherent 3-dimensionality of the turbulence.

Figure 12 shows details of the adopted mesh. In Fig. 12a one can see the total mesh region of 10° of dihedral angle highlighting the central longitudinal plane and 2 transversal planes (in green) at the tip and base of the model. The far-field limit is located about 170 diameters or 14 model lengths. Figure 12b shows detail of a longitudinal plane in the model region, delimiting model wall from its tip at (0.000 m; 0.000 m) and its base at (0.360 m; 0.0146 m), highlighting the point clustering in the fore-body region. Figure 12c shows the detail of the fore-body region showing the point clustering normal to the body wall to represent the boundary-layer, according to the requirement of the turbulent model ($y^+ < 1$), with the first point at 1×10^{-6} m and a stretching of 17%.

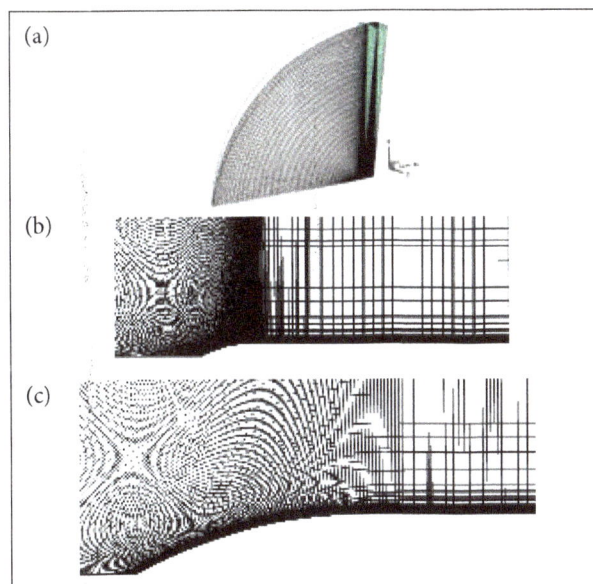

Figure 12. Details of the numerical mesh: (a) Total view with far-field distant 170 diameters; (b) Longitudinal plane detailing the model with 0.36 m of length and showing the regions of calculating point clustering; (c) The model fore-body detailing clustering close to the model wall to capture the boundary-layer.

RESULTS

Table 3 shows 12 of the main tests run in TTP to obtain *schlieren* images, along with the main test parameters with standard deviations related to the test section conditions. For some tests, numerical simulation was performed with the SU2 code aiming comparisons. Since the tests were carried out using supersonic first-throat, the Mach number informed

by the control needed correction, as it is shown in the Table 3. The study cases encompassed the relevant Mach number range in the transonic regime, and they were limited by the good quality of the available *schlieren* images, as allowed by the visualization window.

Figure 13 shows *schlieren* photographs from the tests undertaken in the TTP from free-stream Mach number from 0.8777 to 0.9893. The model was shifted about half diameter

Table 3. Main parameters related to test section condition of the tests performed with VLM-1 in TTP.

	Nominal Mach number	Corrected Mach number	Stagnation pressure (kPa)	Stagnation temperature (K)	Reynolds number (× 10⁻⁹)	Numerical simulation
1	0.90	0.8777 ± 0.0011	91.71 ± 0.14	306.29 ± 0.10	710.8	Yes
2	0.91	0.8869 ± 0.0009	91.93 ± 0.31	308.04 ± 0.09	710.3	Yes
3	0.92	0.8968 ± 0.0008	91.67 ± 0.10	308.59 ± 0.12	709.7	Yes
4	0.93	0.9069 ± 0.0008	91.62 ± 0.12	309.17 ± 0.09	710.7	Yes
5	0.94	0.9164 ± 0.0008	91.57 ± 0.06	309.57 ± 0.08	711.9	Yes
6	0.95	0.9272 ± 0.0008	91.66 ± 0.20	310.18 ± 0.16	713.8	Yes
7	0.96	0.9382 ± 0.0009	91.67 ± 0.12	310.20 ± 0.08	716.6	Yes
8	0.97	0.9475 ± 0.0008	91.72 ± 0.23	310.41 ± 0.08	718.8	No
9	0.98	0.9589 ± 0.0007	91.64 ± 0.06	310.75 ± 0.07	719.9	No
10	0.99	0.9688 ± 0.0007	91.67 ± 0.04	310.70 ± 0.07	722.6	No
11	1.00	0.9789 ± 0.0009	91.73 ± 0.12	310.78 ± 0.07	725.1	No
12	1.01	0.9893 ± 0.0009	91.81 ± 0.29	311.00 ± 0.09	727.3	No

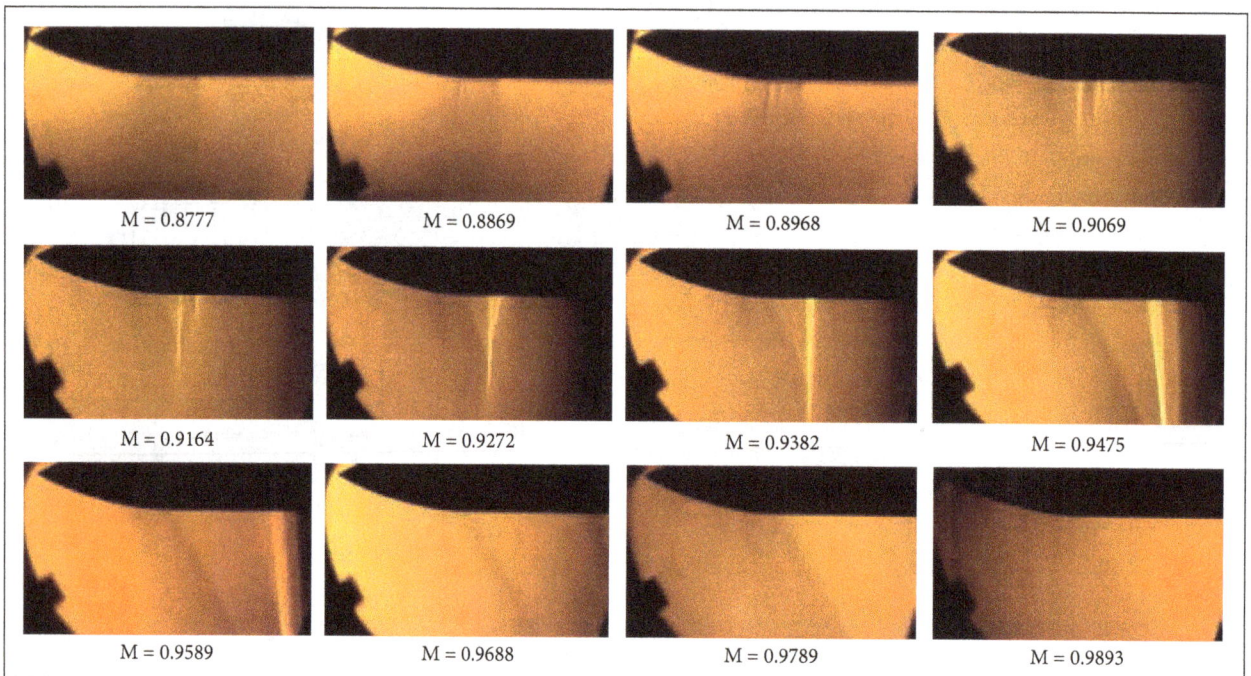

M = 0.8777   M = 0.8869   M = 0.8968   M = 0.9069
M = 0.9164   M = 0.9272   M = 0.9382   M = 0.9475
M = 0.9589   M = 0.9688   M = 0.9789   M = 0.9893

Figure 13. *Schlieren* photographs of tests with VLM-1 in TTP for different free-stream Mach numbers.

up from the test section center-line to attain a wider space of visualization. The proximity from the upper wall of the tunnel was about 3 model diameters which did not significantly influence the effects of the wall presence. It is important to note that, as it is common in *schlieren* images, in order to obtain a better contrast in the picture, the adjustment of the system caused small shadow effects in some regions of the photograph that do not represent density changes in the flow field.

As the vehicle front-body has aerodynamic shape, the *schlieren* photographs did not show interesting aspects until near free-stream Mach number of 0.8777, as it can be seen in Fig. 13. Therefore, the critical Mach number probably occurs a little lower than Mach number of 0.8777. At this Mach number condition, it is possible to distinguish a darker region starting before the nose end and finishing some distance after the cylindrical part of the vehicle. Numerical simulation will help to investigate better this condition in further studies.

For Mach number of 0.8869, the presence of a small shock wave formation near the nose end can be observed. In fact the formation looks like a main shock wave followed by a secondary weaker one, as discussed in the section "Theoretical Approach" for laminar flow. The Reynolds number for this Mach number condition is about 740,000 related to nose length, suggesting the existence of turbulent flow conditions. However, it is difficult to guarantee this condition since it is an expansion region and the local acceleration of the flow with a non-adverse pressure gradient contributes to its stabilization.

For Mach number of 0.8968, the main shock wave becomes stronger and it is followed by 3 secondary shock waves. The shock waves perceived at free-stream Mach numbers of 0.8869 and 0.8968 are normal to the local body surface and their feet are straight, indicating small growth in the boundary-layer thickness. This pattern resembles in some way the prediction shown in Fig. 3.

For Mach number of 0.9069, the pattern changes to a stronger main shock wave followed by a secondary one, which was formed by the collapse of the secondary shock waves, observed at Mach number of 0.8968. The shock waves are located in the cylindrical part of the model, and so they are normal to the free-stream flow. It is important to note that, now, the shock wave feet have lambda shape denouncing compression waves due to the growth of the boundary-layer thickness. This pattern resembles the prediction shown in Fig. 4a. In this case, it is possible that the boundary-layer experienced a recirculation bubble or even detachment. It is

worth noting how the shock wave formation pattern changes from Mach number from 0.8777 to 0.9069 and, even, for higher Mach numbers.

For Mach number of 0.9164, the pattern is repeated but the shock waves are stronger, and, finally, for Mach number of 0.9272, the 2 shock waves collapse in one stronger shock wave with its foot in a lambda shape. For Mach numbers from 0.9382 to 0.9589, the strong and unique shock wave formation progresses and advances downstream. For Mach numbers from 0.9688 to 0.9893 the available window does not show the shock wave formation because it traveled downstream. It is noteworthy, from the results for Mach number of 0.9164 and, repeatedly, for higher Mach numbers, a clear contrast in the pictures, starting before the nose end and finishing just after the nose end, which shows the limits of a supersonic expansion at the same location.

Figure 14 shows the numerical simulation results in terms of Mach number contours for free-stream Mach number of 0.85. In the figure, one should note the iso-Mach line legends being increased by 0.02. The maximum Mach number determined in the field is about 1.00, thus representing the critical free-stream flow condition. The nose has aerodynamic shape, which contributes to smooth flow acceleration along the model body. Starting at this free-stream condition and increasing the Mach number, it is expected a shock wave formation and an expressive rise of the drag coefficient, as the flow experiences transonic regime.

Figure 15 shows the numerical simulation result in terms of Mach number contours for free-stream Mach number of 0.8777. The maximum Mach number observed in this case was 1.067, corresponding to the first supercritical condition of the studied cases. It is already possible to realize that the sonic Mach-line underwent a small disturbance at approximately 0.050 m, which indicates initial formation of a shock wave, too

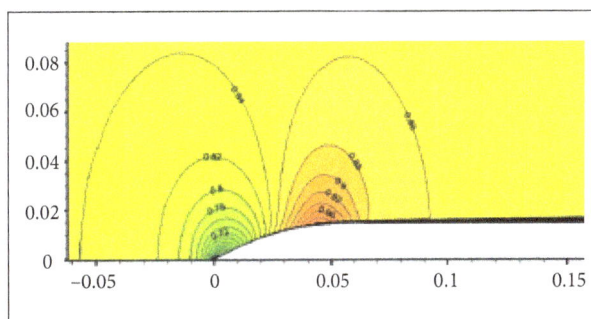

Figure 14. Mach number contours for free-stream Mach number of 0.85, corresponding to critical conditions.

weak to be noticed by *schlieren* images, as shown in Fig. 13. In Figs. 14 and 15, one can see the accumulation of iso-Mach lines close to the model wall due to the deceleration of the flow, and the region grows along the model surface because of the boundary-layer thickness growth.

In Fig. 16, it is shown the pressure field for free-stream Mach number of 0.8869. As the isobaric lines cross the boundary-layer on the model surface, there is no accumulation of lines, as those observed in Mach number contour plots. A shock wave formation at about 0.0500 m from the model tip is clear. A careful observation of the *schlieren* image for Mach number 0.8869 (Fig. 13) will indicate the main shock wave at a distance of 0.0519 m and the secondary at 0.0541 m. As previously discussed, no eddy-viscosity model can predict transition, or relaminarization; therefore, the laminar 2-shock-wave formation observed in the *schlieren* image could not be numerically represented. However, the comparison of the main shock wave location measured from the model tip in both techniques had a disagreement of only 3.8%.

As discussed, it was already expected that the laminar multiple shock wave formation could not be predicted by the present numerical simulation approach. However, observing Fig. 13, one can see that the laminar shock formation will practically end at Mach number 0.9272, when the shock wave collapses into a unique shock wave formation. Figure 17 compares both numerical and *schlieren* image results for free-stream Mach number $M_\infty = 0.9272$, where one can see a good prediction of the shock wave location through the numerical simulation, with an error of 6.2%. In general, it is possible to visualize that the numerical result overpredicts the physical phenomena, such as sonic line, supersonic region, and shock wave location. The shadowed expansion region in the *schlieren* image is followed by a more illuminated area, which would indicate compression wave formation region, but this fact was not demonstrated by the numerical result. Besides, the numerical simulation could not predict the complex formation at the shock wave foot, where the boundary-layer has an important role, most certainly due to the lack of resolution of the boundary layer.

Figure 18 shows the same results for free-stream Mach number of 0.9382. In this case, it is possible to note a better comparison between results, and the error in the shock wave location predicted by numerical calculation was 3.7%. Isobaric line representing the sonic condition (dashed red line) had a noticeable increase in the field. It is interesting to observe that the shock wave in the *schlieren* image is a straight line while the same shock wave in numerical simulation has a curvature, indicating some effect from the boundary-layer.

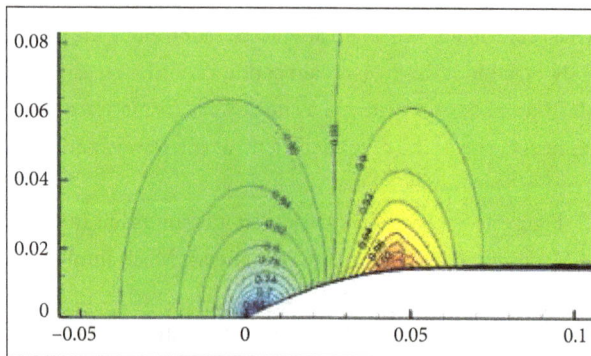

Figure 15. Mach number contours for free-stream Mach number of 0.8777.

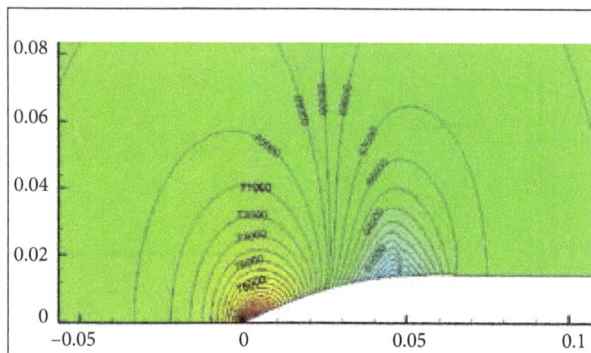

Figure 16. Numerical simulation of pressure field for free-stream Mach number of 0.8869.

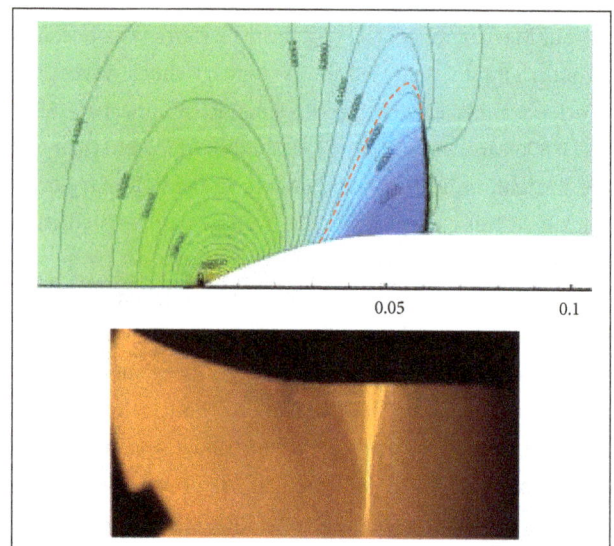

Figure 17. Numerical simulation result of pressure contours compared with *schlieren* image for free-stream Mach number of 0.9272. The isobaric line increases by 1,000 Pa. Dashed red line highlights isobaric line related to sonic condition.

A very important issue is the role of the turbulence in the flow field. Figure 19a shows turbulent viscosity contours. Here, eddy viscosity is made dimensionless by the free-stream viscosity coefficient. The accumulation of lines close to the model wall indicates the boundary-layer development. Although the present approach implies that the flow is fully turbulent from the beginning, it is clear that eddy viscosity generation occurs inside the boundary layer. Hence, close to the model tip, since the

boundary-layer is very thin, turbulent effects are almost negligible. As the boundary layer grows along the body, so does the eddy viscosity coefficient, which reaches a value equal to 320 times the viscosity coefficient of free-stream far-field. This increase occurs just downstream of the shock wave impingement point, hence, indicating a tremendous increase in turbulent effects together with the growth of boundary-layer due to shock wave interaction.

It is also important to note how the turbulent activity decreases when the flow experiences acceleration by expansion in the supersonic region. Figures 19b and 19c show a detailed view of the shock wave foot, highlighting the iso-lines of turbulent viscosity and also plotting the velocity vectors in the flow field. It is notorious the increase of the boundary-layer thickness after the shock wave along with the turbulent viscosity. A detailed inspection of the boundary-layer indicated its thickness to be about 0.5 mm just before (and 0.8 mm just after) the shock wave, using the criterion of 95%.

CONCLUSION

VLM-1 model in scale 1:50 was tested in the TTP of IAE, and very important results obtained with *schlieren*

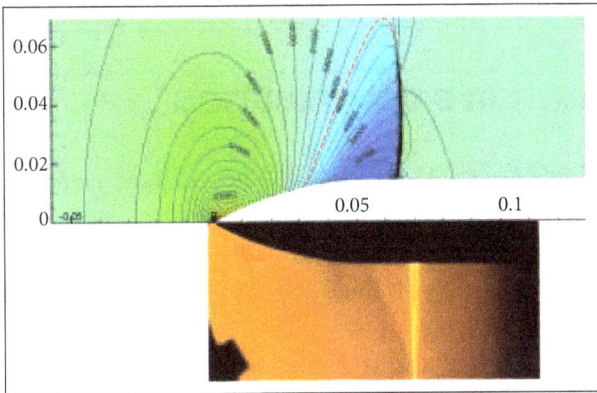

Figure 18. Numerical simulation result of pressure contours compared with *schlieren* image for free-stream Mach number of 0.9382. The isobaric line increases by 1,000 Pa. Dashed red line highlights isobaric line related to sonic condition.

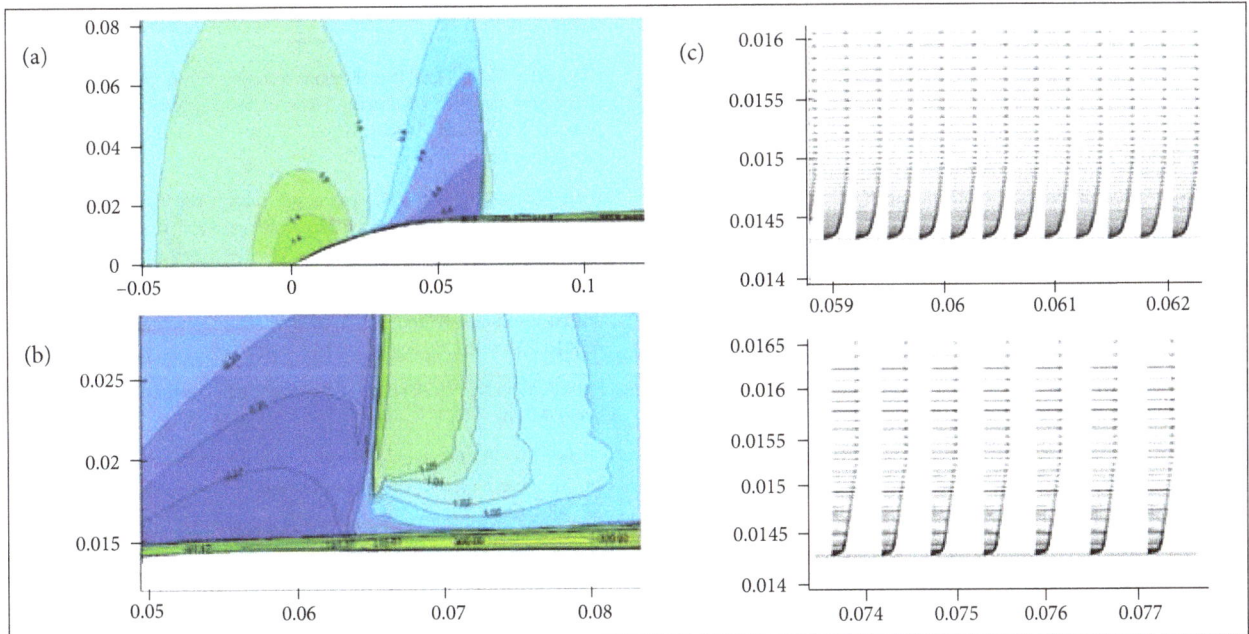

Figure 19. (a) Numerical simulation result of turbulence field for free-stream Mach number of 0.9382; (b) Detail of the base of shock wave with turbulence field (the iso-lines of turbulent viscosity were determined by dividing local values by the value at free-stream condition); (c) Velocity profiles before and after the shock wave (the boundary-layer thickness at x = 0.060 m is 0.00055 m, and at x = 0.076, 0.00094 m).

visualization technique in transonic regimes were presented and discussed. Shock wave formation over the model could be observed as the Mach number increased from 0.8777 to 0.9893. Because of the aerodynamic shape of the model ogive, *schlieren* image showed shock wave formation starting at free-stream Mach number of 0.8869. Typical laminar shock wave/boundary-layer interaction was perceived by multiple shock wave formation.

Numerical simulation performed by SU2 using the same flow conditions for some of the experiments allowed to observe other variables from the physical field, such as pressure, Mach number, turbulent viscosity, and velocity vectors. Numerical search indicated free-stream critical Mach number to be about 0.85. Although numerical simulation could not predict well the laminar and transition behavior of the boundary-layer, it was possible to observe good agreement of the shock wave distance location with a maximum error of about 6% from model tip. The numerical simulations were very helpful to understand *schlieren* images obtained and to quantify some important flow parameters.

ACKNOWLEDGEMENTS

The authors gratefully acknowledge the Agência Espacial Brasileira (AEB) for the support for this research. Partial support was also provided by the Conselho Nacional de Desenvolvimento Científico e Tecnológico (CNPq), under Research Grants No. 309985/2013-7 and 308829/2015-8. The authors are also indebted to the partial financial support received from the Fundação de Amparo à Pesquisa do Estado de São Paulo (FAPESP), under Research Grant No. 2013/07375-0.

AUTHOR'S CONTRIBUTION

Mata HO and Falcão Filho JBP conceived and conducted the experiments; Avelar AC participated in the experiments as well. The numerical simulations were conducted mainly by Falcão Filho JBP and Carvalho LMMO; Azevedo JLF participated in the discussion of the results and in the writing of the main text, which was done mostly by Falcão Filho JBP, with the participation of the other authors.

REFERENCES

Anderson DA, Tannehill JC, Pletcher RH (1984) Computational Fluid Mechanics and heat transfer. Washington: Hemisphere Pub. Corp.; New York: McGraw-Hill.

Barlow JB, Rae WH, Pope A (1999) Low-speed wind tunnel testing. 3rd ed. New York: Wiley.

Da Mata HO (2013) Procedimento Experimental para Análise Aerodinâmica do Veículo Lançador de Microssatélite VLM-1 (Undergraduate thesis). São José dos Campos: Instituto Tecnológico de Aeronáutica.

Davis MW, Gunn JA, Herron RD, Kraft EM (1986) Optimum Transonic Wind Tunnel. Proceedings of the 14th AIAA Aerodynamic Testing Conference; West Palm Beach, USA.

Dolling DS (2001) Fifty years of shock-wave/boundary-layer interaction research: what next?. AIAA J 39(8):1517-1531. doi: 10.2514/2.1476.

Economon TD, Palacios F, Copeland SR, Luckaczyk TW, Alonso JJ (2016) SU2: an open-source suite for multiphysics simulation and design. AIAA J 54(3):828-846. doi: 10.2514/1.J053813

Falcão Filho JBP, Avelar AC, Reis MLCC (2009) Historical review and future perspectives for Pilot Transonic Wind Tunnel of IAE. J Aerosp Technol Manag 1(1):19-27. doi: 10.5028/jatm.2009.01011927

Goethert BH (2007) Transonic Wind Tunnel testing. Mineola: Dover Publications.

Halila GLO, Bigarella EDV, Azevedo JLF (2016) Numerical study on transitional flows using a correlation-based transition model. J Aircraft 53(4):922-941. doi: 10.2514/1.C033311

Houghton EL, Carpenter PW (2003) Aerodynamics for engineering students. 5th ed. Amsterdam: Elsevier.

Kline SJ (1943) Flow Visualization. Internet movie from the National Committee for Fluid Mechanics Films under a grant from the National Science Foundation; [accessed 2017 Jan 13]. http://web.mit.edu/hml/ncfmf.html

Kraft EM (2010) After 40 years why hasn't the computer replaced the wind tunnel?. International Test and Evaluation Association Journal 31:329-346.

Langtry RB, Menter FR (2009) Correlation-based transition modeling for unstructured parallelized Computational Fluid Dynamics codes. AIAA J 47(12):2894–2906. doi:10.2514/1.42362

Pope A, Goin KL (1978) High-speed wind tunnel testing. New York: John Wiley & Sons

Spalart PR (2000) Strategies for turbulence modelling and simulation. Int J Heat Fluid Flow 21(3):252-263. doi: 10.1016/S0142-727X(00)00007-2

Spalart PR, Allmaras SR (1992) An one-equation turbulence model for aerodynamic flows. Proceedings of the 30th Aerospace Sciences Meeting and Exhibit; Reno, USA.

Tropea C, Yarin AL, Foss JF (2007) Springer handbook of Experimental Fluid Mechanics. Berlin: Springer.

Wilcox DC (2006) Turbulence modeling for CFD. 3rd ed. La Cañada: D. C. W. Industries.

6

Assembly and Testing of a Thermal Control Component Developed in Brazil

Marcos Galante Boato[1], Ezio Castejon Garcia[1], Marcio Bueno dos Santos[2], Antonio Fernando Beloto[2]

ABSTRACT: The optical solar reflector is basically a mirror of second surface with low absorptivity/emissivity ratio and negligible degradation in the space environment, which makes it an excellent coating for thermal control of satellites. It works as a radiator and is used in particular parts of the external surfaces of the satellites in order to reject the undesirable heat to the deep space. In the Brazilian Space Programs, the radiators of the satellites are generally painted with special white paints in order to reject heat instead of the use of optical solar reflector. The problem of white-ink radiators is the high degradation of the thermo-optical properties that happen over the useful lives. Thus, a process of manufacturing and assembly of optical solar reflector was developed in Brazil. To validate this process in terms of mechanical and thermal properties, 3 types of optical solar reflector radiators were manufactured, and their absorptivity and emissivity properties at the temperature of 23 °C were measured. Optical solar reflector coupons were mounted on aluminum plates to perform vibration, thermal vacuum and thermal-shock tests. A study was also done to optimize the thickness of the glue to fix the structure of the satellite on the optical solar reflector. It showed an excellent environmental stability and maintained its thermo-optical characteristics after the tests.

KEYWORDS: Solar radiator, Satellite thermal control, Solar absorptivity, Emissivity.

INTRODUCTION

Artificial satellites are equipment sent into space in order to perform a certain task, such as: meteorological studies, telecommunications, data collection for scientific studies, etc. In the space environment, the satellite is exposed to extreme thermal conditions such as solar radiation, radiation that comes from the Earth (albedo and infrared), and a strong heat sink that is deep space, where the fund temperature is 4 K. Therefore, the satellites are supposed to be protected from all thermal loads from space.

To protect the satellites, they are externally coated with thermal-control materials in order to isolate themselves of the external environment. Due to the absence of convective medium, heat exchange between the satellite and the space environment is made exclusively by radiation. One of the materials commonly used to coat the satellite is the thermal blanket known as multi-layer insulation (MLI) (Nagano *et al.* 2011). However, this coating cannot completely insulate the satellite, because of several devices which, by Joule effect, generate heat inside of the satellite. This heat raises the internal temperature, reaching above the acceptable limits, and it is necessary to eliminate the excess of heat through openings in the blanket. These openings act as their heat rejection areas, radiating the heat of the equipment into deep space. Figure 1 shows the satellite CBERS-2B and the MLI with the respective openings for rejection of heat to deep space.

As all environments in satellites, the interaction is by radiation, and there is a high dependency of the thermal coating properties as a function of temperature. Thus, these heat rejection areas, known as radiators, are coated with a material which has

1.Departamento de Ciência e Tecnologia Aeroespacial – Instituto Tecnológico de Aeronáutica – Divisão de Engenharia Mecânica – São José dos Campos/SP – Brazil.
2.Instituto Nacional de Pesquisas Espaciais – Laboratório de Integração e Testes – São José dos Campos/SP – Brazil.

Author for correspondence: Marcos Galante Boato | Departamento de Ciência e Tecnologia Aeroespacial – Instituto Tecnológico de Aeronáutica – Divisão de Engenharia Mecânica | Praça Marechal Eduardo Gomes, 50 – Vila das Acácias | CEP: 12.228-900 – São José dos Campos/SP – Brazil | Email: marcosmhd@hotmail.com

good properties of emissivity and absorptivity (when exposed to direct solar radiation and albedo).

The Instituto Nacional de Pesquisas Espaciais (INPE) and the Instituto Tecnológico de Aeronáutica (ITA) have developed, manufactured, and qualified an optical solar reflector (OSR) to be used for thermal control of satellites. The OSR is a coating to be used as radiator for certain external surfaces of satellites. The aim is to reject heat that occurs in positions where it appears the incidence of solar radiation (direct and/or albedo).

In the design, thermal control coatings have to be resistant to the degradation effects of the space environment. Due to this degradation, there is an increase in the solar absorptivity and, thus, an increase occurs in the internal temperature of the satellite along its orbital life. The temperatures may exceed acceptable values, which can cause serious damage for equipment pieces in the satellite, consequently reducing the life of the mission.

The OSR-type heat radiators, as well as the white-ink radiators, are passive and used for thermal control of satellites. One of their characteristics is to have a lower absorptivity (α_s) in the solar band if compared to the white-ink radiators, as well as a high emissivity (ε), which means that they can reject heat from the interior of the satellite to the deep space. The main feature is to present an extreme low degradation of thermo-optical properties if compared to other thermal control coatings (Gilmore 1994; Marshall and Breuch 1968).

As previously mentioned, the OSR is basically a mirror of second surface. It consists of a layer of thin silver film that is deposited on a surface of high-quality glass coverslip. A high ε in the infrared spectrum is related to the glass coverslip, which is transparent in solar band and substantially opaque in the infrared (spectrum of wave length below 4.5 μm) (Marshall and Breuch 1968). Furthermore, the thickness of the glass layer also influences the ε value. The low α_s in the solar spectrum is related to the thin silver film deposited on the glass coverslip (Greenberg et al. 1967).

What differs the OSR from other passive radiators is its negligible thermo-optical degradation of properties in relation to: atomic oxygen, protons, free electrons, and ultraviolet radiation. In addition, the OSR has a low degradation factor in relation to volatile organic photo-depositions that are generated by the outgassing from the satellite's internal components and products of the satellite's thrusters. that are generated by the outgassing from the satellite's internal components and products of the satellite's thrusters.

In this study, it was used a borosilicate glass with a thickness of 120 μm and dimension of 20 × 40 mm, supplied by the Shanghai Institute of Space Power Sources, in China. This material is doped with Cerium to avoid the appearance of color center in the glass surface due to ultraviolet radiation. These coverslips were originally used as a protective coating of the Brazilian satellite solar cells, and this material has demonstrated a good environmental stability in space flight.

EXPERIMENT

Silver has low adhesion to glass due to its lack of interaction with the active oxygen in the substrate (Benjamin and Weaver 1961). For greater adhesion between the substrate and the silver film, it was deposited an interface layer to increase this adhesion. As the metal interface might change the OSR of the α_s, 3 kinds of sample sets were made, each one with different interfaces. Thus, there was the possibility to analyze which of the 3 interfaces might have the best adhesion for the thin film of silver and the best a_s/ε relation. Then, the adopted interface materials were: chromium (Iacovangelo et al. 2003), aluminum, and magnesium fluoride (MgF_2) (Tanzilli and Gebhardt 1996). In order to avoid the oxidation of silver in the terrestrial environment, a layer of Cr was deposited to protect the silver layer. Thus, the final mirror was composed of 3 layers (Fig. 2).

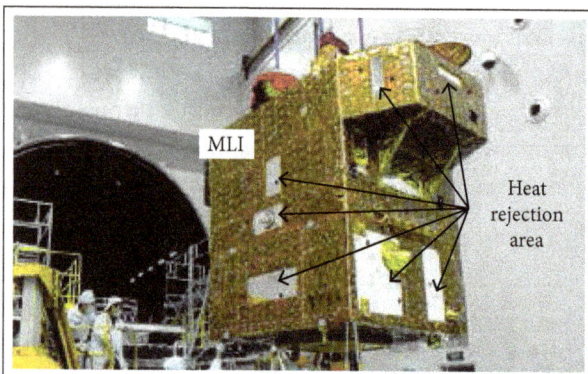

Figure 1. Satellite CBERS-2B and its heat rejection areas in the MLI.

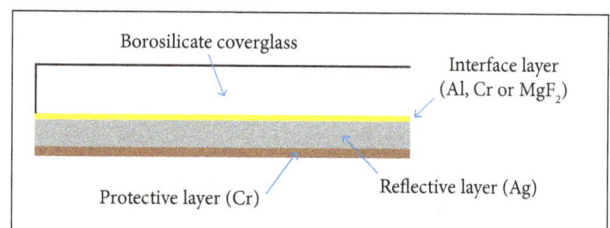

Figure 2. Representation of the layers deposited on the substrate.

Table 1. Borosilicate samples with their respective interfaces.

Substrate	Quantity	Interface material	Reflective metal	Metal protection
Borosilicate SiO_xN_y	14	MgF_2 (100 Å)	Ag (3,000 Å)	Cr (200 Å)
Borosilicate SiO_xN_y	14	Al (60 Å)	Ag (3,000 Å)	Cr (200 Å)
Borosilicate SiO_xN_y	14	Cr (60 Å)	Ag (3,000 Å)	Cr (200 Å)

To conduct the proposed experiment, 3 series of deposits were made on the borosilicate coverslip glass. The study about the deposits was based on the verification of optical and thermo-mechanical properties of the OSR for different interfaces. Therefore, it was deposited on each set of substrates (Fig. 2; the thicknesses are described in Table 1) The layers of the films on the substrate surfaces were made with an apparatus which deposits thin films, called electron beam.

THERMO-OPTICAL PROPERTY MEASUREMENTS

For a thermo-optical surface to be considered a good passive radiator, its emissivity in the infrared spectrum has to be the highest as possible, and its absorption in the solar spectrum, the lowest. In this paper, the data from α_s and ε were obtained from samples of OSR with different interfaces; so, it was possible to identify which sample had the best α_s/ε ratio. To measure these properties, a Gier Dunkle Device was used for: a) normal solar absorptivity α_s, which was done by the MS251 Solar Reflectometer; b) hemispherical emissivity ε, by DB100 Infrared Reflectometer. Table 2 shows the measurements for the 3 types of interfaces in the OSR. Initial measures have highlighted the differences among the values of the α_s for different samples.

Table 3 shows the ε values. In all 3 cases, no significant differences were detected among the samples. What could be observed is the following: the interface layer deposited has high influence on α_s values, but it has low influence on the values of emissivity in infrared spectrum (ε_{IR}). In this test, it can be concluded that the OSR with the best $\alpha_s/\varepsilon_{IR}$ ratio was the aluminum interface, with a value of $\varepsilon_{IR} = 0.810$ and $\alpha_s = 0.026$.

Table 2. Solar absorptivity for 3 interface types on the OSR.

Cover slip interface	Normal solar absorptivity (α_s)
MgF_2	0.051 ± 0.005
Al	0.026 ± 0.005
Cr	0.237 ± 0.006

Table 3. Data emissivity of 3 types of OSR.

Cover slip interface	Hemispherical emissivity (ε)
MgF_2	0.812 ± 0.005
Al	0.810 ± 0.006
Cr	0.814 ± 0.005

APPLICATION METHODS FOR MATERIAL EVALUATIONS

The OSR with each interface type was glued on 3 aluminum plates, each one with dimensions of $150 \times 150 \times 10$ mm and mass of 800 g. Each plate contained 6 coverslips with different types of interface (Fig. 3). The total mass of the 3 coupons was 2.4 kg, approximately.

For the OSR to be fixed on the aluminum plates, it was used the Dow Corning® glue RTV566, a qualified resin-based silicon for space. This glue was used for having a high elasticity, keeping its mechanical properties at high temperature gradients, and presenting low rates of outgassing at low pressures.

Each thickness of the glue, used to join the OSR to the aluminum plates, was calculated. The development of an equation was based on Volkersen's formulation, which was expanded to Goland and Reissner (Chen and Nelson 1979). This equation takes into account the thermal expansion, elastic modulus of each material (glass and aluminum), and the shear modulus of the adhesive.

Another important consideration is that the shear stress is equal to 0 at the center of the assembly and increases gradually to the edge. Typically, the maximum shear stress occurs at the end of the bond and it is the item of most concern. Figure 4 shows how the expansion of the materials joined by bonded joints would be studied.

Equation 1 provides the maximum tension applied to the glue due to the thermal expansion of the materials (glass + aluminum). If this stress is higher or equal to the shear stress of the adhesive, it can break.

$$\tau_{m\acute{a}x} = \frac{(\omega_2 - \omega_1)\Delta T G}{a \cdot \left[\frac{G}{a}\left(\frac{1}{E_1 b_1} + \frac{1}{E_2 b_2}\right)\right]^{\frac{1}{2}}} \cdot \tanh\left\{x\left[\frac{G}{a}\left(\frac{1}{E_1 b_1} + \frac{1}{E_2 b_2}\right)\right]^{\frac{1}{2}}\right\} \quad (1)$$

where: τ_{max} is the glue maximum stress; ω is the linear dilatation coefficient; ΔT represents the temperature gradient; G is the glue shear module; a means glue thickness; E_1 is the Young's modulus for glass; b_1 is the glass thickness; E_2 is the Young's modulus for aluminum; b_2 is the plate thickness; x is the maximum distance from the center to the edge (where the diagonal of the square center is).

Figure 5 shows the curve that relates the maximum stresses for each thickness of imposed glue. It was obtained by a Matlab program, developed in this study by using the shear strength data from the manufacturer. In the case of RTV566 glue, the limit pressure of maximum rupture is up to 3 MPa (provided by theoretical calculation). Therefore, in Fig. 5, it is observed that, for borosilicate, the theoretical thickness is extremely small, which shows to be highly-elastic adhesive, suitable for the space requirements.

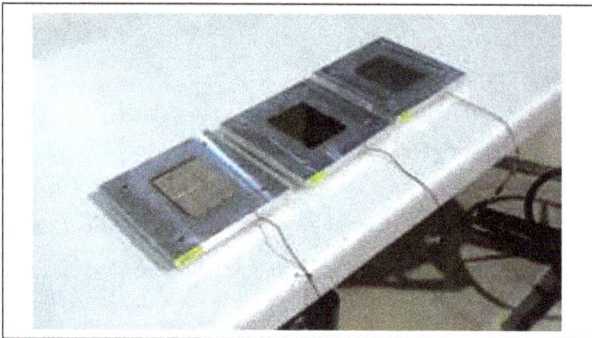

Figure 3. OSR coupons used in the environmental tests.

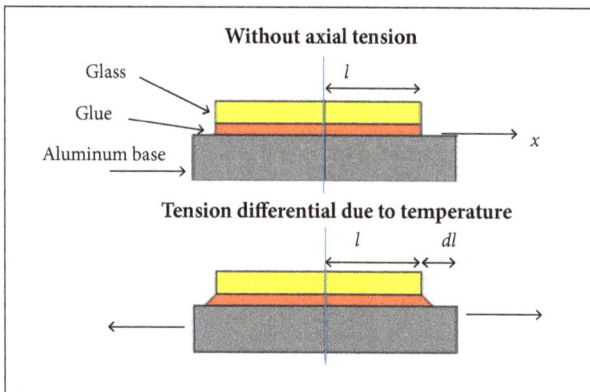

Figure 4. Differential expansion in plates. dl: gap relative to the distance l for dilations between the glass and the aluminum base.

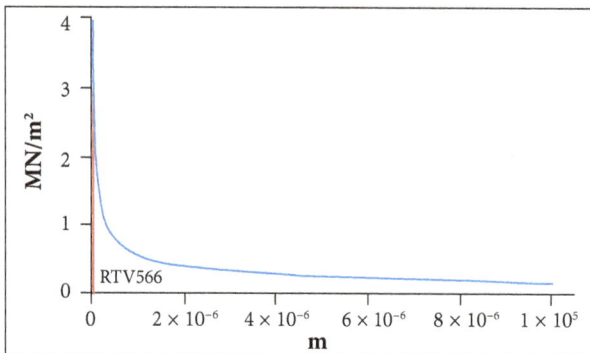

Figure 5. Maximum tension to which the glue was requested in relation to its thickness.

The thickness of adhesive used was between 0.04 and 0.06 mm (measurement made through a device, with ±0.01 mm of error). In order to ensure uniformity in the glue thickness, a fabric screen of 44 wires/cm was used. Figure 6 illustrates the bonding process that ensures the glue thickness. To ensure a good uniformity of the glue under OSR, a weight was evenly placed on the OSR after gluing.

Figure 6. Printing screen being used to apply the RTV566 on aluminum plates.

ENVIRONMENTAL TESTS
RANDOM VIBRATION TEST

The samples were subjected to random vibration according to the levels specified by the MIL-STD 883 — Method 2026 — K condition (Table 4). This test is intended to check the performance and mechanical strength of the bonding of the coverslips on OSR aluminum plate as well as the strength of adhesion of thin films of metal deposited on the borosilicate. This level was imposed for qualification of the OSR as a component.

The sets of sample (OSR + aluminum plate) were installed on the electromechanical shaker (Fig. 7). To monitor the test, acceleration sensors were installed. One sensor was installed on the underside of the center of each coupon, 3 in total, focused on measuring the direction normal to the OSR. An additional sensor was installed in the circular adapter plate test of the set to the head of the electromechanical vibrator which belongs to

Table 4. Specification of random vibration test for the OSR.

Frequency range (Hz)	Level
5 to 100	+6 dB/ILO
100 to 1,000	1.5 g²/Hz
1,000 to 2,000	−6 dB/ILO
Effective acceleration	44.8 g$_{rms}$
Duration	15 min
Direction	Normal to the ORSs plane

the Laboratório de Integração e Testes (LIT) at INPE in order to control the applied excitation function.

The maximum excitation applied was 46.2 g_{rms}. The responses have reached in 53.1 g_{rms} for Cu; 55.5 g_{rms} for Al and 56.6 g_{rms} for MgF_2, considering from the bottom plate. It is possible to verify that the 3 interface layers showed excellent adhesion of the silver film to the borosilicate glass after the vibration tests. There was no delaminating of the silver from the glass surface. Furthermore, visual inspection was done, and it was found that there were no other damages, neither in the OSR, nor in the bonding.

Figure 7. Set of coupons installed for vibration testing.

THERMAL-VACUUM TEST

The test in the thermal-vacuum chamber simulates the space environment, in which the satellite will be exposed. For this test, it was used the 250 L # 1 thermal-vacuum chamber, which has a thermal shroud with temperature control operating from −180 to +150 °C and vacuum up to 10^{-7} Torr. This chamber is part of the Thermal-Vacuum Laboratory (TVL) at the LIT/INPE. The infrastructure and test setup are presented in Fig. 8.

The boundary conditions of the test were taken from the specification for the solar cells, from the classification of the AMAZON satellite program (Nagano *et al.* 2011). These conditions present the exposure time to the maximum and minimum pressure levels as well as the quantity of temperature cycles. The specification of the test is shown in Table 5.

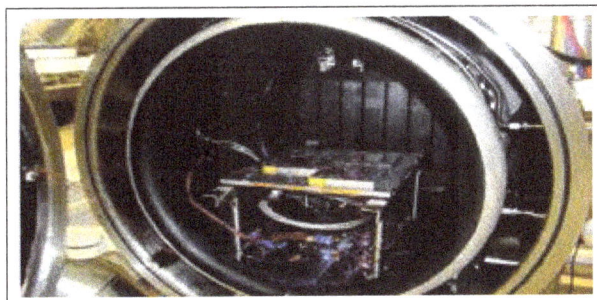

Figure 8. Coupons inside the LIT Thermal-Vacuum Chamber #1.

Table 5. Parameters for the test in thermal-vacuum chamber.

Test parameters	Data
Number of cycles	4
Soak time	4 h
Maximum temperature	90 ± 5 °C
Minimum temperature	−90 ± 5 °C
Vacuum chamber pressure	Below 10^{-5} Torr

The use of these coupons previously used in vibration tests served as a method to check for potential problems that could arise from the vibration test (satellite launch simulation). Thus, any problem could be magnified during the thermal vacuum test and then could be checked after this.

This test aimed to verify possible cracks caused by OSR differential thermo-dilation, delamination of thin films of the metal borosilicate coverslip glass, as well as problems related to assembly with air bubbles in the glue, which could expand due vacuum and damage the radiator.

As the thermal-vacuum test setup, 2 thermocouples (type T) were installed on each coupon in order to provide the individual temperatures. These thermocouples worked to guaranty the parameters for the temperature control system of the chamber during the test. After installation of the thermocouples, coupons were installed inside the thermal-vacuum chamber.

What differs the OSR from other passive radiators is its negligible thermo-optical degradation of properties in relation to: atomic oxygen, protons, free electrons, and ultraviolet radiation. In addition, the OSR has a low degradation factor in relation to volatile organic photo-depositions

Figure 9 shows the simulated curve profile during the thermal-vacuum test. The test began at the hot level in order to take advantage of the temperature of the coupons. This procedure aimed at reducing the duration of the test, facilitating the chamber operation (Almeida *et al.* 2006).

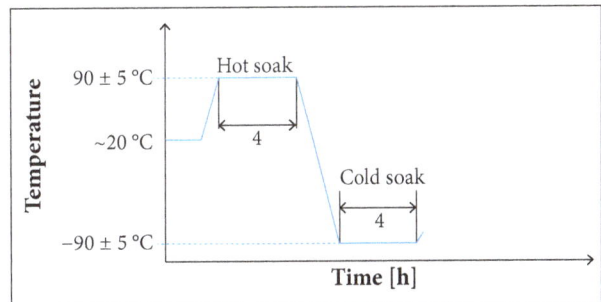

Figure 9. Theoretical curve for test in thermal-vacuum chamber.

Figure 10 shows the graph of the temperature in each specimen during the hot and cold cycles. At the end of each cycle, only a merely operational stop occurred in the thermal cycling, without compromising the vacuum pressure.

At the end of the thermal vacuum test, a visual inspection was performed, and no damage was observed, neither in the OSRs, nor in the bonding. With this test, one can conclude that the OSR was successfully qualified according to the AMAZON program.

THERMAL-SHOCK TEST

The thermal-shock test is an experimental method to check possible cracks on the glass and/or delaminating problems on the metal film due to high gradients of temperature imposed in the samples. For this test, it was used a thermal-shock chamber Thermotron Model ATS-320-V-10-705 LIT/INPE.

The minimum and maximum thermal-shock limits were imposed according to predicted temperature levels for the OSR, once in orbit. Temperatures about 90 °C were set for the hot cases and −90 °C for the cold ones. The number of cycles for test was 80 thermal profiles. These cycles were based on the qualification process for solar cells of the AMAZON satellite program (Instituto Nacional de Pesquisas Espaciais 2013) as well as on MIL-STD-1540B military standard. The qualification parameters for the thermal-shock test are presented in Table 6.

For the thermal-shock test, all thermocouples of the thermal-vacuum test described before were used. With these thermocouples, it was possible to measure the extreme temperatures and thermal transients of the coupons during the test.

The control of the thermal-shock chamber was directly connected with the measurements of the thermocouples. Thus,

such temperatures were utilized to monitor the operation of the thermal-shock chamber elevator. When the coupons reached in a determined temperature level, the elevator was activated to take the coupons to the next compartment, and so on. Figure 11 shows the thermal-shock chamber with the coupons exposed in its elevator.

Figure 12 shows the 80 cycles performed in the coupon qualification test. In this figure, the developed temperature profile in the coupons is presented. After testing, visual inspection was performed on the samples, and no failure was observed.

Furthermore, visual inspections were done. It was observed there was not any damage, neither in the OSR, nor in the bonding.

Table 6. Parameters of the thermal-shock test.

Test parameters	Data
Number of cycles	80
Maximum temperature	90 ± 5 °C
Minimum temperature	−90 ± 5 °C

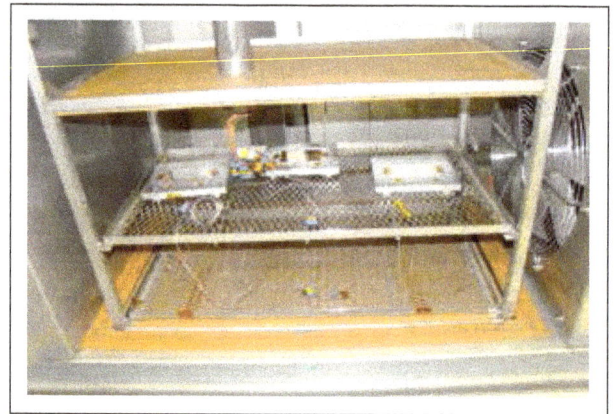

Figure 11. Coupons placed in the thermal-shock chamber elevator.

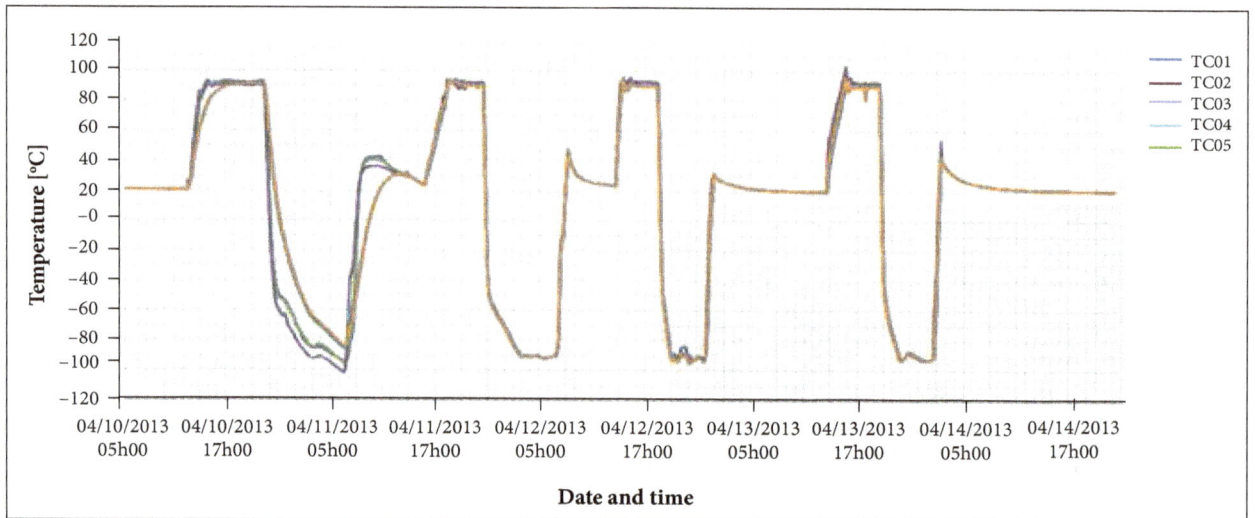

Figure 10. Developed temperature profiles of the coupons in the thermal-vacuum test.

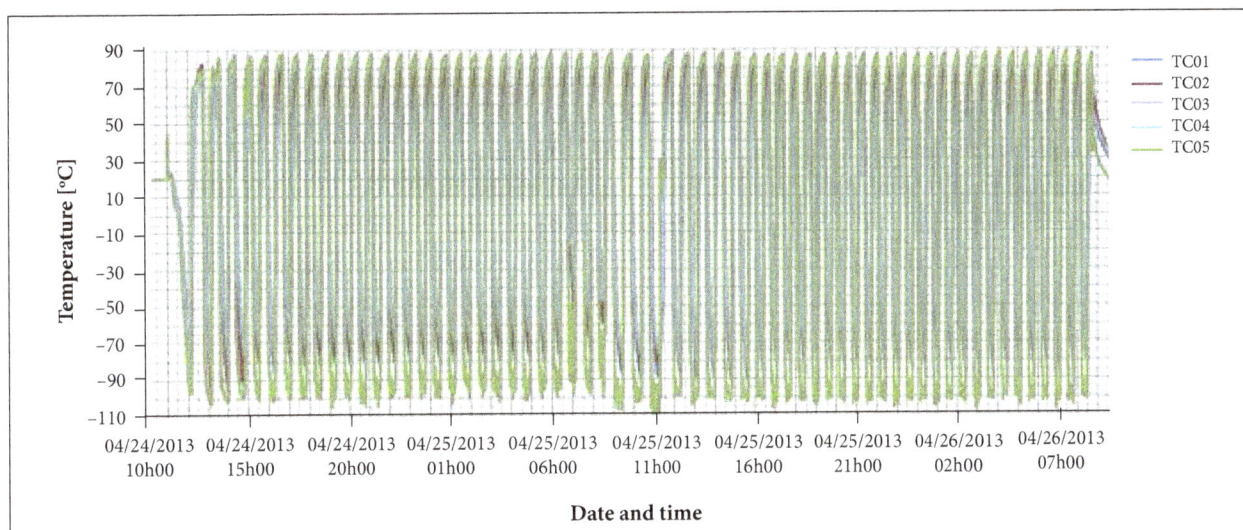

Figure 12. Results of the developed temperature profile in one of the coupons during the thermal-shock test.

By this test, it was concluded that the OSR has supported thermal-shock transient of 200 °C/min, and no damage was verified in its structure.

RESULT ANALYSIS

Once completed the package of environmental and mechanical tests, the final visual inspection and measurements of IR emissivity and solar absorptivity were made. For the final qualification, 3 main topics were considered:

Evaluation of the bonding process for OSR adhesion on base: it was taken into consideration the detachment of OSR aluminum plate in environmental and mechanical testing, and no damage was observed.

Evaluation of the process of films deposited on glass cover slip: aspects of delamination of the silver film from the glass surface were taken into consideration. For this case, no damage was observed too.

Evaluation of measurements of α_s and ε, before and after the environmental and mechanical tests: there were not significant changes in the measurements.

CONCLUSION

The OSR, manufactured and tested at the LIT/INPE, proved to be an excellent tool for thermal control of satellites, due to

its characteristics of low degradation and excellent α_s/ε ratio if compared with white-ink radiators.

Other kinds of OSR, such as substrate of kapton or polytetrafluoroethylene, currently used by the Brazilian Space Program, could be developed too. In terms of mass, or maybe cost, these might be better; however, certainly in terms of degradation and α_s/ε relation, the OSR developed in this study is one of the best.

The developed manufacturing process was very reliable: the glass was already qualified for space flight, and the deposition process has maintained the same characteristics during the manufacturing of each unit.

The interface layers showed the adhesion needed to withstand the stresses that the space environment demands. In addition, the manufacturing process has been studied to improve the adhesion of the silver film on the surface of borosilicate, introducing 3 kinds of interfaces among them. The 3 interface layers showed an improvement in the silver film adhesion on the glass, as could be seen in the vibration and climate tests: there was not delamination of the silver in glass surface. All thermo-optical measurements applied to the OSR showed the aluminum interface layer to have the lowest solar absorptivity; the others have resisted to all environmental tests.

The OSR coupons have demonstrated good stability of the thermo-optical properties after being tested in the vibration, thermal-vacuum, and thermal-shock experiments. The stability in the large temperature range ensures their use in space vehicles, once predicted for conditions similar to the AMAZON Project.

The chosen adhesive (RTV566) has demonstrated mechanical and thermal stability due to differential expansion that occurred between the OSR and aluminum. The thermo-optical tests have shown that the deposited adhesion layer interferes with the values of α_s: the best option is the aluminum interface, which provides the best α_s/ε relation.

Thus, it can be concluded that the process of OSR development can be a good option in the manufacture of thermal radiators,

being used as a tool for the thermal control of satellites. This qualification process has been well done at the LIT/INPE.

AUTHOR'S CONTRIBUTION

Boato MG and Garcia EC conceived the idea and co-wrote the main text; Beloto AF, Boato MG, and Santos MB performed the experiments. All authors discussed the results and commented on the manuscript.

REFERENCES

Almeida JS, Santos MB, Panissi DL, Garcia EC (2006) Effectiveness of low-cost thermal vacuum tests of a micro-satellite. Acta Astronaut 59(6):483-489. doi: 10.1016/j.actaastro.2006.03.003

Benjamin P, Weaver C (1961) The adhesion of evaporated metal films on glass. Proc Roy Soc Lond Math Phys Sci 261(1307):516. doi: 10.1098/rspa.1961.0093

Chen WT, Nelson CW (1979) Thermal stress in bonded joints. IBM J Res Dev 23(2):179-188. doi: 10.1147/rd.232.0179

Gilmore DG (1994) Satellite thermal control handbook. Vol. 1: Fundamental technologies. 2nd ed. El Segundo: The Aerospace Corporation Press.

Greenberg SA, Vance DA, Streed ER (1967) Low solar absorptance surface with controlled emittance: a second generation of thermal control coatings. In: Heller GB, editor. AIAA Progress in Astronautics and Aeronautics: thermophysics of spacecraft and planetary bodies. Vol. 20. New York: Academic Press. p. 297-314.

Iacovangelo CD, Pan Y, Wei C, Chen M, inventors; Lockheed Martin Corporation, assignee. 2003 Jul 1. Optical solar reflector. United States Patent US 6,587,263.

Instituto Nacional de Pesquisas Espaciais (2013) AMAZONIA 1. Satellite Environmental Specification. A820000-SPC-009/04.

Marshall KN, Breuch RA (1968) Optical solar reflector: a highly stable, low α_s/ε spacecraft thermal control surface. J Spacecraft Rockets 5(9):1051-1056.

Nagano H, Ohnishi A, Nagasaka Y (2011) Development of a lightweight deployable/stowable radiator for interplanetary exploration. Appl Therm Eng 31(16):3322-3331. doi: 10.1016/j.applthermaleng.2011.06.012

Tanzilli RA, Gebhardt JJ, inventors; The United States of America as represented by the Secretary of the Air Force, assignee. 1996 Jul 30. Optical solar reflector. United States Patent US 5,541,010.

7

Formation of Composite Polyaniline and Graphene Oxide by Physical Mixture Method

Ludmila Resende Vargas[1], Anne Karoline Poli[1], Rita de Cássia Lazzarini Dutra[1], Camila Brito de Souza[2,3], Maurício Ribeiro Baldan[1,4], Emerson Sarmento Gonçalves[1,2]

ABSTRACT: The development of polyaniline and graphene oxide composites aims to join the unique properties of each material for aerospace applications. The present paper demonstrates an easy and quick method, compared to the ones found in the literature, to obtain a composite made with polyaniline doped with dodecylbenzenesulfonic acid, a combination commonly called polyaniline, and graphene oxide. Nowadays, the most common studied methods are electrochemistry and *in situ* chemical polymerization. Differently from these methods, the films were obtained by a physical mixture of equimolar suspension of graphene oxide (4 mg/mL) with 3 concentrations of polyaniline powder: 25; 50 and 75%, being compared to pure graphene oxide and polyaniline. The morphology and structure behavior of all the films were studied, besides the bonding nature between both materials. The films were analyzed by scanning electron microscopy, X-ray diffraction, Fourier transform infrared spectroscopy, and differential scanning calorimetry. The apparent interaction between graphene oxide corrugated sheets and polyaniline grains was verified by scanning electron microscopy images. It can be noticed, as the concentration of polyaniline increases, that more polymer was entrapped. To prove the formation of polyaniline/graphene oxide composite, X-ray diffraction and Fourier transform infrared spectroscopy techniques demonstrated the changes on graphene oxide crystallographic plans and on the chemical bonding between polyaniline and graphene oxide, suggesting an interaction between polyaniline and graphene oxide, especially in the composite with 50% polyaniline/50% graphene oxide. Differential scanning calorimetry was used to highlight this effect through the increase in thermal stability. The method of physical mixture was efficient to obtain the polyaniline/graphene oxide composites.

KEYWORDS: Graphene oxide, Polyaniline, Structure, Morphology, FT-IR, DSC.

INTRODUCTION

The search for innovative technologies in the area of electronics and electromagnetic absorbing materials has grown in the last few years (Singh *et al.* 2011; Mi *et al.* 2008; Kim *et al.* 2010). Materials that combine good electric properties, thermal stability (Feng *et al.* 2011), and low cost of production are used as initial requirements for the development of new materials.

The development of polyaniline (PAni) and graphene oxide (GO) composites aims to join the unique properties of each material and to obtain a composite with better performance than the original materials. These composites can be easily applied in aerospace industry devices, such as sensors, radar absorbing systems (RAS), and energy storage elements. Both materials present special properties that can be changed according to their manufacturing parameters.

As described by Wang *et al.* (2009), graphene (GE) and materials based on it (GO and reduced GO — rGO) have attracted even more attention thanks to their applications as energy storage devices (Wei *et al.* 2015; Yang *et al.* 2013). GO can be considered the most common graphene-based material produced in large scale and at a lower cost.

Differently from GE, GO presents oxygenated, hydrophilic functional groups and can be easily dispersed in aqueous solution (Dreyer *et al.* 2010), facilitating the interaction between GO and PAni. Gupta *et al.* (2013) describe the GO as a material that presents a planar structure of oxidized carbon with sp^2 hybridization, rich in oxygen groups, such as hydroxyl, carboxyl,

1.Departamento de Ciência e Tecnologia Aeroespacial – Instituto Tecnológico de Aeronáutica – Programa de Pós-Graduação em Ciência e Tecnologia Espaciais – São José dos Campos/SP – Brazil. **2.**Departamento de Ciência e Tecnologia Aeroespacial – Instituto de Aeronáutica e Espaço – Divisão de Materiais – São José dos Campos/SP – Brazil. **3.**Universidade Federal de São Paulo – Instituto de Ciência e Tecnologia – Campus Parque Tecnológico – São José dos Campos/SP – Brazil. **4.**Instituto Nacional de Pesquisas Espaciais – Laboratório Associado de Sensores – São José dos Campos/SP – Brazil.

Author for correspondence: Emerson Sarmento Gonçalves | Departamento de Ciência e Tecnologia Aeroespacial – Instituto de Aeronáutica e Espaço – Divisão de Materiais | Praça Marechal Eduardo Gomes, 50 – Vila das Acácias | CEP: 12.228-904 – São José dos Campos/SP – Brazil | Email: sarmgon@yahoo.com

and epoxides. Dreyer *et al.* (2010) consider that the presence of these groups can be used as a good site for surface modification.

Regarding PAni and its special chemical property of doping and dedoping (ranging from acidic to basic medium), Feng *et al.* (2011) and Rakic *et al.* (2011) showed its suitable environmental stability and application versatility. This polymer can be produced in several oxidation states (leucoemeraldine, emeraldine, pernigraniline), each one characterized by a different structure, with several electrical properties and the ability to be produced by chemical or electrochemical synthesis (Tamburria *et al.* 2011; Maia *et al.* 2000; Sapurina and Stejskal 2010; Ćirić-Marjanović 2013). The formation of conductive structures in the polymer is shown in Fig. 1.

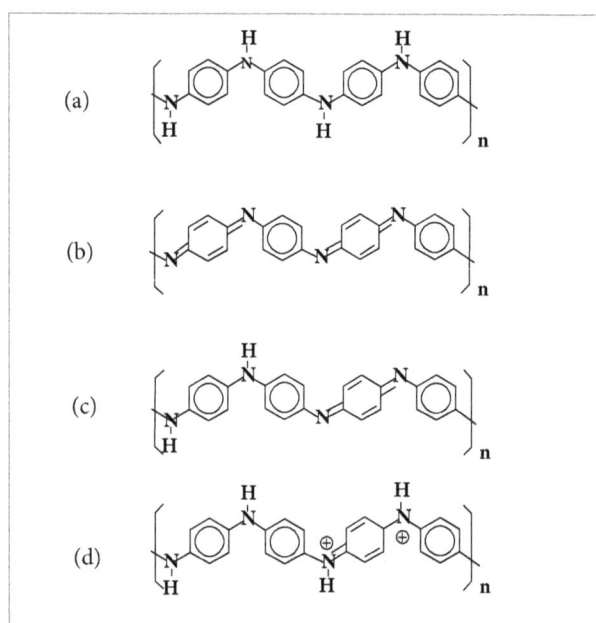

Figure 1. Linear structure of PAni and its several oxidation states. [a] Leucoemeraldine; [b] Pernigraniline; [c] Emeraldine; [d] Protoned emeraldine (Padilha 2011).

When the emeraldine base is doped in an acid environment, a polaron or bipolaron can be formed through successive formations of positive species. Bipolaron structures are thermodynamically more stable and conductive (Bockris and Reddy 2004). As mentioned by Bhadra *et al.* (2009), these polarons are responsible for electrical conduction through a jumps mechanism in the crystalline polymer region. The adjacent nitrogen electron (neutral) moves to a vacant spot and neutralizes it. Consequently, this spot moves, creating new spaces in the nitrogen structure and in the polarons structures, resulting in electron transportation and, thus, electrical conductivity along the chain (Bockris *et al.* 2004).

Recently, the PAni/GO composition was successfully synthesized by electrochemical and chemical methods applied to obtain supercapacitor materials (Wang *et al.* 2009). In this paper, a composite was obtained through a physical mixture of PAni and GO in suspension. The aim is to achieve, through this simple method, the formation of a PAni/GO composite and to study the potential interaction between PAni polar groups and the graphenic structure. The morphological, structural, chemical, and thermal aspects were analyzed. Furthermore, low-cost factors to develop a high-quality composite using innovative methods and processes were considered in the present study.

METHODOLOGY
POLYANILINE SYNTHESIS

The PAni synthesis was carried out through aniline monomer oxidation (1.0 mol/L aqueous solution) with ammonium persulfate (1.9 mol/L), in 1.0 mol/L solution of hydrochloric acid and sodium chloride. The solution was kept in a reactor at −5 °C under constant stirring, immersed in a bath of 1.0 L capacity with ethylene glycol in a temperature around −40 °C, while the ammonium persulfate solution was added during 50 min.

A dark green precipitate in the solution was observed, indicating the formation of PAni doped with HCl. After the complete addition of oxidizer, the system was left under mechanical stirring for 2 h. The polymer was removed from the reactor and washed several times with an 50% v/v ethanol/distilled water solution and a basic solution of NH_4OH (1.0 mol/L).

After dedoping, the material was washed with a 50% v/v ethanol/distilled water solution several times and filtered to obtain a particulate powder (base form of pure PAni).

It was added a quantity of 5 g of PAni in an aqueous solution containing 9 g of dodecylbenzenesulfonic acid (DBSA); this solution was left under constant stirring for 24 h at room temperature. The doped PAni-DBSA was washed with distilled water and ethyl alcohol, filtered, and the green powder was dried under vacuum at 60 °C until constant weight. The doping with functionalized DBSA was carried out by a solution method in the molar ratio 1:8 of PAni/DBSA. The PAni doped with this acid is referred to as "PAni" throughout this paper.

POLYANILINE/GRAPHENE OXIDE COMPOSITE

The cited method aims to verify the impact of solvent action in the composite formation. Furthermore, it is not previously described in the literature, and the interaction effects need to be confirmed.

Composites were prepared with doped PAni (2 h; 200 mesh/ 325 mesh) and GO from SIGMA-ALDRICH (4 mg/mL in water dispersion). Initially, the desired amount of PAni powder was transferred into a 10 mL-beaker and the mixture with GO; 5 mL of this mixture were transferred to a Petri dish and dried in a vacuum oven at 40 °C, resulting in a dried film. The films were obtained after 5 h of drying using 3 different PAni concentrations (25; 50 and 75%) that were compared with the original materials — GO and PAni.

CHARACTERIZATION METHODS

The morphological characterization of the films was performed using Zeiss® Leo 440 scanning electron microscopy (SEM), and the crystallographic structures of the films were analyzed using Panalytical® X-ray diffractometer (XRD), a rotating anode X-ray generator working at $5° \leq 2\theta \leq 90°$, with Cu monochromatic radiation (0.154 nm). Functional groups were determined by Fourier transform infrared (FT-IR) spectroscopy using PerkinElmer® Spectrum One spectrometer. The samples were prepared as KBr pressed pellet samples, and the FT-IR spectra were obtained by transmission mode from 4000 to 400 cm^{-1}. Differential scanning calorimetry (DSC) data were obtained with PerkinElmer® Pyris 1, at a heating rate of 5 °C/min in nitrogen atmosphere.

RESULTS AND DISCUSSION
SCANNING ELECTRON MICROSCOPY

The material morphology was studied by SEM. Figure 2 shows the original materials and the films that were obtained with the different ratios of PAni. The corrugated structures refer to the GO (Fig. 2a), and the granular ones, to the PAni (Fig. 2b). The corrugated structures are microscopically-oriented and grouped with a certain degree of parallelism, which was broken as the polymer was added. For all the different ratios of GO/PAni, the PAni was homogeneously surrounded by GO fibers. This morphology remains similar, in greater or lesser intensity, for all GO/PAni ratios.

This encapsulation effect appears more evident in the proportions of 50 and 75% of PAni, as shown in Figs. 2d and 2e. The interaction between PAni and corrugated sheets of GO can be easily noticed, as well as corrugated structures are localized between PAni particles, which can be partially encapsulated

as demonstrated in Fig. 3. In all cases, the GO acts as a placental structure, very thin and transparent, and involves most of the PAni particles. This fact may be explained by the presence of

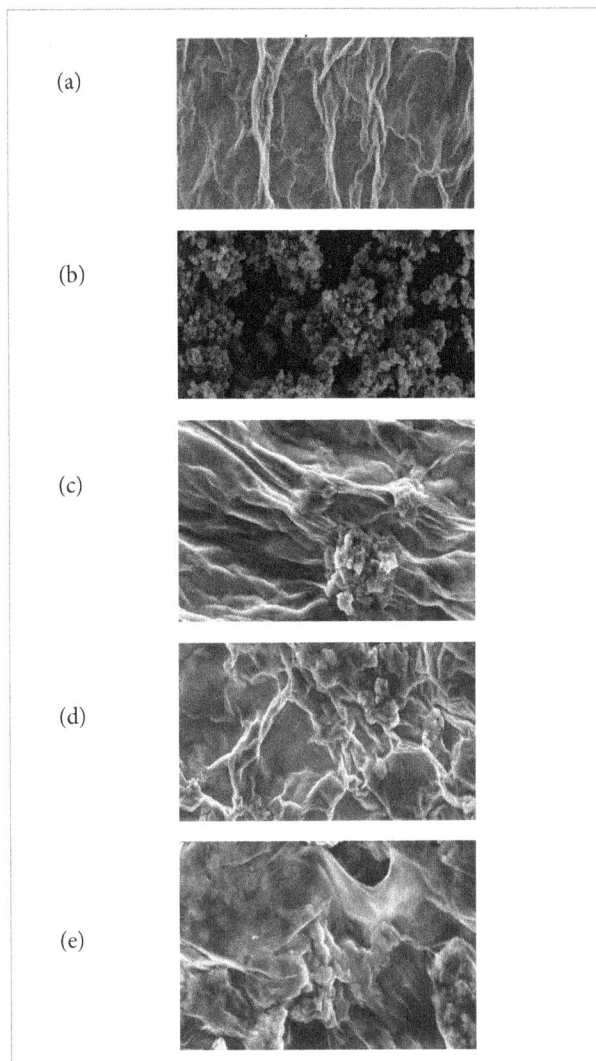

Figure 2. SEM of PAni, GO and their mixture (scale 10 μm). (a) GO (1,800X); (b) PAni (1,000X); (c) PAni 25%/GO 75% (2,700X); (d) PAni 50%/GO 50% (2,700X); (e) PAni 75%/GO 25% (3,000X).

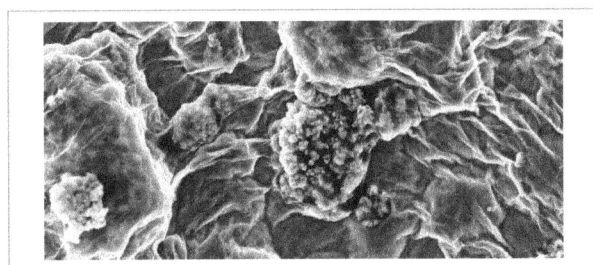

Figure 3. SEM of PAni 50%/GO 50% composite.

a solvent action (deionized water) to connect and encapsulate the PAni particles.

There is evidence of interactions between PAni and GO structures, so the nature of their bonds is not conclusive. Wang *et al.* (2010) made a relevant contribution and raised possible chemical bonds and electrostatics interactions between PAni and GO. They suggested a chemical synthesis of PAni/GO composites and its interactions using XRD and FT-IR techniques. With a different synthesis method compared with that of Wang *et al.* (2010), the present paper intends to analyze the existence of chemical bonds between PAni and GO, proving the composite formation.

As shown in Fig. 4, some interactions are noticed. By combining the results of UV-vis absorption, Raman spectra, and XPS analysis, Wang *et al.* (2010) proposed that the interaction of GO/PAni can be divided into 3 major interactions: (a) π-π stacking; (b) electrostatic interactions; and (c) hydrogen bonding. In order to get the best of both materials, it is important to understand the interaction between GO and PAni to exploit the potential for supercapacitor applications. The interaction between polar groups (polymer charge carriers) and oxygenated groups of GO results in the possible formation of ionic or coordinated complexes. In addition, hydrogen bonds may be formed between the non-protonated amine groups with hydroxyl hydrogen atoms. This

can also occur between epoxide groups and hydrogen atoms attached to electrically-charged nitrogen (Wang *et al.* 2010).

X-RAY DIFFRACTION

The structure of the composites was investigated by powder XRD measurements. The XRD spectra were measured in a range of 2θ from 5° to 70°, showing an intense and sharp peak at $2\theta = 8.7°$ (GO) corresponding to the interplanar space of GO sheets. The reflection peak can be assigned to (001). According to Blanton and Majumdar (2013), the 2θ peak of GO can shift from 7° to 12°, depending on the amount of residual water that can be intercalated between the basal planes. Zhang *et al.* (2010) reported that the reflection peak might depend on 2 major contributions: the method of preparation and the number of layers of water below the GO surface film. Michell *et al.* (2015) observed that the characteristic peak of the graphite used to synthesize the graphite oxide shifts from $2\theta = 26.4°$ to $2\theta = 10.61°$. This shift was attributed to oxygenated groups and water molecules inserted in the interlayer of graphite. It is important to remember that GO has a 2-D form of graphite and presents oxygen function groups in its basal planes and edges providing compatibility with polymer matrices. The GO surface acts as a nucleation site for the PAni due to oxygen functional groups on GO surface. The tuning of the amount of PAni in GO introduced some interesting aspects in XRD and in the morphology (Fig. 2). After the formation of GO/PAni, it is observed that the GO peak height intensity shows a very strong decrease. Furthermore, the peak broadening is very apparent for all the 3 compositions of GO/PAni (inset of Fig. 5). Another interesting aspect is that the peaks shift to

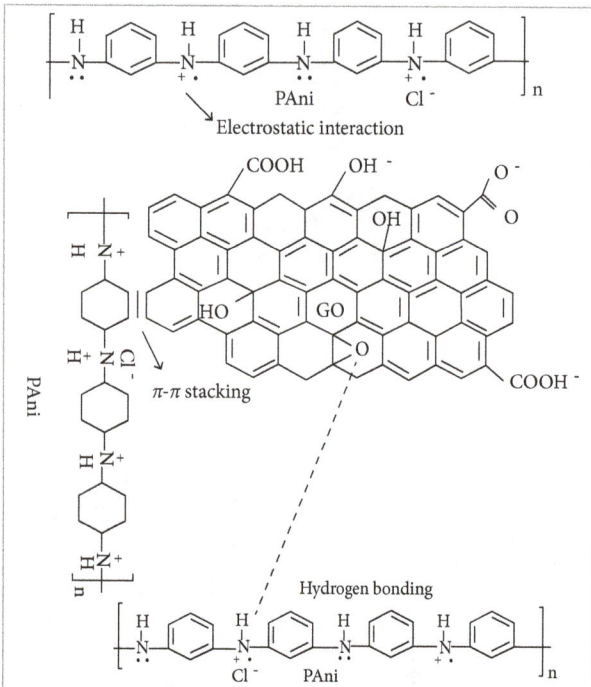

Figure 4. Possible interaction between PAni and GO structures, adapted from Wang *et al.* (2010).

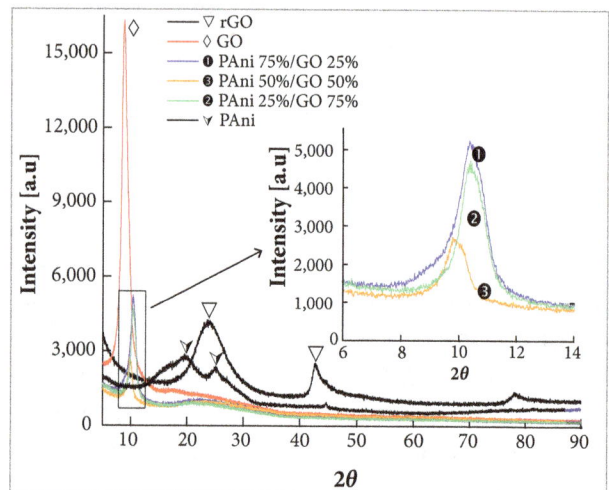

Figure 5. XRD of PAni/GO composites.

a higher 2θ. Singh *et al.* 2011 observed the decrease in GO height peak. According to them, the GO peak decreases its height intensity when residual water is removed from the GO surface, which suggests a possible interaction between PAni and GO that may be influenced by water content or oxygen function groups on the GO surface. The pure PAni exhibits 2 important reflection peaks, $2\theta = 2\theta°$ and $2\theta = 25°$, that can be associated to some crystalline order in the bulk PAni. The PAni reflection peaks decrease drastically during the formation process of GO/PAni, which was revealed by the shallow reflection in just one peak around $2\theta = 21°$. This change is associated with PAni/GO interaction and crystallographic sheets changes. The combination of the substantial shift of the PAni, in the formation process of PAni/GO, and the decrease in GO peak height intensity justify the formation of a hybrid composite. By analyzing the XRD of the 3 compositions (PAni 25%/GO 75%, PAni 50%/GO 50% and PAni 75%/GO 25%), a new, broad, and minor peak appears indicating the intercalation of PAni between GO sheets.

The broadening of the peaks is evident for all the PAni/GO compositions studied in this paper. Due to the interaction between PAni and GO, the material structure may change and affect their thermal and mechanical stability. In order to confirm the nature of these interactions, the FT-IR and DSC techniques are discussed next.

FOURIER TRANSFORM INFRARED SPECTROSCOPY

FT-IR spectroscopy was used to elucidate the nature of the interaction between PAni and GO. Figure 6 shows the spectra of PAni, GO, and of their mixtures. It is possible to observe the greatest similarity between the spectra of PAni 25%/GO 75% and the spectra of GO. The samples containing PAni 50%/GO 50% and PAni 75%/GO 25% demonstrated similarity with the PAni, especially PAni 50%/GO 50%, suggesting that the interaction effect achieved an optimized level.

All contributions were grouped in 10 different items as shown in Table 1. The main spectra changes in the medium infrared region (MIR) are included in Table 1 as well. In order to analyze and compare all the spectra, they were divided into different intervals in this region. It is very important to notice that, in MIR region, from 4000 down to 1800 cm^{-1}, the spectra showed a broad and intense band related to stretching O-H and N-H in hydroxyl and amine groups, respectively. In addition, the presence of moisture in KBr pads, used to perform the

experiment by transmittance or humidity from GO, could contribute to this. The band may be associated to hydrogen bond between NH from PAni and possible oxygenated group from GO. The behavior of the OH group can suggest electrostatic interaction (Smith 1979), and the relevant FT-IR bands in the MIR region are shown in Table 1 (item 1).

In the range from 1,800 to 1,600 cm^{-1}, it is possible to analyze the GO spectrum and its consequences in the presence of different PAni/GO compositions (Table 1, items 2 and 3). Probably, π-π stacking is a predominant effect, which moves electrons preferably from GO to PAni structure. The main bands at 1720 and 1630 cm^{-1} in the spectrum of GO are related to C=O stretching in COOH group and C=C in aromatic rings, respectively. However, 1630 cm^{-1} band may be associated to vibrations of the residual water mainly coming from KBr pads. This effect certainly masks the analysis of this band. It is possible to notice the constant presence of these bands for all PAni/GO compositions. Furthermore, a noticeable blueshift at 1720 and 1630 cm^{-1} regarding PAni 50%/GO 50% was observed (Fig. 7), which may be related with 2 phenomena: electrostatic attraction of carboxylic oxygen and π-π stacking from C=C supplied from graphene sheets to benzenoid structure of PAni, once the

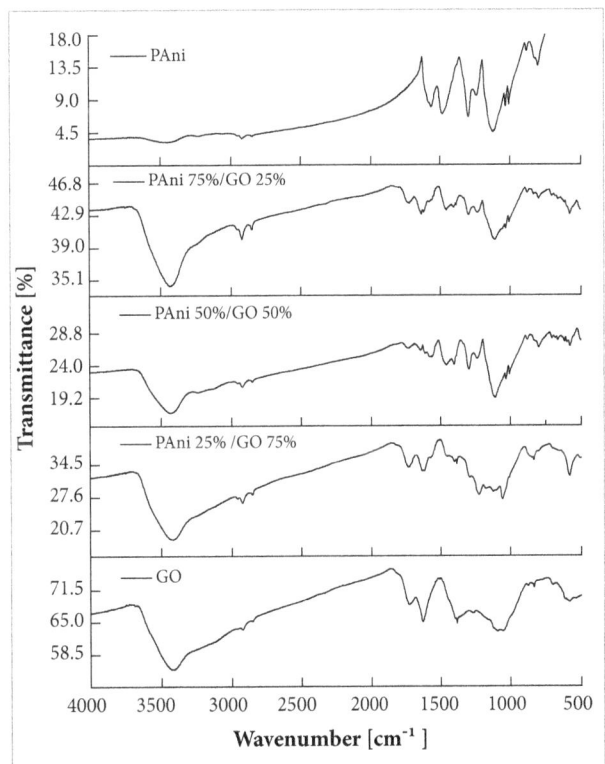

Figure 6. Transmission MIR spectra of PAni/GO, prepared as KBr pellets.

Table 1. More relevant FT-IR bands to the obtained composites.

Item	GO (cm⁻¹)	PAni (cm⁻¹)	25% PAni (DBSA)/ 75% GO (cm⁻¹)	50% PAni (DBSA)/ 50% GO (cm⁻¹)	75% PAni (DBSA)/ 25% GO (cm⁻¹)	Wavenumber shift	Changing on form and intensity	Attribution and probable effects (Wang et al. 2009; Smith 1979)
1	3413	3442	3422	3431	3431	In relation to GO, it increased with the increasing in PANI ratio; in relation to PAni spectrum, a decrease was observed	Widening observed for PAni 50%/GO 50% and band higher than others for PAni 75%/GO 25%	- ν OH (GO) and ν NH (PAni/DBSA), referring to hydrogen bond between NH from PAni and possible oxygenated group from GO, except hydroxyl (possibly carboxyl or epoxyde). The behavior of OH group can suggest electrostatic interaction. Furthermore, there is a possible overlap with humidity from KBr pad (Smith 1979)
2	1720	----	1727	1734	1717	Highest value for PAni 50%/GO 50%	The lowest intensity for PAni 50%/GO 50%	νC=O in COOH from GO, which may pass to C=O similar to ester (Smith 1979). Suggested changing on mode of electrostatic interaction. Possible action of DBSA (PAni dopant)
3	1630	-----	1616	1641	1632	The lowest value for PAni 25%/GO75% and the highest for PAni 50%/GO 50%	Intensity decreases, especially for PAni 50%/GO 50%	π–π stacking, possibly masked to PAni 50%/GO 50% by KBr humidity
4	----	1559	Shoulder at 1559	1564	Shoulder at 1556	No variation	The highest value for PAni 50%/GO 50%	Quinoid structure (N=Q=N) in aniline oligomers (Wang et al. 2009; Tang 1988)
5	----	1481	Shoulder at 1465	1457	1457	Decrease	It is visible for PAni 50%/GO 50% and PAni 75%/GO 25%	π (C=C) in benzenoid ring (N-B-N) (Tang 1988; Dmitrieva and Dunsch 2011). Wang et al. (2009) suggest π–π stacking
6	1385	---	1385	1384	1385	No variation	Apparently, intensity decreased until maximize PAni ratio	δ OH acid and δ OH phenol (GO) (Smith 1979)
7	---	1297	Shoulder at 1297	1296	1296	No changes	The highest value for PAni 50%/GO 50%	υ C-N in QB$_{cis}$Q, QBB, BBQ (Wang et al. 2009) and υ C-N secondary aromatic amine (Smith 1979)
8	----	1240	1232	1239	1234	The lowest value for PAni 25%/GO 75%	Apparently, the highest value for PAni 50%/GO 50%	ν C-N in BBB (Wang et al. 2009). Probably performing on π-π stacking
9	---	1122	1125	1108	1108	Decrease	The highest intensity for PAni 50%/GO 50%	1,170 – 1,140 cm⁻¹ refers to vibrational mode Q=NH⁺-B or B-N⁺H-B (related to electrical conductivity). At 1,115 cm⁻¹, substituting 1,2,4 of ring (Zeng and Ko 1998; Tang 1988; Dmitrieva and Dunsch 2011). It can be related to π-π stacking and hydrogen bonds with oxygen groups from GO
10	---	800	Shoulder at 800	798	797	No variation	The highest value for PAni 50%/GO 50%	Substituting 1,4 of ring (Wang et al. 2009 and Smith 1979)

blueshift of this band may be related to decreased availability of electrons π in graphenic rings.

Concerning the second region (Fig. 8), in the range from 1,600 to 1350 cm^{-1}, the bands at 1481 and 1559 cm^{-1} are assigned to C=C in benzenoid and quinoid structures, respectively. The 1481 cm^{-1} mode in the PAni spectrum was redshifted from its position for all PAni/GO combinations. The blueshift for the mode 1559 cm^{-1} in combination with a decrease in the benzoid structure may be associated to the greater electron availability from GO in the π-π stacking. Besides, the electron availability contributes to the reduction of quinoid structures. This behavior was described in Table 1 (items 4 and 5).

The band at 1296 cm^{-1} (Fig. 9), observed in the PAni spectrum, was attributed to C-N stretching of secondary amine in QB$_{cis}$Q, QBB, and BBQ units (Table 1, item 7). It can be observed that this band is more evident for the PAni 50%/GO 50%. This effect is related to the alternation of benzenoid and quinoid structures. The next band, at 1240 cm^{-1} (Fig. 9), is related to C-N stretching in BBB units (and it is a result of the reduction from quinoid to benzenoid structures, as discussed next, coherent with the proportion between the bands 1,564 and 1457 cm^{-1}) and is decreased for this composition.

The band at 1122 cm^{-1} (Fig. 9) may be associated to p electron cloud in PAni chain (Table 1, item 9). No relative increase was observed for the charged polymer unit Q=NH^{+}-B or B-N^{+}H-B. This structure may be suited to evaluate the degree of doping. As can be observed, this effect is more pronounced in PAni 50%/GO 50%, indicating that this composition favors a conductivity greater than the others.

According to Table 1, all effects noticed in the literature are present, but with predominance of π-π stacking. Concerning GO bands, shifts are observed for higher values of wavenumber, while, for PAni, values decrease in contrast to those outlined by Wang et al. (2009). However, the techniques and orders of magnitude of ratios are radically different. The effects observed in the interaction are more highlighted for the composite PAni 50%/GO 50%.

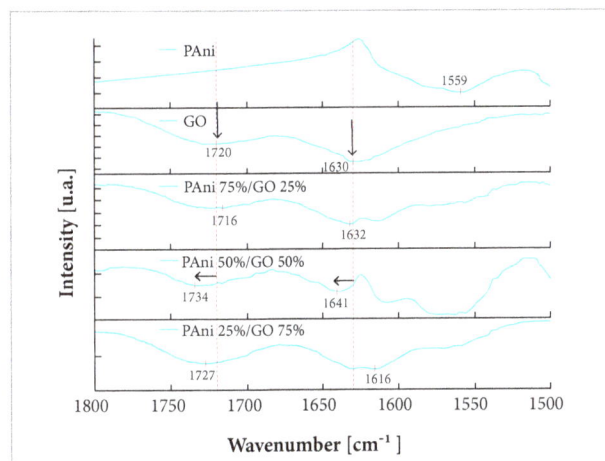

Figure 7. FT-IR partial spectra: first significant region of MIR for PAni/GO composites.

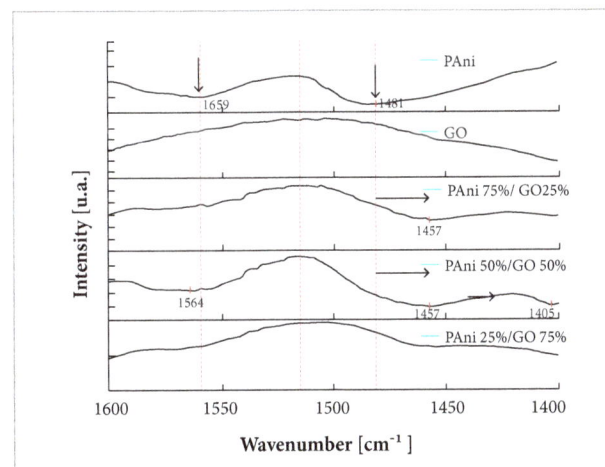

Figure 9. FT-IR partial spectra: third significant region of MIR for PAni/GO composites.

DIFFERENTIAL SCANNING CALORIMETRY

The thermal analysis was made in order to study the thermal stability of the PAni/GO composites. Energy variation behavior can be associated to possible links between groups of PAni and GO.

According to DSC analysis, PAni exothermic peaks are associated to cross-linking reactions during the annealing process, as observed by Cheng (2002). This author attests that

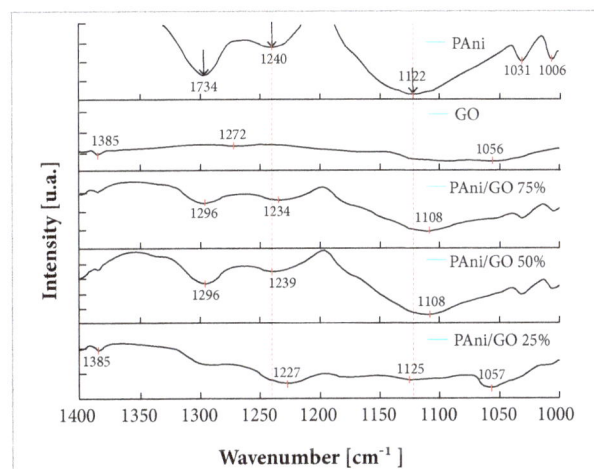

Figure 8. FT-IR partial spectra: second significant region of MIR for PAni/GO composites.

PAni may suffer a change in conformation or reticulation reaction above 100 °C. An increase from 107 °C (Fig. 10) up to around 180 °C (Fig. 11) was observed for this peak temperature when PAni was combined with GO.

These data indicate the formation of PAni cross-links. However, in the presence of GO, there is an engagement of PAni p electrons in a stacking with GO aromatic structures, decreasing their availability to obtain s bonds of cross-linking. Due to this fact, the cross-linking reaction is thermodynamically disfavored at temperatures below 170 °C, probably because these interactions. This effect is more evident in PAni 50%/GO 50%, which showed a cross-linking temperature higher than that of PAni alone. In all compositions of PAni/GO, a very strong increasing in the process enthalpy was noticed. The jumping from −94.3 J/g (Fig. 10) up to values between 2 and 7 times more than that, inversely proportional to the concentration of PAni (Fig. 11), highlights the effect of entrapment of PAni by GO and its difficulty to form cross-linkings, when trapped by GO. This effect indicates that the stability of PAni/GO is possibly valid to space application because of normative limitation. For example, the European Aerospace

Norms EN 4644-011:2015-8, EN 4644-013:2015-08, and EN 4644-131:2015-8 determine that aerospace materials must resist temperatures below 175 °C. For aeronautic applications, MIL-STD-810 determines that the materials must be stable until 85 °C. So, for aerospace application, PAni/GO seems to be more suitable than PAni.

The GO does not have an exothermic peak, as illustrated in Fig. 12, but an endothermic one can be noted, which merely indicates loss of water and, therefore, does not interfere in the composite thermal stability.

The evidence of electronic interactions of π-π stacking on composites was strongly confirmed by DSC, and the method of physical mixture has proven relatively easy and successful to obtain PAni/GO composites.

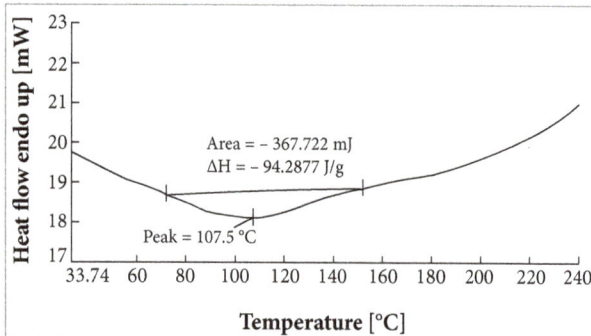

Figure 10. DSC thermogram of doped PAni.

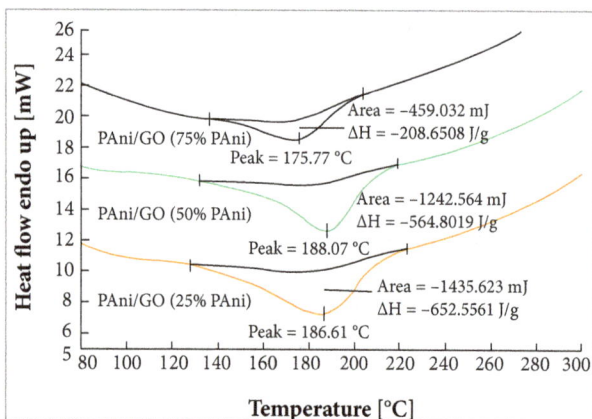

Figure 11. DSC of PAni/GO composites in different ratios.

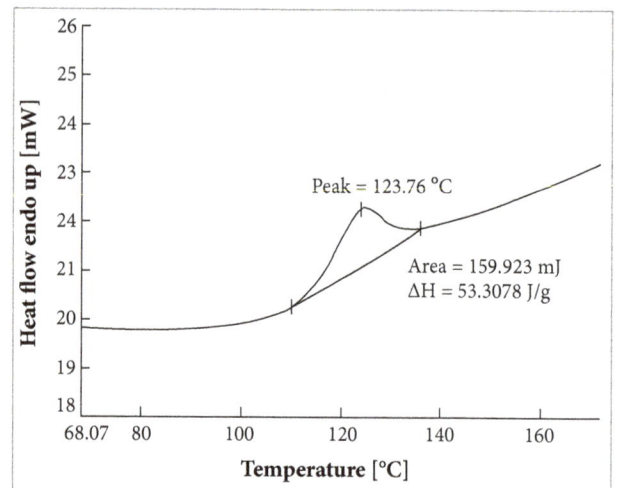

Figure 12. DSC of GO.

CONCLUSION

Through the experiments conducted with the used composite formulation, it is noticed that there is an increase in the attraction intensity of the corrugated GO and PAni particles through solvent action. However, this behavior is not linear.

It can be seen, within the examined range, that this effect is most pronounced in the PAni 50%/GO 50% composite. XRD tests show important interactions that occur between PAni structure and oxygen groups from GO, provoking a possible crystallographic rearrangement. This occurs, according to FT-IR data set, because of the important interaction between PAni and graphene structure, due to predominant p-p stacking; however, other effects may be observed (electrostatic interactions and, in lower incidence, hydrogen bonds). The thermal effect of this interaction is the formation of a more stable conformed PAni, disfavoring cross-linking reactions and improving the composite thermal stability, which, around 175 °C, is desirable for aerospace applications.

ACKNOWLEDGEMENTS

The authors would like to thank Andreza de Moura Cardoso, Maria Aparecida Miranda de Souza, Rogério Duque Gonçalves, Aline Fontana Batista, and João Marcos Kruczinski, from the Departamento de Materiais of the Instituto de Aeronáutica e Espaço, for their assistance and support in this project.

Special thanks also go to Milton Faria Diniz, from the Divisão de Química of the Instituto de Aeronáutica e Espaço, and to Coordenação de Aperfeiçoamento de Pessoal de Nível Superior/ Instituto Tecnológico de Aeronáutica, for the research grant PVS process that made this research possible.

AUTHOR'S CONTRIBUTION

Vargas LR, Dutra RCL, Baldan MR, and Gonçalves ES conceived the idea and co-wrote the main text; Vargas LR, Poli AK, and Souza CB performed the experiments; Vargas LR, Baldan MR, Souza CB, and Gonçalves ES prepared the figures. All authors discussed the results and commented on the manuscript.

REFERENCES

Bhadra S, Khastgir D, Singha NK, Lee JH (2009) Progress in preparation, processing and applications of polyaniline. Prog Polymer Sci 34(8):783-810. doi: 10.1016/j.progpolymsci.2009.04.003

Blanton TN, Majumdar D (2013) Characterization of X-ray irradiated graphene oxide coatings using X-ray diffraction, X-ray photoelectron spectroscopy, and atomic force spectroscopy. Powder Diffr 28(2):68-71. doi: 10.1017/S0885715613000109

Bockris JO, Reddy AKN (2004) Modern electrochemistry. 2nd edition. New York: Kluwer Academic Publishers.

Cheng CH (2002) Thermal studies of polyaniline doped with dodecyl benzene sulfonic acid directly prepared via aqueous dispersions. J Polymer Res 9(3):195-200. doi: 10.1023/A:1021395726060

Ćirić-Marjanović G (2013) Recent advances in polyaniline research: polymerization mechanisms, structural aspects, properties and applications. Synthetic Met 177:1-47. doi: 10.1016/j.synthmet.2013.06.004

Dmitrieva E, Dunsch L (2011) How linear is "linear" polyaniline? J Phys Chem B 115(20):6401-6411. doi: 10.1021/jp200599f

Dreyer DR, Park S, Bielawski CW, Ruoff RS (2010) The chemistry of graphene oxide. Chem Soc Rev 39:228-240. doi: 10.1039/B917103G

Feng XM, Li RM, Ma YW, Chen RF, Shi NE, Fan QL, Huang W (2011) One-step electrochemical synthesis of graphene/ polyaniline composite film and its applications. Adv Funct Mater 21(15):2989-2996. doi: 10.1002/adfm.201100038

Gupta RK, Alahmed ZA, Yakuphanoglu F (2013) Graphene oxide based low cost battery. Mater Lett 112:75-77. doi: 10.1016/j.matlet.2013.09.011

Kim H, Abdala AA, Macosko CW (2010) Graphene/polymer nanocomposites. Macromolecules 43(16):6515-6530. doi: 10.1021/ma100572e

Maia DJ, De Paoli MA, Alves OL, Zarbin AJG, Neves S (2000) Síntese de polímeros condutores em matrizes hospedeiras. Quím Nova 23(2):204-215. doi: 10.1590/S0100-40422000000200011

Mi H, Zhang X, Yang S, Ye X, Luo J (2008) Polyaniline nanofibers as the electrode material for supercapacitors. Mater Chem Phys 112(1):127-131. doi: 10.1016/j.matchemphys.2008.05.022

Michell E, Candler J, De Souza F, Gupta RK, Gupta BK, Dong LF (2015) High performance supercapacitor based on multilayer of polyaniline and graphene oxide. Synthetic Met 199:214-218. doi: 10.1016/j.synthmet.2014.11.028

Padilha RMA (2011) Estudo de transporte de cargas de polímeros de polianilina (Master's thesis). Rio de Janeiro: Pontifícia Universidade Católica.

Rakic A, Bajuk-Bogdanović D, Mojović M, Ćirić-Marjanović G, Milojević-Rakić M, Mentus S, Marjanović B, Trchová M, Stejskal J (2011) Oxidation of aniline in dopant-free template-free dilute reaction media. Mater Chem Phys 127(3):501-510. doi: 10.1016/j.matchemphys.2011.02.047

Sapurina IY, Stejskal J (2010) The effect of pH on the oxidative polymerization of aniline and the morphology and properties of products. Russian Chemical Review 79(12):1123-1143. doi: 10.1070/RC2010v079n12ABEH004140

Singh V, Joung D, Zhai L, Das S, Khondaker SI, Seal S (2011) Graphene based materials: past, present and future. Progr Mater Sci 56(8):1178-1271. doi: 10.1016/j.pmatsci.2011.03.003

Smith AL (1979) Applied infrared spectroscopy: fundamentals, techniques, and analytical problem-solving. New York: Wiley-Interscience.

Tamburria E, Orlanducci S, Guglielmotti V, Reina G, Rossi M, Terranova ML (2011) Engineering detonation nanodiamond — polyaniline composites by electrochemical routes: structural features and functional characterizations. Polymer 52(22):5001-5008. doi: 10.1016/j.polymer.2011.09.003

Tang J (1988) Infrared spectra of soluble polyaniline. Synthetic Met 24:231-238. doi: 10.1016/0379-6779(88)90261-5

Wang H, Hao Q, Yang X, Lu L, Wang X (2009) Graphene oxide doped polyaniline for supercapacitors. Electrochem Comm 11(6):1158-1161. doi: 10.1016/j.elecom.2009.03.036

Wang H, Hao Q, Yang X, Lu L, Wang X (2010) Effect of graphene oxide on the properties of its composite with polyaniline. ACS Appl Mater Interfaces 2(3):821-828. doi: 10.1021/am900815k

Wei P, Fan M, Chen H, Yang X, Wu H, Chen J, Li T, Zeng L, Zou Y (2015) High-capacity graphene/sulfur/polyaniline ternary composite cathodes with stable cycling performance. Electrochimica Acta 174:963-969. doi: 10.1016/j.electacta.2015.06.052

Yang JE, Jang I, Kim M, Baeck SH, Hwang S, Shim SE (2013) Electrochemically polymerized vine-like nanostructured polyaniline on activated carbon nanofibers for supercapacitor. Electrochim Acta 111:136-143. doi: 10.1016/j.electacta.2013.07.183

Zeng XR, Ko TM (1998) Structures and properties of chemically reduced polyanilines. Polymer 39(5):1187-1195. doi: 10.1016/s0032-3861(97)00381-9

Zhang K, Zhang LL, Zhao XS, Wu J (2010) Graphene/polyaniline nanofiber composites as supercapacitor electrodes. Chem Mater 22(4):1392-1401. doi: 10.1021/cm902876u

8

NSGA-II-Based Multi-objective Mission Planning Method for Satellite Formation System

Xiaowei Shao[1], Zehao Zhang[1], Jihe Wang[1], Dexin Zhang[1]

ABSTRACT: This study proposes a non-dominated sorting genetic algorithm-II-based multi-objective optimization method to solve the multi-objective mission planning problem for satellite formation flying system which has the ability to obtain both digital elevation map and ground moving target indicator information at the same time when certain conditions are satisfied. The 2 objectives considered in this study are maximizing total profits and maximizing numbers of completed acquisitions. Thus, the multiple-objective satellite scheduling optimization problem is formulated and solved by the proposed method. Its validity and effectiveness are verified by numerical simulations, and the results show that it can achieve better performance with respect to 2 different objectives in an overall perspective than the traditional scheduling optimization, which can consider only 1 objective.

KEYWORDS: NSGA-II method, Multi-objective mission planning, DEM and GMTI, Formation flying system.

INTRODUCTION

With the development of small satellite technology, formation flying becomes a new enabling technique for many space missions, such as virtual synthetic aperture radar (SAR), space surveillance system, etc. In this study, we focus on the interferometric SAR (InSAR) mission by using new proposed formation flying technology. Differently from traditional space mission, by satisfying certain conditions, formation flying-based InSAR mission can perform both digital elevation map (DEM) and ground moving target indicator (GMTI) missions at the same time. Hence, to increase the system effectiveness, it is essential to propose a multiple-objective satellite scheduling optimization method to deal with the new formation flying-based InSAR mission planning problem. In order to propose the new multi-objective mission optimization method, the relevant literature review regarding to satellite mission planning is conducted here.

In relation to traditional single satellite mission scheduling problem, Bensanna *et al.* (1996) studied the daily scheduling problem associated with the SPOT5 satellite and generalized the satellite image collection planning as a version of the knapsack model, in which the objective was to maximize a profit function. They proposed a Tabu search algorithm to determine the image collection schedule. Lemaître *et al.* (2002) applied 4 methods to solve the simplified scheduling problem, which are greedy algorithm, dynamic programming, constraint programming, and local search method. Jang *et al.* (2013) provided a heuristic solution approach to solve the image collection planning problem of a Korean satellite, KOMPSAT-2.

With the development of multiple satellite system, many researchers have proposed various methods to solve optimal

1.Shanghai Jiao Tong University – School of Aeronautics and Astronautics – Distributed Spacecraft System Technology Laboratory – Shanghai – China.

Author for correspondence: Xiaowei Shao | Shanghai Jiao Tong University – School of Aeronautics and Astronautics – Distributed Spacecraft System Technology Laboratory | Shanghai 200240 – China | Email: shaoxwmail@163.com

mission scheduling for multiple satellite system. Xiaolu *et al.* (2014) partitioned the problem into 2 sub-problems: task assignment and task merging. In task assignment, they proposed an adaptive ant colony optimization algorithm to select specific time window for each task and create a task list for each satellite. In task merging, they proposed the concept of task combination and developed a dynamic programming algorithm to find the best merging plan for each satellite. Bianchessi *et al.* (2007) studied the selecting and scheduling requests for the multi-satellite, multi-orbit, and multi-user. An upper bounding procedure based on column generation is used to evaluate the quality of the solutions. Kim and Chang (2015) performed the optimization of constellation operation by minimizing the system response time. He *et al.* (2011) formulated the model of satellites observation scheduling problem with task merging and developed the simulated annealing algorithm to solve the mission scheduling problem. Gao *et al.* (2013) designed a framework of ant colony algorithm for remote satellite and ground integration scheduling problem in the parallel environment. Hao *et al.* (2013) proposed a combination of genetic and ant colony algorithms to solve the mission scheduling problem. The performance of the proposed algorithm was compared with the genetic one. Tangpattanakul *et al.* (2015) presented an indicator-based multi-objective local search (IBMOLS) to solve a multi-objective optimization problem. The objectives are to maximize the total profit of the selected acquisitions and simultaneously to ensure the fairness of resource sharing by minimizing the maximum profit difference between users.

As aforementioned, there are various methods to deal with single task mission planning problem with 1 satellite; however, due to the unique features brought by the new developed formation flying technology, these methods cannot solve the multiple tasks scheduling problem of formation flying mission considered in this study. Hence, the multiple objective optimization method to handle satellite formation flying system with multiple task abilities is proposed here.

In this study, we focus on the mission scheduling problem for formation flying system which has multiple task ability. Instead of single objective mission planning, 2 objectives are considered, which are total profit maximization and completed mission numbers maximization. The contribution of this study is that, to solve the unique formation flying SAR mission planning which can perform 2 different tasks at the same time, a non-dominated sorting genetic algorithm (NSGA)-II-based multi-objective mission planning optimization method with

several practical constraints is proposed. The new proposed method can achieve better performance with respect to 2 different objectives in an overall perspective than traditional genetic algorithm method, which can optimize only 1 objective.

PROBLEM DESCRIPTION

This section describes the imaging process of InSAR satellite, introduces the mission scheduling problem of InSAR formation flying system, and formulates the mathematical model of the multi-objective scheduling problem.

OBSERVATION CHARACTERISTICS OF SYNTHETIC APERTURE RADAR SATELLITE

SAR satellites collect image data by its SAR sensor, which observes the Earth by sweeping mode. The observation range is composed of multiple observation strips. The length of the strip depends on observing time and its width depends on the inner and outer half angle, as illustrated in Fig. 1. For simplicity, we assume the widths of all strips are equal, and there is a small amount of overlap between different observing strips.

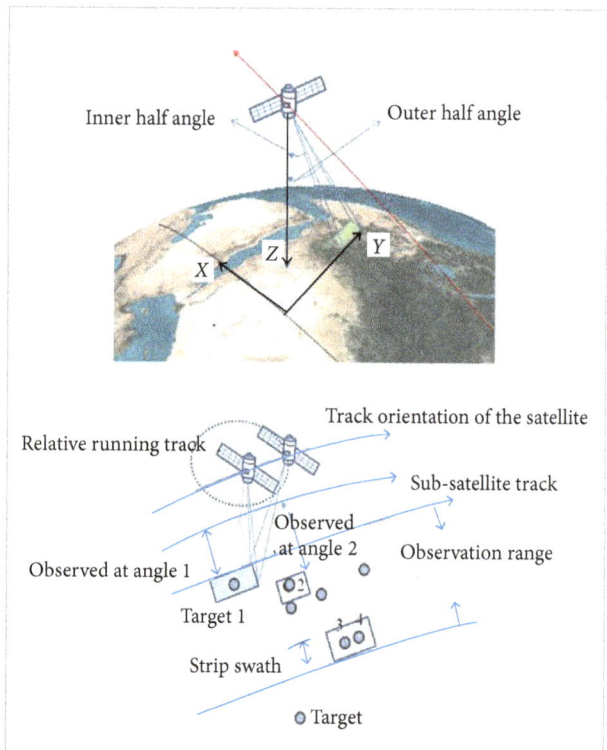

Figure 1. Observation process of the SAR satellite.

Because the observation ranges can overlap, some targets can be observed by 2 strips, as illustrated in Fig. 2. Hence, each target can have more than 1 observation chance. With the help of orbit predicting and satellite-ground visibility computing software (Satellite Tool Kits — STK), the observation time windows between satellites and targets can be obtained as the input for the scheduling optimization algorithm, as illustrated in Table 1.

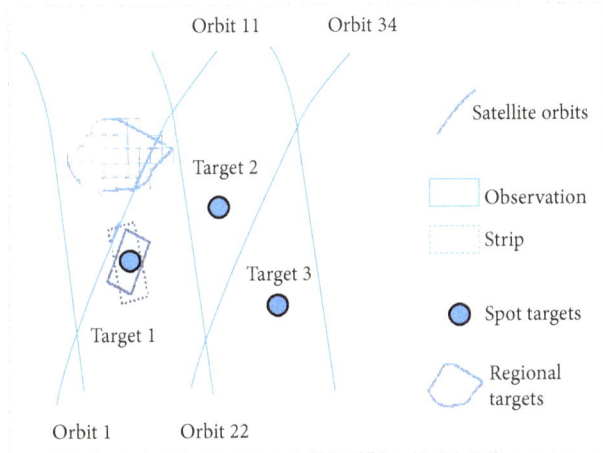

Figure 2. Satellite orbits which can observe targets.

Table 1. Observation opportunity of targets.

Target	Satellite orbits which can observe targets
1	1 11 34
2	1 11 22
3	1 11 22 34

OBSERVATION TARGET

Depending on the size, all the observation areas can be grouped into 2 types: point targets and regional targets. Point targets can be observed by 1 strip and the regional one needs more strips to complete region observation. Therefore, the method to decompose regional targets into multiple point ones is shown in Fig. 3, and the detail procedure of decomposition will be introduced in the next section.

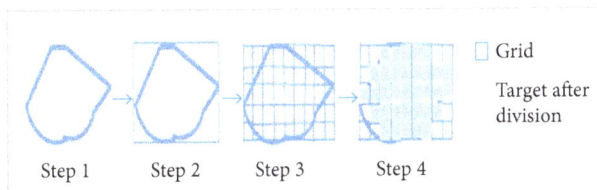

Figure 3. The steps of regional target decomposition.

MULTI-OBJECTIVE DESCRIPTION

By satisfying certain geometrical conditions, formation flying-based InSAR system can obtain DEM and GMTI images at the same time. For instance, in Fig. 1, by using strip 1, the satellite can observe target 1.The observation of targets 1 and 2 can satisfy angle switching constraints. The observing angle of targets 1 and 2 can obtain images satisfying 2 tasks (DEM and GMTI). But there is time window conflict between target 2 and targets 3 and 4; the observing angle of targets 3 and 4 can only obtain 1 image satisfying 1 task (DEM). If we choose to observe targets 3 and 4, we can complete 2 targets observation. If we choose to observe target 2, we can only obtain 1 GMTI image. Thus, we need to choose the target to be observed on the basis of objective and constraints. This study studies the multi-objective mission scheduling optimization problem of InSAR formation flying system. The constraints considered here include the maximum number of observation in 1 orbit, the longest observation time in 1 orbit, and the time to switch from a strip to another; the 2 optimization objectives are maximizing the total profit of the selected acquisitions and the number of finished acquisitions.

MATHEMATICAL FORMULATION

Based on the above discussion, the mathematical model of multi-objective mission planning is established and represented by the operator (O, A, T, C, F), where O is the set of orbit numbers of satellites, $O = \{1, 2, ..., Num\_O\}$; A is the set of strips of each orbit, $A = \{1, 2, ..., Num\_a\}$; C is the set of observing chances; F is the set of objective functions. For simplicity, we assume that the number of strips for each orbit is equal, and the satellite can perform 2 imaging tasks (DEM and GMTI) by using the first 3 strips (1, 2, and 3); $T = \{t_1, t_2, L, t_m\}$ is the set of observation targets including point and regional targets, where denotes the total number of targets. For each target, we define $t_i = \{p_i, chance_i\}$, where p_i is the profit when target i is observed, whose value varies from 1 to 10, and $chance_i$ is defined as $chance_i\{c_1, c_2, L, c_n\}$, where n is the total number of observation chances of target i, $c_i = \{a_i, w_i, we_i, o_i\}$, where a_i denotes the observation strip and satisfies $a_i \in A$, ws_i and we_i stand for start and end times of observation time window, respectively, and o_i stands for the sequence number of observation orbit. Apart from the aforementioned variables, there are other parameters defined as follows:

Time_start: observation mission start time.

Time_end: observation mission end time.

min_ont: the shortest observation time for 1 strip.

max_ont: the shortest observation time for 1 strip.

g_onum: maximum observation numbers in 1 orbit.

g_maxt: the longest observation time in 1 orbit.

time_onf: the shortest time to restart an observation since the previous observation is stopped.

bt: switching time from one strip to another.

Mg: satellite memory storage capacity in 1 orbit.

Num_O: total number of observation time (unit: orbits).

Num_a: total number of observation strips.

n^g: total number of strip switches in the gth orbit.

ont_k^g: k^{th} observing start time in the gth orbit.

$offt_k^g$: k^{th} observing stop time in the gth orbit.

After defining the variables used to formulate the multi-objective optimization problem, the multi-objective mission planning problem can be formulated as follows.

OPTIMIZATION VARIABLES

$$x_{ik} = \begin{cases} 1 & \text{Target } t_i \text{ is arranged in the } k^{th} \text{ observation chance} \\ 0 & \text{Target } t_i \text{ is not implemented} \end{cases} \quad (1)$$

$$y_{ij} = \begin{cases} 1 & \text{when } O_j = O_i, \text{ implemented after } t_j \\ 0 & \text{otherwise} \end{cases} \quad (2)$$

$$xn_i = \begin{cases} 1 & \text{implemented after } t_j \text{ and select the current} \\ & \text{chance to complete two images } (x_{ik} = 1, a_k \le 3) \\ 0 & \text{otherwise} \end{cases} \quad (3)$$

OBJECTIVE FUNCTIONS

$$F_1 = \max \sum_{i=1}^{m} \sum_{k=1}^{len} p_i(x_{ik}), len = length(chance_i) \quad (4)$$

$$F_2 = \max \sum_{i=1}^{m} xn_i \quad (5)$$

CONSTRAINTS

$$offt_k^g - ont_k^g \le max\_ont \quad (6)$$

$$offt_k^g - ont_k^g \le min\_ont \quad (7)$$

$$n^g \le g_{onum} \quad (8)$$

$$\sum_{k=1}^{n^g} \left(offt_k^g - on_k^g \right) \le g_{maxt} \quad (9)$$

$$ws_j \ge we_i + bt + Time\_start, \\ y_{ij} = 1, \text{and } a_i \ne a_j \quad (10)$$

$$ont_k^g \le ws_i < we_i \le offt_k^g, \quad (11)$$

$$ont_{k+1}^g \le ws_j < we_j \le offt_{k+1}^g \\ \text{when} \quad y_{ij} = 1, \ O_i \ne O_j \ \text{ and } \ ws_j \ge we_i + time_{onf} \quad (12)$$

$$\sum_{k=1}^{len} x_{ik} \le 1, Len = length(chance_i) \quad (13)$$

The multi-objective mission planning problem is formulated by using the aforementioned expressions. In details, constraints 6 and 7 indicate that the time of single observation task has the lower and upper bounds. Constraint 8 regulates the longest time of observation in single orbit. Constraint 9 gives the maximum observation number in single orbit. Constraint 10 points out that the time for switching the observation strip from one strip to another has an upper bound. Constraints 11 and 12 present the constraints regarding to the shortest time to restart an observation. Constraint 13 restricts that only 1 observation strip is activated at the same time, and each point target can only be observed once. The optimal objectives contain 2 parts: the first one is the total profit, which is described as objective function (4); the other is the completed image acquisitions expressed by objective function (5). Therefore, the optimal multi-objective mission planning problem can be described as to find the optimal x_{ik}, xn_i to maximize F_1, F_2 under the constraints from 6 to 13.

OPTIMAL SCHEDULING BASED ON NSGA-II ALGORITHM
REGIONAL TARGET DECOMPOSITION

To observe the regional target, we need to divide it into several point targets. Here, we divide the regional target into several squares which have the same longitude and latitude length. The regional target decomposition process is described as follows:

- Step1: determine the maximal and minimal values of the latitude and longitude of the regional target.
- Step 2: after obtaining 4 extreme values from step 1, find the rectangle to surround the regional target, as illustrated in Fig. 3.

- Step 3: divide the rectangle into several squares based on the latitude and longitude sample intervals.

ENCODING

Encoding is the key step to apply genetic algorithm. The encoding methods affect the crossover and mutation operator, which determines the efficiency of the genetic evolution. Here, we use the integer coding in which each gene represents a point target whose values range between $0 \sim n$, representing the number of observation opportunities. If the value of this gen is 0, the corresponding point target has not been observed. As shown in Fig. 4, targets 1 and 140 have not been observed.

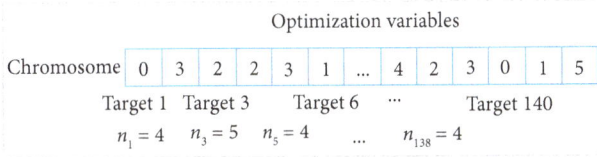

Figure 4. The structure of the chromosomes.

CROSSOVER AND MUTATION OPERATOR

First, we use uniform crossover to select 2 individuals from parent generation by binary tournament, then generate a binary string which has the same length with parent individual randomly, in which 0 represents non-exchange and 1, exchange. Based on the template of binary string to cross 2 parent solutions, new solutions can be obtained, as shown in Fig. 5.

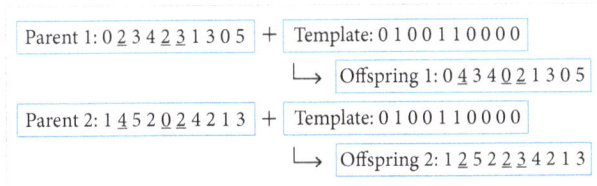

Figure 5. Template of binary string.

The offspring creation process is given by:
- Step 1: select parent solutions $xg_i^{(1,l)}$ and $xg_i^{(2,l)}$ by binary tournament, where l is the length.
- Step 2: choose a random number $r \in [0,1]$, if $r < Pc_1$ (Pc_1 is the crossover probability); go to step 3, otherwise, non-exchange.
- Step 3: generate a binary string randomly and create offspring solutions $xg_i^{(1,l+1)}, xg_i^{(2,l+2)}$ by using crossover operator.

We use uniform mutation as mutation operator to improve the local search ability and prevent premature convergence.

ADJUSTMENT SOLUTION CONSIDERING CONSTRAINTS

Solutions generated by the crossover and mutation processes may not satisfy constraint conditions, hence, these solutions must be filtered to delete unfeasible solutions. For the potential solutions which conflict with the constraints, we change the corresponding gene value of these solutions to 0 (Fig. 6). As shown in Fig. 6, point target 3 does not satisfy constraint conditions, hence, its gene value is changed to 0.

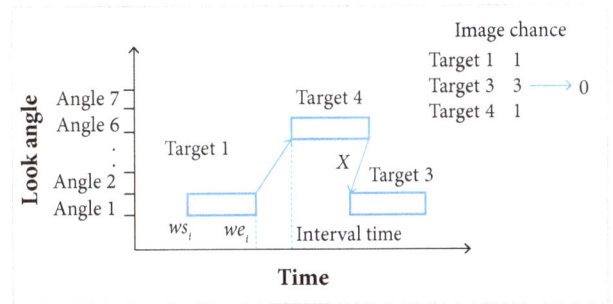

Figure 6. Illustration of observation interval time.

NSGA-II ALGORITHM

NSGA-II has been demonstrated an efficient algorithm to solve multi-objective optimization problems. Therefore, it has been chosen to solve the proposed multi-objective mission planning optimization problem. Due to the unique feature of this mission planning problem, basic NSGA-II algorithm has been properly modified (Fig.7):
- Step 1: if there are regional targets, they should be decomposed into several point ones.
- Step 2: determine observation opportunity of each target during the mission planning time span.
- Step 3: set proper NSGA-II algorithm parameters, such as population, maximum generation number, crossover, and mutation probabilities.
- Step 4: generate initial population P_0 randomly.
- Step 5: adjust initial solutions and make them satisfy constraint conditions; evaluate objective functions and perform non-dominated sorting, then calculate crowding distance of individuals and sort them according to crowding distance. Get the first generation P_1, the generation count number N_gen is set to 1, and record P_1.
- Step 6: perform uniform crossover and uniform mutation for population P_1, then obtain offspring generation Q_1, adjust solution set, and evaluate objective functions (fitness) for individuals in Q_1.

- Step7: merge parent population P_1 and offspring population Q_1, perform non-dominated sorting, calculate crowding distance between individuals, and sort individuals according to crowding distance in each non-dominated layer.
- Step 8: select individuals which locate on the front and obtain new parent population P_2.
- Step 9: set P_1 equals to P_2, increase generation count number N_gen, and record P_1.
- Step 10: repeat steps from 5 to 8 until N_gen reaches the maximal number.

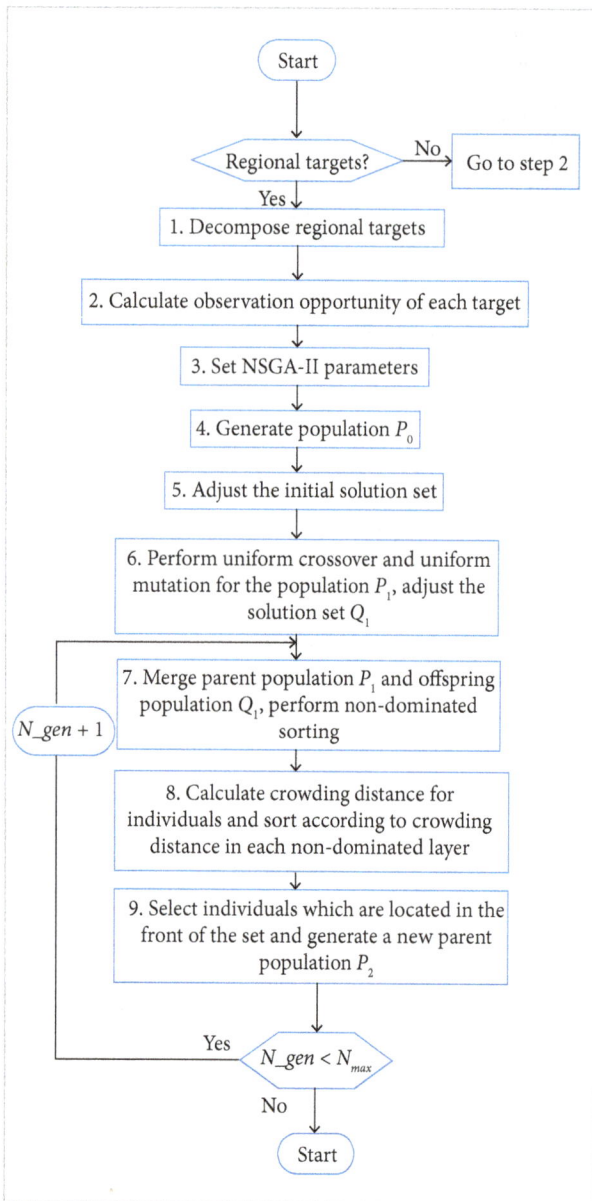

Figure 7. Flowchart for NSGA-II algorithm.

NUMERICAL SIMULATION RESULTS

In order to verify the validity of the proposed multi-objective mission planning method, several numerical simulations are conducted. The parameters of the numerical simulation are:

- The orbital elements of the chief satellites in the formation are shown in Table 2, where a is orbital semi-major axis, e is eccentricity, i is orbital inclination, ω is argument of perigee, Ω is right ascension of ascending node, and M is mean anomaly. The satellite is in Sun synchronous repeating orbit, whose revisit period and orbit repeating cycle are 15 days and 227 orbits, respectively.
- The mission planning start and end times are 2015/07/01 12:00:00 and 2015/07/16 12:00:00, respectively.
- The mission planning mission is conducted, whose point target number is 200. For single planning mission, the targets are generated and distributed randomly inside Chinese main land (Fig. 8). The priority values of targets vary from 0 to 10. There is 1 regional target, whose range is 1° in both latitude and longitude directions, and the sample interval to decompose the regional target is 0.1° in both latitude and longitude directions. Table 3 presents the point target numbers and the corresponding NSGA-II algorithm parameters.
- The minimal sensor half angle is 19° and the maximal one is 43°. The whole sensor range is divided into 12 strips (which is abbreviated as st in Table 4). Only the first 3 strips (1, 2, and 3) of each orbit can perform the 2 tasks (DEM and GMTI) at the same time, which is highlighted in Table 4. The minimal half angles (α^{min}) and the maximal ones (α^{max}) of these strips are listed in Table 4.

To verify the validity and effectiveness of the proposed multi-objective mission planning optimization method proposed in this study, comparison simulations have been conducted. For the same observation planning missions, 2 different methods are utilized: one is the new proposed NSGA-II method and the other is the general genetic algorithm one. For the latter, 2 different objectives are considered, which are profits and completed image acquisitions.

Table 2. Orbital elements of chief satellite.

a (km)	e	i (°)	ω (°)	Ω (°)	M (°)
6,898.224109	0.001075	97.482440	0	9.312805	180

Table 3. Simulation parameters of scenario 1.

Scenario	Target numbers	N_pop	N_top	Crossover probability	Mutation probability
1	200	100	1,200	0.9	0.1

N_pop: Population size; N_top: Maximal generation number.

Table 4. Inner and outer half angles of different strips.

	st 1	st 2	st 3	st 4	st 5	st 6	st 7	st 8	st 9	st 10	st 11	st 12
α^{min} (°)	19.7	22.3	24.8	27.2	29.5	31.8	33.9	36.0	37.9	39.8	41.6	43.4
α^{max} (°)	23.1	25.6	27.9	30.2	32.4	34.6	36.6	38.5	40.4	42.2	43.9	45.5

Figure 8. Distribution of observation targets in the map.

For target number equal to 200, Fig. 9 shows the results of the proposed NSGA-II method and of the general algorithm one. The 3 lines in Figs. 9a and 9b show the total profits and completed image acquisitions of the NSGA-II method, as well as genetic algorithm method with total profits as the optimization function (abbreviated as GA_Income) and genetic algorithm method with total completed image acquisitions as the optimization function (abbreviated as GA_Num). In Fig. 9a, the green line shows the profit values by using genetic algorithm method with the total profits as the optimization function, the blue line indicates the profit value by using genetic algorithm method with the completed image acquisitions as the optimization function, and the red line shows the profit values by utilizing NSGA-II methods. Figure 9a shows that the total profit values obtained by GA_Income method are larger than those of the proposed NSGA_II method, which are larger than those of the GA_Num method. Figure 9b shows that the total completed image acquisitions obtained by GA_Num method are larger than those of the proposed NSGA_II method, which are larger than those of the GA_Income method. From Figs. 9a and 9b, the conclusion is that, by using the proposed NSGA-II method, although the single objective value is less than the values obtained by GA_Num

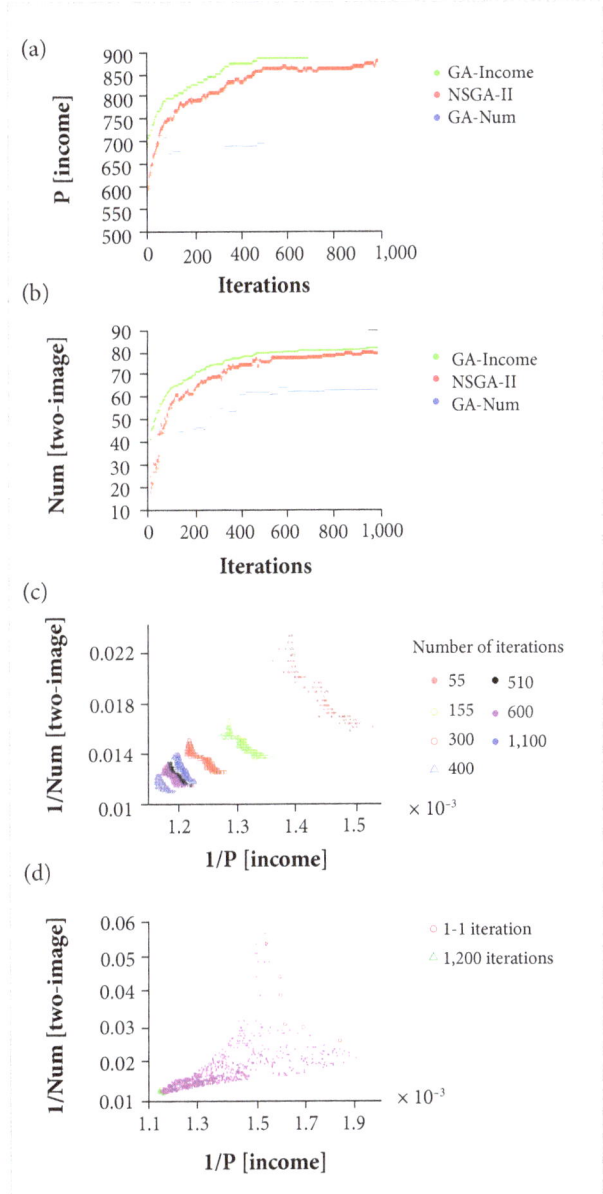

Figure 9. Multi-objective mission planning results (targets number equal to 200).

or *GA_Income* methods, one can obtain better total profits and completed image acquisitions objectives in an multi-objective optimization sense. Figures 9c and 9d demonstrate that the proposed NSGA-II method is converged after 1,200 iterations. In Fig. 9d, the pink indication means that the middle interactions show the coverage process of solution.

CONCLUSION

In this study, a NSGA-II-based multi-objective mission planning optimization method is proposed to deal with the unique SAR formation flying image mission planning problem, whose unique feature is that the formation system can perform 2 different tasks (DEM and GMTI) at the same time compared to the traditional single satellite mission planning problem, which can perform only a single task. The 2 different objective functions are defined as total profits and completed image acquisitions numbers, respectively. Several practical constraints are firstly considered when formulating the multi-objective mission planning problem, such as the longest observation time in 1 orbit and strip switching time from one strip to another.

Numerical simulations show that, although the single objective value is less than the values obtained by genetic algorithm method with single objective function, the proposed multi-objective mission planning optimization method can obtain better performance with respect to 2 mission planning objectives in an overall perspective.

ACKNOWLEDGEMENTS

The authors are grateful for the support provided for this study by the National Natural Science Foundation of China (No. 11502142).

AUTHOR'S CONTRIBUTION

Shao X and Zhang Z conceived the idea and co-wrote the main text; Shao X, Zhang Z, and Wang J performed the numerical simulation and analyzed the results; Zhang D prepared the tables and figures. All authors discussed the results and commented on the manuscript.

REFERENCES

Bensanna E, Verfaillie G, Agnèse JC, Bataille N, Blumstein D (1996) Exact and inexact methods for the daily management of an earth observation satellite. Proceedings of the 4th International Symposium on Space Mission Operations and Ground Data Systems; Munich, Germany.

Bianchessi N, Cordeau JF, Desrosiers J, Laporte G, Raymond V (2007) A heuristic for the multi-satellite, multi-orbit and multi-user management of Earth observation satellites. Eur J Oper Res 177(2):750-762. doi: 10.1016/j.ejor.2005.12.026

Gao P, Tan YJ, Li JF, He RJ (2013) An ant colony algorithm for remote satellite and ground integration scheduling problem in parallel environment. Adv Mater Res 791:1341-1346. doi: 10.4028/www.scientific.net/AMR.791-793.1341

Hao HC, Jiang W, Li YJ (2013) Mission planning for agile earth observation satellites based on hybrid genetic algorithm. Science Technology and Engineering 13(17):4972-4978.

He RJ, Gao P, Bai BC (2011) Models, algorithms and applications to the mission planning systems of imaging satellites. Systems

Engineering – Theory and Practice 31(3):411-422.

Jang J, Choi J, Bae HJ, Choi IC (2013) Image collection planning for KOrea Multi-Purpose SATellite-2. Eur J Oper Res 230(1):190-199. doi: 10.1016/j.ejor.2013.04.009

Kim H, Chang YK (2015) Mission scheduling optimization of SAR satellite constellation for minimizing system response time. Aero Sci Tech 40:17-32. doi: 10.1016/j.ast.2014.10.006

Lemaître M, Verfaillie G, Jouhaud F, Lachiver JM, Bataille N (2002) Selecting and scheduling observations of agile satellites. Aero Sci Tech 6(5):367-381. doi: 10.1016/S1270-9638(02)01173-2

Tangpattanakul P, Jozefowiez N, Lopez P (2015) A multi-objective local search heuristic for scheduling Earth observations taken by an agile satellite. Eur J Oper Res 245(2):542-554. doi: 10.1016/j.ejor.2015.03.011

Xiaolu L, Baocun B, Yingwu C, Feng Y (2014) Multi satellites scheduling algorithm based on task merging mechanism. Appl Math Comput 230:687-700. doi: 10.1016/j.amc.2013.12.109

9

Reducing the Effects of Inaccurate Fault Estimation in Spacecraft Stabilization

Rouzbeh Moradi[1], Alireza Alikhani[1], Mohsen Fathi Jegarkandi[2]

ABSTRACT: Reference Governor is an important component of Active Fault Tolerant Control. One of the main reasons for using Reference Governor is to adjust/modify the reference trajectories to maintain the stability of the post-fault system, especially when a series of actuator faults occur and the faulty system can not retain the pre-fault performance. Fault estimation error and delay are important properties of Fault Detection and Diagnosis and have destructive effects on the performance of the Active Fault Tolerant Control. It is shown that, if the fault estimation provided by the Fault Detection and Diagnosis (initial "fault estimation") is assumed to be precise (an ideal assumption), the controller may not show an acceptable performance. Then, it is shown that, if the worst "fault estimation" is considered, it will be possible to reduce the effects of fault estimation error and delay and to preserve the performance of the controller. To reduce the effects of this conservative assumption (worst "fault estimation"), a quadratic cost function is defined and optimized. One of the advantages of this method is that it gives the designer an option to select a less sophisticated Fault Detection and Diagnosis for the mission. The angular velocity stabilization of a spacecraft subjected to multiple actuator faults is considered as a case study.

KEYWORDS: Active Fault Tolerant Control, Fault estimation error and delay, Reference Governor, Angular velocity stabilization.

INTRODUCTION

Active Fault Tolerant Control (AFTC) is an important field in automatic control that has attracted a large amount of attention. The main responsibility of an AFTC is to tolerate component malfunctions while maintaining desirable performance and stability properties of the faulty system (Zhang and Jiang 2008). Latterly, a review paper published recent developments of the spacecraft AFTC system (Yin *et al.* 2016).

One of the main components of any AFTC is the Fault Detection and Diagnosis (FDD) module. There are several challenges that FDD designs have in common (Zhang and Jiang 2008). Among them, fault estimation error and delay are considered in this paper. These challenges have destructive effects on the stability and performance (Zhang and Jiang 2008).

Reference Governor (RG) is one of the components of the general AFTC structure (Zhang and Jiang 2008). The terms Command Governor (CG) and Reference Trajectory Management (RTM) have been also used in the literature. The main responsibility of RG is to adjust/modify the reference trajectories, so the post-fault model of the system remains stable, even after the occurrence of multiple actuator faults (Garone *et al.* 2016). There are several papers in the literature that have studied the effects of RG on the performance and stability of the post-fault model (Boussaid *et al.* 2010; Boussaid *et al.* 2011; Boussaid *et al.* 2014; Almeida 2011). According to these papers, RG has been able to deal with the actuator faults/failures efficiently.

To the authors' best knowledge, reducing the effects of fault estimation error and delay using the concept of RG still remains an open problem. This is the main subject that is pursued in this paper. It is shown that, as long as the estimated fault

1.Ministry of Science, Research and Technology – Aerospace Research Institute – Astronautics Department – Tehran/Tehran – Iran. 2.Sharif University of Technology – Engineering College – Department of Aerospace Engineering – Tehran/Tehran – Iran.

Author for correspondence: Alireza Alikhani | Ministry of Science, Research and Technology – Aerospace Research Institute – Astronautics Department | PO box: 14665-834 – Tehran/Tehran – Iran | Email: aalikhani@ari.ac.ir

reported by the FDD (initial "fault estimation") is assumed to be precise (an ideal assumption), the controller may not show an acceptable performance.

However, if the maximum fault estimation error is considered (worst "fault estimation"), RG can be used to reduce the effects of FDD errors and preserve the performance of the closed-loop system. To reduce the effects of this conservative assumption (considering maximum fault estimation error), a quadratic cost function is defined and optimized.

In order to validate the results, the angular velocity stabilization of a spacecraft subjected to multiple actuator faults is considered. It is shown that, if the initial "fault estimation" (the fault estimation reported by the FDD) is considered accurate, the response will not converge to the origin. However, if RG is designed based on the worst "fault estimation", AFTC will be able to asymptotically stabilize the faulty spacecraft in a wide range of actuator fault and despite FDD errors. This paper consists of the following sections: firstly, the modeling of the proposed RG is described. Then, the spacecraft dynamics and controller are shown. Finally, results obtained and the discussions are presented.

MODELING THE REFERENCE GOVERNOR

The structure of the considered AFTC is shown in Fig. 1. It is assumed that the FDD block provides "an estimation of" the post-fault model of the system. The RG block uses the proposed methodology to find the most suitable reference trajectories for the post-fault model, despite the presence of fault estimation error and delay. The signals ω and ω_d are the plant output (angular velocity) and the desired reference trajectory vectors, respectively.

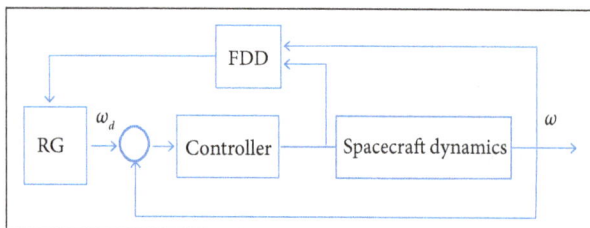

Figure 1. Structure of the AFTC.

In this paper, the mission of the controller is to make the origin an asymptotically stable equilibrium for the post-fault system, i.e. $\omega \to 0$ as $t \to t_f$ (final time).

It is assumed that the actuator fault/failure occurs at $t = t_{fault}$ and the FDD determines $\hat{t}_{fault}$ (estimated t_{fault}) with a fault estimation delay equal to:

$$\delta t_{fault} = \hat{t}_{fault} - t_{fault} \tag{1}$$

which is a positive value, since $\hat{t}_{fault}$ is always bigger than t_{fault}.

Fault estimation error is another property of the considered FDD block. The control inputs are bounded according to the following saturation function:

$$sat(u_i) = \begin{cases} u_i & if \quad -u_{max} \le u_i \le u_{max} \\ u_{max} & if \quad u_i \ge u_{max} \\ -u_{max} & if \quad u_i \le -u_{max} \end{cases} \tag{2}$$

where u_{max} is the maximum torque that can be produced by the actuators.

The reduction in the actuator region is considered as the actuator fault and is modeled according to Eq. 3 (Miksch and Gambier 2011):

$$sat_{p-f}(u_i) = \begin{cases} u_i & if \quad -u_{max_{p-f}} \le u_i \le u_{max_{p-f}} \\ u_{max_{p-f}} & if \quad u_i \ge u_{max_{p-f}} \\ -u_{max_{p-f}} & if \quad u_i \le -u_{max_{p-f}} \end{cases} \tag{3}$$

The subscript p-f shows the post-fault condition. The relation between pre- and post-fault actuator region is given according to:

$$u_{max_{p-f}} = a u_{max} \quad a \in (0\ 1) \tag{4}$$

where a is the actuator effectiveness coefficient (Sobhani-Tehrani and Khosravi 2009), a real value between 0 and 1; u_{max} is the pre-fault actuator region. FDD determines the estimated value of a (shown by $\hat{a}$). It is assumed that the FDD provides $\hat{a}$ with an estimation error given by:

$$\delta_{a/\hat{a}} = \frac{a}{\hat{a}} \tag{5}$$

where $\delta_{a/\hat{a}}$ is a value between 0 and 1. The larger/smaller values of $\delta_{a/\hat{a}}$ show better/worse fault estimation, respectively.

According to the considered mission, the goal of RG is to determine ω_d such that the faulty model of the system remains asymptotically stable, even after the occurrence of multiple actuator faults and in the presence of fault estimation error and delay in the FDD module. The RG flowchart is presented in Fig. 2. The consecutive steps are explained in the following paragraphs.

According to Fig. 3, $\omega_d(t_1) \ldots \omega_d(t_n)$ are initialized by the solver, which is the Genetic Algorithm (GA), as will be explained in the results section.

Note 1: although the GA is used to solve the problem, other numerical solvers can be also employed. However, the main concern of this paper is to find a method to decrease the consequences of fault estimation error and delay. Therefore, any numerical solver (possibly faster than GA) that solve the problem can be considered as well.

Note 2: as will be seen in the simulation section, GA can find a solution within a reasonable time.

When these points are determined, a cubic spline is passed through them, similarly to Fig. 4. A detailed analysis about cubic

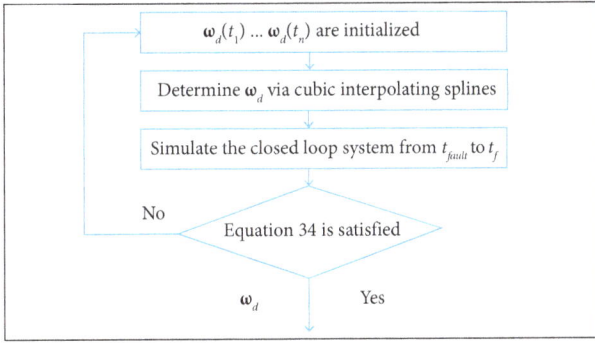

Figure 2. RG flowchart.

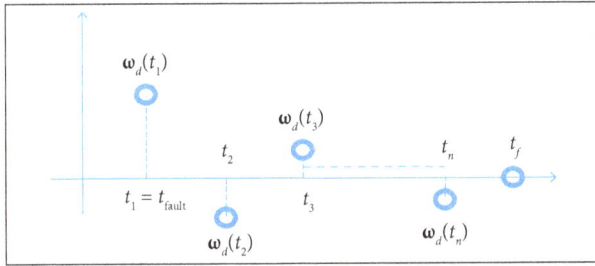

Figure 3. Initializing $\omega_d(t_1) \ldots \omega_d(t_n)$.

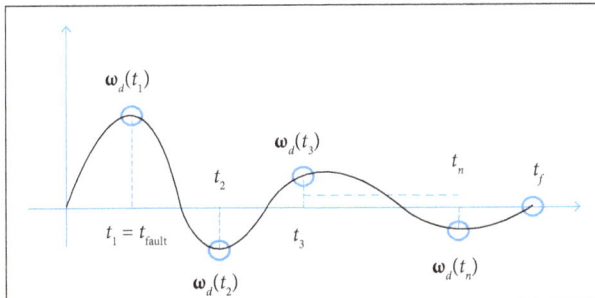

Figure 4. ω_d produced by cubic spline.

spline interpolation can be found in de Boor (1978). One of the main advantages of cubic splines is their smoothness (they are twice continuously differentiable). This will prevent the controller inputs from being discontinuous (refer to Eqs. 25 – 27).

According to the FDD information, an estimation of the post-fault model of the system is known. The faulty closed-loop system is simulated from t_{fault} to t_f. This simulation is a part of the flowchart shown in Fig. 2 and several simulations may be needed to obtain ω_d.

After simulation, the value of $\omega(t_f)$ is checked to see whether the following equality is satisfied or not:

$$\omega(t_f) = 0 \tag{6}$$

Such a final state constraint is well-known in the literature and is introduced to ensure asymptotic stability (Fontes 2001). Since this equality will never hold numerically, Eq. 34 will be considered in simulations.

Note 3: to ensure that ω_d approaches the origin before $t = t_f$, its value is set to 0 as t passes t_s (settling time). In other words:

$$\omega_d = 0 \ \forall \ t \geq t_s, t_s < t_f \tag{7}$$

To give the solver more flexibility, another variable (k_s) is introduced, satisfying Eq. 8:

$$t_s = k_s t_f, 0 < k_s < 1 \tag{8}$$

In addition to $\omega_d(t_1) \ldots \omega_d(t_n)$, k_s is another variable that should be found by the solver.

SPACECRAFT DYNAMICS AND CONTROLLER STRUCTURE
SPACECRAFT DYNAMICS

The rigid body spacecraft rotational dynamics in the principal coordinate system is described by the following equations (Sidi 2000):

$$\dot{\omega}_1 = \alpha_1 \omega_2 \omega_3 + u_1' \quad \alpha_1 = \left(\frac{J_2 - J_3}{J_1}\right) \tag{9}$$

$$\dot{\omega}_2 = \alpha_2 \omega_1 \omega_3 + u_2' \quad \alpha_2 = \left(\frac{J_3 - J_1}{J_2}\right) \tag{10}$$

$$\dot{\omega}_3 = \alpha_3\omega_1\omega_2 + u'_3 \quad \alpha_3 = \left(\frac{J_1 - J_2}{J_3}\right) \tag{11}$$

where ω_1, ω_2, ω_3 are the angular velocities; u'_1, u'_2, u'_3 are the normalized control inputs; J_1, J_2, J_3 are the principal moments of inertia of the rigid body. The relation between control torques and inputs are given by Eqs. 12 – 14:

$$u'_1 = u_1/J_1 \tag{12}$$

$$u'_2 = u_2/J_2 \tag{13}$$

$$u'_3 = u_3/J_3 \tag{14}$$

where u_1, u_2, u_3 are the control moments acting on the spacecraft.

CONTROLLER STRUCTURE

The error signal is defined as:

$$\boldsymbol{\omega}_e = \boldsymbol{\omega} - \boldsymbol{\omega}_d \tag{15}$$

where $\boldsymbol{\omega}_d$ and $\boldsymbol{\omega}_e$ are the desired and error angular velocity vectors, respectively.

Inserting the scalar form of Eq. 15 into Eqs. 9 – 11 and eliminating $\boldsymbol{\omega}$, one has:

$$\dot{\omega}_{1e} = \dot{\omega}_1 - \dot{\omega}_{1d} = \alpha_1(\omega_{2e} + \omega_{2d})(\omega_{3e} + \omega_{3d}) + u'_1 - \dot{\omega}_{1d} = u''_1 \tag{16}$$

$$\dot{\omega}_{2e} = \dot{\omega}_2 - \dot{\omega}_{2d} = \alpha_2(\omega_{1e} + \omega_{1d})(\omega_{3e} + \omega_{3d}) + u'_2 - \dot{\omega}_{2d} = u''_2 \tag{17}$$

$$\dot{\omega}_{3e} = \dot{\omega}_3 - \dot{\omega}_{3d} = \alpha_3(\omega_{1e} + \omega_{1d})(\omega_{2e} + \omega_{2d}) + u'_3 - \dot{\omega}_{3d} = u''_3 \tag{18}$$

Canceling the non-linear terms using feedback linearization, the closed-loop system will change into the following simple linear time invariant form:

$$\dot{\omega}_{1e} = u''_1 \tag{19}$$

$$\dot{\omega}_{2e} = u''_2 \tag{20}$$

$$\dot{\omega}_{3e} = u''_3 \tag{21}$$

and the following form of control inputs

$$u''_1 = -k_1\omega_{1e} \quad k_1 \in R^+ \tag{22}$$

$$u''_2 = -k_2\omega_{2e} \quad k_2 \in R^+ \tag{23}$$

$$u''_3 = -k_3\omega_{3e} \quad k_3 \in R^+ \tag{24}$$

will lead to the exponential stabilization of $\boldsymbol{\omega}_e$ to 0; consequently, $\boldsymbol{\omega}$ will converge to $\boldsymbol{\omega}_d$ exponentially. The numerical values of k_1, k_2 and k_3 determine the exponential convergence rate of $\boldsymbol{\omega}_e$ to 0. Therefore, larger values of k_1, k_2 and k_3 mean a faster response and vice-versa.

Considering Eqs. 16 – 18 and Eqs. 22 – 24, the following relations will be obtained:

$$u'_1 = -\alpha_1(\omega_{2e} + \omega_{2d})(\omega_{3e} + \omega_{3d}) + \dot{\omega}_{1d} - k_1\omega_{1e} \tag{25}$$

$$u'_2 = -\alpha_2(\omega_{1e} + \omega_{1d})(\omega_{3e} + \omega_{3d}) + \dot{\omega}_{2d} - k_2\omega_{2e} \tag{26}$$

$$u'_3 = -\alpha_3(\omega_{1e} + \omega_{1d})(\omega_{2e} + \omega_{2d}) + \dot{\omega}_{3d} - k_3\omega_{3e} \tag{27}$$

For feedback purposes, it is better to rewrite u'_1, u'_2 and u'_3 as a function of the original variables:

$$u'_1 = -\alpha_1(\omega_2)(\omega_3) + \dot{\omega}_{1d} - k_1(\omega_1 - \omega_{1d}) \tag{28}$$

$$u'_2 = -\alpha_2(\omega_1)(\omega_3) + \dot{\omega}_{2d} - k_2(\omega_2 - \omega_{2d}) \tag{29}$$

$$u'_3 = -\alpha_3(\omega_1)(\omega_2) + \dot{\omega}_{3d} - k_3(\omega_3 - \omega_{3d}) \tag{30}$$

According to Eqs. 28 – 30, for the control inputs to be continuous, the desired reference trajectory ($\boldsymbol{\omega}_d$) should be continuously differentiable. As stated previously, this is one of the main reasons for using cubic spline interpolation to find $\boldsymbol{\omega}_d$. These are the desired control inputs that will lead to the exponential convergence of $\boldsymbol{\omega}$ to $\boldsymbol{\omega}_d$.

If $\boldsymbol{\omega}_d = 0$, the equations of closed-loop system will be:

$$\dot{\omega}_1 = -k_1\omega_1 \tag{31}$$

$$\dot{\omega}_2 = -k_2\omega_2 \tag{32}$$

$$\dot{\omega}_3 = -k_3\omega_3 \tag{33}$$

Clearly, as long as there is no saturation and the actuators can produce the required control inputs, will remain globally exponentially stable (GES). However, after the occurrence of severe actuator faults, GES will not be guaranteed.

RESULTS

The system/controller parameters and initial conditions are given in Table 1. The values chosen for the moments of inertia are taken from Wang *et al.* (2013), and the range of variables is presented in Table 2.

Table 1. System/controller parameters and initial conditions

Controller parameters	Initial conditions (deg/s)	Moments of inertia (kg·m²)
$k_1 = 0.1$	$\omega_1(0) = 10$	$J_1 = 449.5$
$k_2 = 0.1$	$\omega_2(0) = -10$	$J_2 = 449.5$
$k_3 = 0.1$	$\omega_3(0) = 5$	$J_3 = 449.5$

Table 2. Range of variables.

Optimization variable	Range
ω_d	[−100 100] deg/s
k_s	[0.5 0.9]

In order to satisfy the final state constraint given by Eq. 6, the following inequality is defined:

$$\textstyle\sum_{i=1}^{3} \omega_i^2 \left(t_f\right) \leq 0.0001 , \quad \omega : deg/s \tag{34}$$

As already mentioned, to determine $\boldsymbol{\omega}_d$, GA (Goldberg 989) is used as the solver; $[\boldsymbol{\omega}_{1d}(t_1) \dots \boldsymbol{\omega}_{1d}(t_n)]$, $[\boldsymbol{\omega}_{2d}(t_1) \dots \boldsymbol{\omega}_{2d}(t_n)]$ and $[\boldsymbol{\omega}_{3d}(t_1) \dots \boldsymbol{\omega}_{3d}(t_n)]$ are initialized every 10 s ($\Delta t = 10$ s or equivalently, $n = 10$) from the beginning of the fault time (t_{fault}). Therefore, considering k_s, the total number of decision variables will be 31. The considered parameters for GA are presented in Table 3. Other GA parameters are the default values considered in MATLAB® (MathWorks® 2011).

The actuation system consists of 6 thrusters (without considering hardware redundancy), that are placed in opposite directions, and each thruster can produce maximum 50 N variable thrust. The effective moment arm of all thrusters is 1 m along the principal body axis. However, the configuration of the thrusters is such that $(T_1 - T_2)$, $(T_3 - T_4)$ and $(T_5 - T_6)$ produce net moments about the first, second and third principal axes,

Table 3. GA parameters.

Parameter	Value
Cross-over fraction	0.8
Elite count	2
Population size	5 × number of decision variables = 5 × 31 = 155
Initial population	$\omega_{d,initial} = 0$, $k_{s,initial} = 0$

respectively. The direction of the arrows shows the direction of the forces produced by the thrusters (Fig. 5). Therefore, the relation between control torques (u_1, u_2, u_3) and $T_1 - T_6$ can be obtained according to the following equations:

$$u_1^+ = T_1, u_1^- = -T_2 \tag{35}$$

$$u_2^+ = T_3, u_2^- = -T_4 \tag{36}$$

$$u_3^+ = T_5, u_3^- = -T_6 \tag{37}$$

where the superscripts + and – show the positive and negative control torques, respectively.

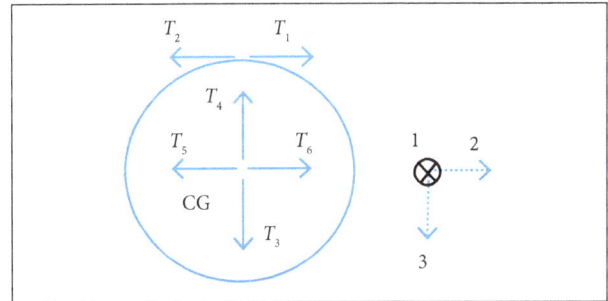

Figure 5. Thruster configuration.

Note 4: it seems that the thrusters T_3, T_4, T_5 and T_6 pass through the center of gravity. However, as indicated before, they have a moment arm of 1 m along the first body axis. Three important concepts are introduced:

- Initial "fault estimation": the fault estimation reported by the FDD.
- Worst "fault estimation": the biggest error of the FDD in providing the fault information. Its value is determined from the initial "fault estimation", according to the experience or the FDD specifications.
- Real fault: the fault that happens in reality (unknown).

The fault scenario that FDD reports is:

- Initial "fault estimation": T_5 and T_6 have lost 99% of their effectiveness ($\hat{a}_5 = \hat{a}_6 = 0.01$) and the remaining thrusters are at a good health ($\hat{a}_1 = \hat{a}_2 = \hat{a}_3 = \hat{a}_4 = 1$). The fault occurs at $\hat{t}_{fault} = 10$ s.
- Worst "fault estimation": based on the experience or the FDD specifications; in the worst case, the following parameters are given: $\delta t_{fault} = 5$ s and $\delta_{a/\hat{a}} = 0.01$. Therefore, it can be concluded that, in the worst case, $a_5 = a_6 = 0.0001$, i.e. T_5 and T_6 can produce a maximum 0.05 N thrust and the fault occurrence time is $t_{fault} = 5$ s.

Note 5: it is assumed that the real fault is less severe than the one reported by the worst "fault estimation". In this case, the controller will show an acceptable performance for less severe, and therefore, a wide range of faults.

Qualitatively, it is assumed that the severity of the faults satisfies the following inequalities:

$$S_{w.f.e} > S_{r.f} > S_{i.f.e} \tag{38}$$

where S is a quality that represents the severity of the fault; the subscripts w.f.e, r.f and i.f.e stand for worst "fault estimation", real fault and initial "fault estimation", respectively.

According to the previous discussion, the proposed method is very conservative, because it considers the worst "fault estimation". To reduce the adverse effects of this assumption, the following quadratic cost function is introduced:

$$J = \int_{t_{fault}}^{t_{final}} \boldsymbol{\omega}^T \boldsymbol{\omega} dt + \int_{t_{fault}}^{t_{final}} \mathbf{u}^T \mathbf{u} dt \tag{39}$$

Minimizing this cost function will decrease the adverse effects of considering the worst fault estimation. The considered sample time for integration is 0.1 s. The problem consists of 2 phases: first, GA tries to satisfy the constraint given by Eq. 34. Then, the result is used as an initial solution to optimize Eq. 39. The following penalty on cost function is considered:

$$J = \begin{cases} \int_{t_{fault}}^{t_{final}} \boldsymbol{\omega}^T \boldsymbol{\omega} dt + \int_{t_{fault}}^{t_{final}} \mathbf{u}^T \mathbf{u} dt & \text{Eq. 34 is satisfied} \\ \infty & \text{Eq. 34 is not satisfied} \end{cases} \tag{40}$$

It was verified that 1,000 s elapsed time is considered as the stopping criterion for the second phase — Intel(R) Core™

2 CPU, T7200@2.00 GHz; MATLAB® (MathWorks® 2011). To observe the consequences of employing the proposed method, 2 different cases are considered and summarized in Table 4.

Table 4. Cases considered.

Case	Fault estimation
1	Considering the initial "fault estimation"
2	Considering the worst "fault estimation"

CASE 1

If the initial "fault estimation" is considered (FDD is assumed to report the precise fault information), the results shown in Figs. 6 and 7 will be obtained.

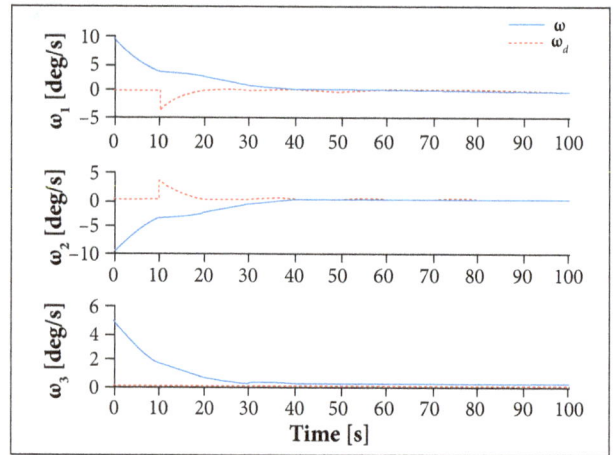

Figure 6. Angular velocities, initial "fault estimation" (case 1).

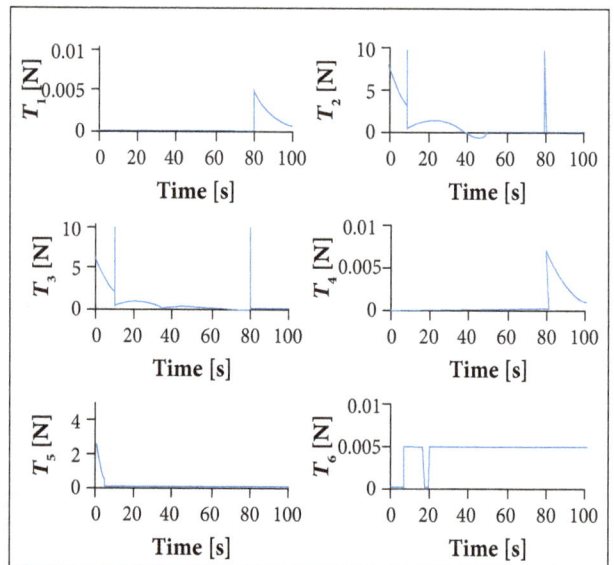

Figure 7. Control inputs, initial "fault estimation" (case 1).

Figure 6 shows that RG can not make the closed-loop system asymptotically stable, because it assumes the fault scenario reported by the FDD (initial "fault estimation"), which is precise. However, since the real fault is worse than the fault reported by the FDD (initial "fault estimation"), does not converge to the origin. This simulation shows the consequences of considering the initial "fault estimation". The main conclusion of this simulation is: if the FDD is assumed to report the precise fault information, the response of the controller may not be acceptable.

CASE 2

The result of considering the worst "fault estimation" is illustrated in Fig. 8. The control inputs are illustrated in Fig. 9.

According to Fig. 8, RG can asymptotically stabilize the closed-loop system, when the worst "fault estimation" is considered. A comparison of Figs. 6 and 8 shows the consequences of considering the worst "fault estimation" in the RG design. Clearly, considering the initial "fault estimation" (case 1) can lead to the poor performance of the controller and even to a non-convergent response. On the other hand, if RG is designed for the worst "fault estimation" (case 2), it can cover less severe faults and stabilize the faulty system for a wide range of faults (Note 5).

Since the assumption of worst "fault estimation" is conservative, the response is optimized via minimizing the cost function (Eq. 39). The GA performance is illustrated in Fig. 10. As stated previously, the quadratic cost function has been introduced to reduce the adverse consequences of considering the worst "fault estimation" (maximum fault estimation error). According to Fig. 10, after 14 generations (1,000 s elapsed time), the cost function is reduced from 8,758 to 5,944

(about 32%). This reduction in the cost function decreases the adverse consequences of considering the worst fault estimation.

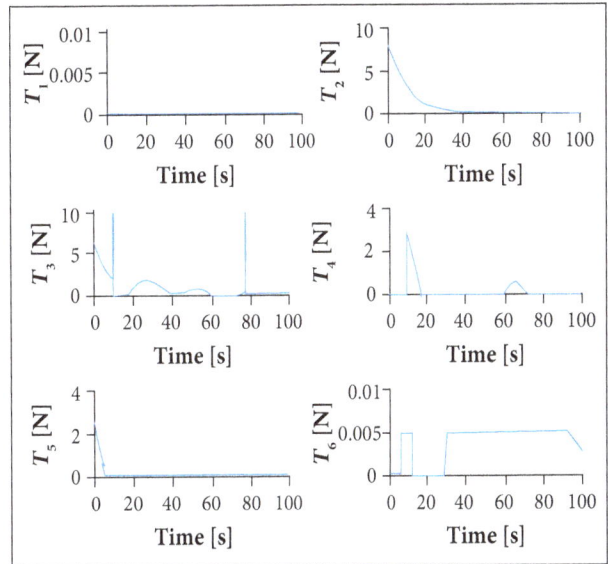

Figure 9. Control inputs, worst "fault estimation" (case 2).

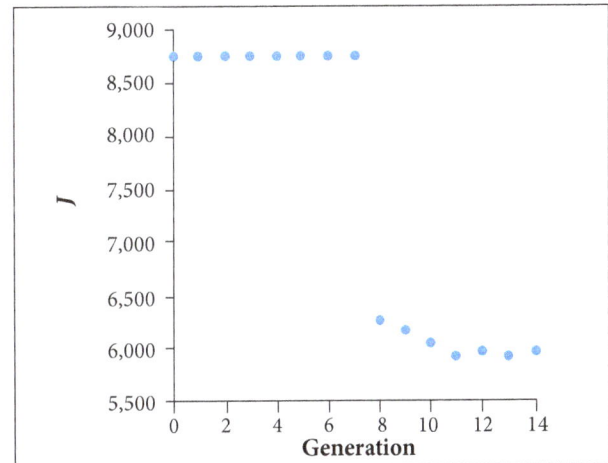

Figure 10. Cost function versus generations (1,000 s elapsed time).

DISCUSSION

Fault estimation error and delay are important characteristics of FDD schemes. RG is a method to adjust/modify the reference trajectories to handle actuator fault/failure. It was shown that, if the initial "fault estimation" was assumed to be precise (an ideal assumption), the controller might not be able to show an acceptable performance. On the other hand, if the worst "fault estimation" was considered, it

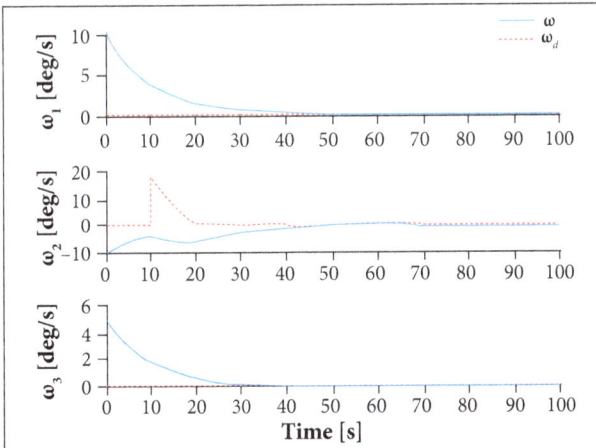

Figure 8. Angular velocities, worst "fault estimation" (case 2).

would be possible to reduce the destructive effects of fault estimation error. A quadratic cost function was defined to reduce the adverse consequences of this conservative assumption (assuming maximum fault estimation error). Therefore, a less sophisticated FDD can be used to satisfy the mission objectives.

AUTHOR'S CONTRIBUTION

Conceptualization, Moradi R; Methodology, Moradi R, Alikhani A, and Fathi Jegarkandi M; Writing – Original Draft, Moradi R and Alikhani A; Writing – Review & Editing, Moradi R, Alikhani A, and Fathi Jegarkandi M.

REFERENCES

Almeida FA (2011) Reference management for fault-tolerant model predictive control. J Guid Control Dynam 34(1):44-56. doi: 10.2514/1.50938

Boussaid B, Aubrun C, Abdelkrim MN (2010) Fault adaptation based on reference governor. Proceedings of the Conference on Control and Fault-Tolerant Systems; Nice, France.

Boussaid B, Aubrun C, Abdelkrim MN (2011) Two-level active fault tolerant control approach. Proceedings of the 8th International Multi-Conference on Systems, Signals and Devices; Sousse, Tunisia.

Boussaid B, Aubrun C, Jiang J, Abdelkrim MN (2014) FTC approach with actuator saturation avoidance based on reference management. International Journal of Robust and Nonlinear Control 24(17):2724-2740. doi: 10.1002/mc.3020

De Boor C (1978) A practical guide to splines. Berlin: Springer.

Fontes FACC (2001) A general framework to design stabilizing nonlinear model predictive controllers. Systems and Control Letters 42(2):127-143. doi: 10.1016/S0167-6911(00)00084-0

Garone E, Di Cairano S, Kolmanovsky IV (2016) Reference and command governors for systems with constraints: A survey on theory and applications. Automatica 75:306-328. doi: 10.1016/j.automatica.2016.08.013

Goldberg DE (1989) Genetic algorithms in search, optimization & machine learning. Reading: Addison-Wesley.

MathWorks® (2011) MATLAB® and SIMULINK®. Natick: MathWorks®.

Miksch T, Gambier A (2011) Fault-tolerant control by using lexicographic multi-objective optimization. Proceedings of the 8th Asian control conference (ASCC); Kaohsiung, Taiwan.

Sidi MJ (2000) Spacecraft dynamics and control: a practical engineering approach. Cambridge: Cambridge University Press.

Sobhani-Tehrani E, Khosravi KH (2009) Fault diagnosis of nonlinear systems using a hybrid approach. Lecture Notes in Control and Information Sciences. Dordrecht; New York: Springer.

Wang D, Jia Y, Jin L, Xu S (2013) Control analysis of an underactuated spacecraft under disturbance. Acta Astronautica 83:44-53. doi: 10.1016/j.actaastro.2012.10.029

Yin S, Xiao B, Ding S, Zhou D (2016) A review on recent development of spacecraft attitude fault tolerant control system. IEEE Trans Ind Electron 63(5):3311-3320. doi: 10.1109/TIE.2016.2530789

Zhang Y, Jiang J (2008) Bibliographical review on reconfigurable fault-tolerant control systems. Ann Rev Contr 32(2):229-252. doi: 10.1016/j.arcontrol.2008.03.008

10

Multi-Disciplinary System Design Optimization of a Launch Vehicle Upper-Stage

Mostafa Zakeri[1], Mehran Nosratollahi[1], Alireza Novinzade[2]

ABSTRACT: The design method presented in this paper is related to the upper-stage system and its instrumentation, expedition and facilitation so as to transfer the satellite from the destination orbit to the target orbit. We used an integrated design method with a structure based on multi-disciplinary system design optimization and developed a simple systematic interference method for designing aerospace products. The subsystems' convergence in an optimized environment, matrix relationship, and integration of the subsystems' parameters and presentation of design give results while meeting all requirements and considering the limitations of the design were the main aims of the research. Instead of a merely mathematical optimization design, in the present study a new design method with a systematic multipurpose optimization approach was designed. In this context, the optimization means the parameters are optimized as a result of the design convergence coefficients. Validation of the design method was not only obtained through comparison with a specific product but also with the systematic parameters of all upper-stage systems with a similar operation through the results of statistical design graphs. The approximate similarities of the results indicate an acceptable and genuine design with a quite systematic approach which is better than an unreal and merely optimized design.

KEYWORDS: Upper stage, Systems design, Multidisciplinary design optimization, Systems integration.

INTRODUCTION

In a article named "Technologies for future precision strike missile systems - missile design technology" (Fleeman 2001), there is a survey of missile technology concepts, influential parameters in design, and balance among subsystems, using new technologies with lower weight and cost communicating with the launcher. Overall configuration as well as missile simulation results from such a design method. The detailed explanations for the study are available in his book of Tactical Missiles Design. It needs to be mentioned that design depth is limited in the method yet the functional area is high while lack of integration and systemic communications implementation are the biggest weak points in this method which he explains and completes in his 2012 edition of the book. It is claimed in the study that all main parameters of a missile at conceptual design are taken into consideration while the missile has operational capability. Operational capability as well as systemic relation integration is the first act in the current study.

The history of transition from classic design to modern design is not accountable in this study, however, some developed countries have been able to take advantage of system design and multidisciplinary optimization methods to improve conceptual design process through considerable savings in design time and costs (Olds 1993).

Brown and Olds (2006) surveyed multi-objective optimization techniques of collaborative optimization (CO), modified collaborative optimization (MCO), bi-level integrated system synthesis (BLISS), and all at once (AAO) on a reusable satellite. The study claims that the best design method cannot be chosen since such activities are for research

1.Space Research Institute – System Engineering Department – Space System Design Laboratory – Tehran/Tehran – Iran. 2.Khaje Nasir Toosi University of Technology – Faculty of Aerospace – Systems Division – Tehran/Tehran – Iran.
Author for correspondence:Mostafa Zakeri | Space Research Institute – System Engineering Department – Space System Design Laboratory | Postal Address: 1316943551 – Tehran/Tehran – Iran | Email: mostafa.zakeri5@gmail.com

purposes only and require various studies on the results. Systemic design activities are briefly mentioned in this study while most of the activities are focused on comparing 3 design techniques. In the final conclusion, a comparison is made among design methods based on running time as well as quality comparisons, which could be said that BLISS designed outputs have higher quality.

Balesdent *et al.* (2011) surveyed various multipurpose optimization methods in space systems design quantitatively. Various MDO methods to design satellite missiles are surveyed in the study and some features, such as strength, price calculations, flexibility and convergence speed and problem implementation are taken into consideration so as to select the most suitable design method in designing launch vehicle. Mathematical equations of optimization of every method as well as main profile of the algorithm with optimization activities of every method are mentioned briefly. Selection of the optimal method to design space system based on the mission and various situations such as implementation time, cost, complexity, etc. is the final conclusion of the study (Balesdent *et al.* 2011).

Riddle (1998) states that using MDO in designing complex systems comes with 2 obstacles. One of them originates from disconnected and nonlinear essence of design process most of mathematical optimization methods are facing. One other unattractive point of MDO methods is design teams' unwillingness to use it and similarity of automatic decision-making with creative process of innovative design.

Tsuchiya and Mori (2002) claimed that in spite of the higher speed of MDO with parametric methods, and based on which studies are reported to improve system and destination optimization for reusable launch vehicles (RLV), they are still recognized unsuitable for space systems design, especially satellite-carrying missiles which are essentially more complicated in configuration steps and trajectory design.

The need to design with a systematic approach in addition to design implementation based on physics of an aerospace product attracts some researchers to optimized systematic design. Aldheeb *et al.* (2012) tried to create an optimized design for a Micro Air Launch Vehicle.

In the present study, optimization and trajectory design are done through a design algorithm with systematic approach to reduce payload mass. The look on functional design physics in the main algorithm of the article and model of subsystems is clear. Villanueva *et al.* (2013) used a systematic approach in an article to design solid fuel engine.

Conceptual design in the current study is optimized through a genetic algorithm.

It could be claimed that creation of a systematic approach and transforming MDO to multi-disciplinary system design optimization (MSDO) includes development of MDO methods which develops operational capability based on MDO. MSDO approach is available in a limited number of the articles.

According to Wronski and Gray (2004), one can verify a comprehensive MSDO implementation in a specific case in the Massachusetts Institute of Technology (MIT). The specific importance of the study is that it gives a true expression of MSDO systematic design, multipurpose optimization and an obvious process of the algorithm.

Additionally, design and multipurpose optimization of an aircraft was studied through implementation and integrated design tools so as to predict and optimize the implementation and related expenditures of commercial aircrafts design and production with the aim of reducing the noise through accurate selection of configuration and mission parameters (Diedrich *et al.* 2006). After surveying some design samples, a brief look to upper stage activities are made.

Engine and trajectory design in Casalino and Pastrone (2010) is optimized simultaneously. Systematic design is brief in the study and it could be included among the optimization articles with systematic approach for propulsion engine.

An upper stage activity is presented with a brief look at upper stage engine of solid fuel engine (Casalino and Pastrone 2010). Adami *et al.* (2015) designed an upper stage performed through three forms of MDO. Mathematically, a detailed comparison among the three design methods is presented. The optimal proof of choice is surveyed mathematically.

After considering the aforementioned articles in addition to their quantitative and qualitative analyses, Table 1 presents a summary of their features.

METHODOLOGY

Designing an optimized multidisciplinary system is a modern example of aerospace product design. MSDO can be complex product design process and multidisciplinary engineering systems. In this method, the subsystems are related to each other and to a system in an optimized and converging space. Also the main feature of this method is the

Table 1. Distribution of mass components.

Design methods	Physics-based methods	Mathematical-based methods	Full configuration	Balanced subsystems (converge)	Requirement meeting between subsystems	Subsystems design depth	Imbedded baseline statistics data
Conceptual design classic	×	–	×	–	–	×	×
System design optimization	×	×	–	×	–	×	×
MDO	–	×	×	–	×	–	×
MSDO (this study)	×	×	×	×	×	×	×

presence of human expert in the Designing tool environment and integration of all designing subdivisions. The aim of this approach is the creation of sophisticated and advanced engineering systems that are competitive not only in terms of performance but also in terms of value of life cycle.

The most important properties of MSDO method can be stated as follows:

- Deal with design models of realistic size and fidelity that will not lead to erroneous conclusions.
- Reduce the tedium of coupling variables and results from disciplinary models.
- Allow for creativity while leveraging rigorous, quantitative tools in the design process. Hand-shaking: qualitative *versus* quantitative.
- Data visualization in multiple dimensions.
- Incorporation of higher-level upstream and downstream system architecture aspects in early design: staged deployment, safety and security, environmental sustainability, platform design, etc.

In this procedure, design algorithm similar to other design algorithms is not of tree type or merely a mathematical optimization. The MSDO algorithm is designed to link all the subdivisions directly to each other and the best convergence is applied according to physics and subdivisions. In this way, the designer can easily put all the limitations and restrictions of design to work. In this paper, the concepts of design and multistep optimization are used in a strategic computing environment to design upper-stage which transfers from parking orbit to the target orbit. Presented parameters in this procedure are classified as:

- Design or independent variables: including fuel mass ratio, engine structural mass ratio to the whole engine, etc.
- Simple limitations: including mass ratios, I_{sp}, etc.

- Restrictions: diameter, orbital altitude, payload.
- Combined merit functions: aiming at minimizing the total mass and designing the best way of sending the satellite to the target orbit.

The most important specifications of the MSDO algorithm are as follows:

- Offering new approaches in systematic design derived from MSDO designing method.
- Using statistical processing in the design process (increasing accuracy and rapid convergence).
- Offering innovative convergence methods.
- Optimizing system and subsystems' design parameters by using communication matrixes.
- Convergence of designing upper stage with previous stages of the rocket.
- Integrating all the design parameters and meeting all the restrictions and requirements.
- Ability to enter any new special requirement in the way of designing.

Algorithm design of upper stage includes the following:

- Statistical designing and analysing statistical data.
- Designing the layout of subsystems and subdivisions.
- Dynamics and trajectory design.
- Propulsion system design and tank design.
- Feeding system design.
- Analysis of mass-dimensions and mass associated with the previous stage of launch vehicle design.
- Structural design and stiffener.
- Systematic analysis (configuration, integrating and optimization).

All the requirements and limitations of designing the upper stage is done based on the objectives, bottlenecks, and administrative constraints. These constraints are applied in all the phases with the presence of the designer in the design environment. All the requirements and restrictions can be classified as follows:

- The requirements of trajectory.
- The requirements of the launch vehicle and launch.
- The requirements of subsystems and subdivisions.
- The requirements of construction and assembly.
- The limitations in choosing the hardware.

Basic hypotheses regarding the design of upper stage are as follows:

- Payload.
- Parking orbit and target.
- Mechanical properties.
- Helium mechanical properties.
- The characteristics of the chosen fuel.
- Safety factors.
- Temperature of tanks and the flame.

The main body of the communication among subsystems, within each other, and the system is created according to the design matrix. Design matrixes are the designer's guide for displaying design communications and the effects of the parameters on each other (Peoples and Schuman 2003). The communicative matrix between the components of propulsion system is shown in Fig. 1 due to the importance

of the communication of the propulsion system components. The most important design matrix is the comprehensive design matrix which is shown in Fig. 2. Only the main parameters of design are mentioned in the matrixes.

In designing the MSDO algorithm, several optimization and convergences were used. The goal of optimization is to achieve the least amount of goal parameters; however, the goal of convergence is to converge all design parameters within each other as well as meeting all the requirements and limitations. Optimization includes:

- Optimizing comprehensive design matrix based on the MSDO and through genetic algorithm.
- Optimizing trajectory through genetic algorithm.
- Optimizing propulsion system through genetic algorithm.
- Optimizing total mass through genetic algorithm.
- Optimizing the thickness of the crust and stiffener based on the buckling test and through genetic algorithm.

In Table 2, one can see the above-mentioned optimization properties. Design convergence items in MSDO algorithm include the acceleration of design process according to the statistical equations (reducing the time while increasing accuracy).

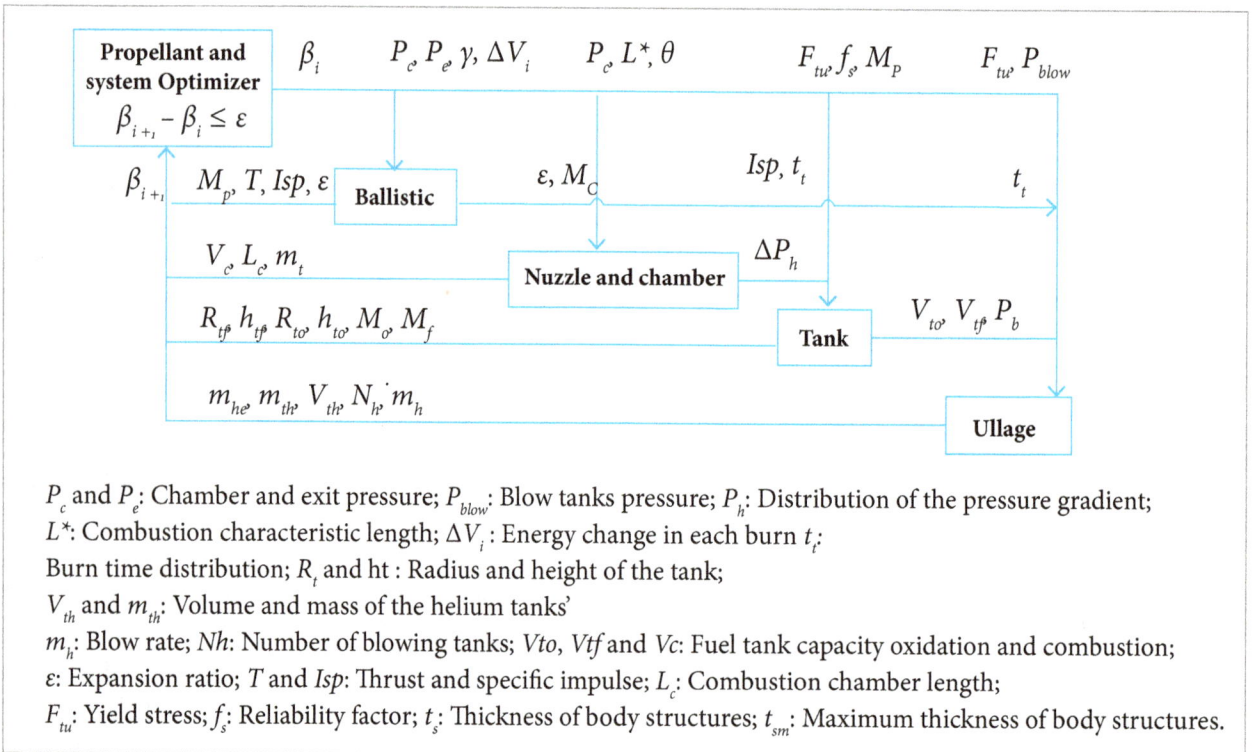

P_c and P_e: Chamber and exit pressure; P_{blow}: Blow tanks pressure; P_h: Distribution of the pressure gradient; L^*: Combustion characteristic length; ΔV_i: Energy change in each burn t_t: Burn time distribution; R_t and ht : Radius and height of the tank; V_{th} and m_{th}: Volume and mass of the helium tanks' m_h: Blow rate; Nh: Number of blowing tanks; Vto, Vtf and Vc: Fuel tank capacity oxidation and combustion; ε: Expansion ratio; T and Isp: Thrust and specific impulse; L_c: Combustion chamber length; F_{tu}: Yield stress; f_s: Reliability factor; t_s: Thickness of body structures; t_{sm}: Maximum thickness of body structures.

Figure 1. Propulsion system matrix.

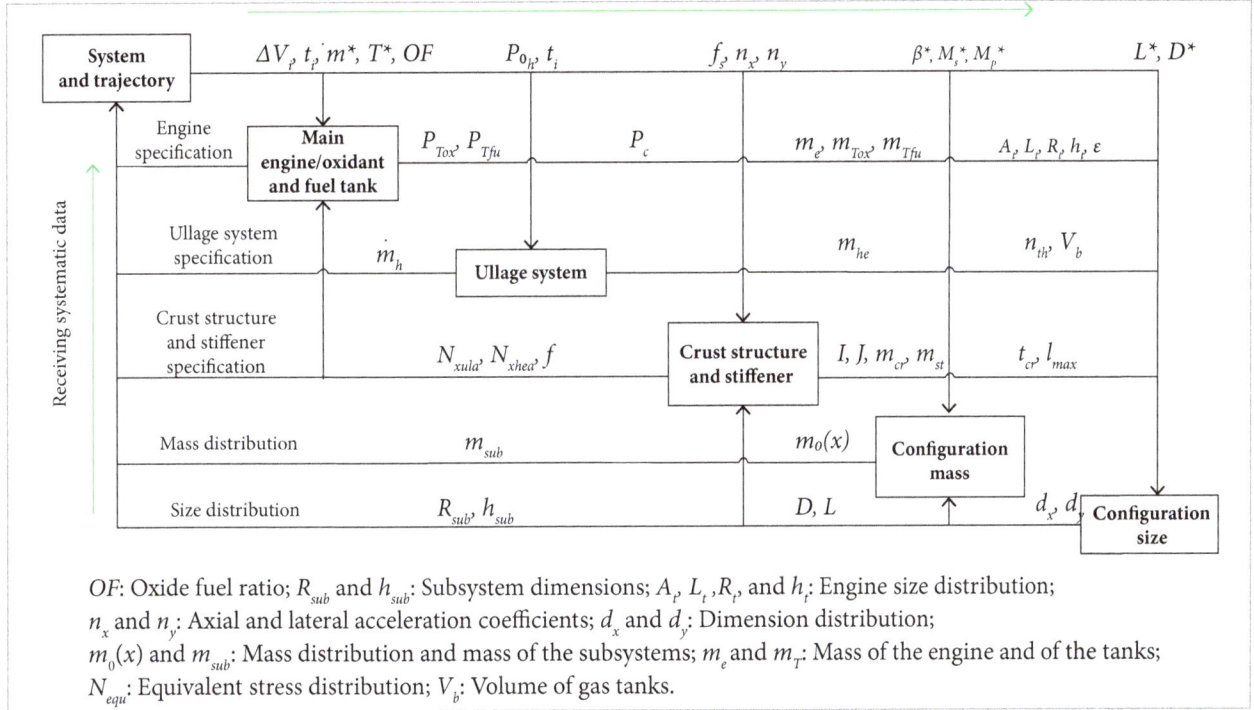

OF: Oxide fuel ratio; R_{sub} and h_{sub}: Subsystem dimensions; A_t, L_t, R_t, and h_t: Engine size distribution; n_x and n_y: Axial and lateral acceleration coefficients; d_x and d_y: Dimension distribution; $m_0(x)$ and m_{sub}: Mass distribution and mass of the subsystems; m_e and m_T: Mass of the engine and of the tanks; N_{equ}: Equivalent stress distribution; V_b: Volume of gas tanks.

Figure 2. Design correlations matrix.

Primary values for design are obtained with statistical equations. Propulsion system convergence through propulsion system mass factor is defined as (Motlagh *et al.* 2013):

$$\beta = M_s / (M_s + M_p) \qquad (1)$$

$$\beta_n = \beta_{n-1} < \varepsilon \qquad (2)$$

The convergence of upper stage compared with the previous stage launch vehicle is:

$$\alpha = M_{p1} / M_{p2} \qquad (3)$$

$$\alpha_n - \alpha_{n-1} < \varepsilon \qquad (4)$$

In Fig. 3, it is presented the interference between convergence

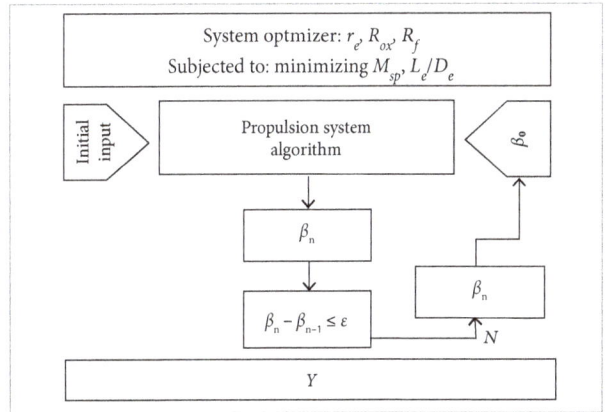

Figure 3. Interference between convergence and optimization.

Table 2. Optimization properties.

System optimizer	Subjected to	Generations	Population	Crossover	Mutation
MSDO optimization	Min: M_t and L/D	100	100	Scattered	Uniform (0.2)
Orbital optimization	Min: M_p	100	20	Scattered	Uniform (0.2)
Propulsion optimization	Min: M_{sp} and L_e/D_e	100	100	Scattered	Uniform (0.2)
Total mass optimization	Min: M_t	100	20	Scattered	Uniform (0.2)
Structure optimization	Min: M_{st}	100	20	Scattered	Uniform (0.2)

and optimization of the algorithm (Fig. 4). Design variables are described in Table 3.

Design output includes the following:

- Systematic parameters of upper stage.
- Subdivisions' mass-dimension distribution.
- Systematic data of subdivisions.

Table 3. Design variables.

System optimizer	Subjected to	Design variables	Name	Unit	Limitation	Description
MSDO Optimization	$Min: M_t$ and L/D	P_c	Chamber pressure	bar	$5 < P_c < 15$	–
		OF	Oxide to fuel ratio	–	$3.5 < OF < 4.5$	According to fuel
		N_h	Number of helium tanks	–	$2 < N_h < 12$	
		N	Thrust to weight ratio	–	$1.5 < n < 4.5$	According to configuration
Orbital Optimization	$Min: M_{sp}$	r_{pt}	Transfer orbit Perigee	km	200 ...36000	According to orbit design
		r_{at}	Transfer orbit Apogee			
Propulsion Optimization	$Min: M_{sp}$ and L_e/D_e	r_e	Nozzle exit		$r_e < D$	
		R_{ox}	Oxidizer tank profile	m	$R_{ox} < D - t_s$	–
		R_f	Fuel tank profile		$R_f < D - t_s$	
Total mass Optimization	$Min: M_t$	–	Selection of components	–	–	According to data feasibility
Structure Optimization	$Min: M_{st}$	t_s	Body thickness	m	$t_{sm} - 2 < t_s < t_{sm}$	–
		J	Stiffener rigidity	m^4	–	Stiffener selection
		–	Structural Materials	–	–	–
Propulsion system convergence	$\beta_n - \beta_{n-1} < \varepsilon$	β	Propulsion Mass factor	–	$0.8 < \beta < 1.5$	According to statistical data
Stages fuel mass ratio convergence	$\alpha_n - \alpha_{n-1} < \varepsilon$	α	Stages mass factor	–	$0.08 < \alpha < 0.3$	According to statistical data

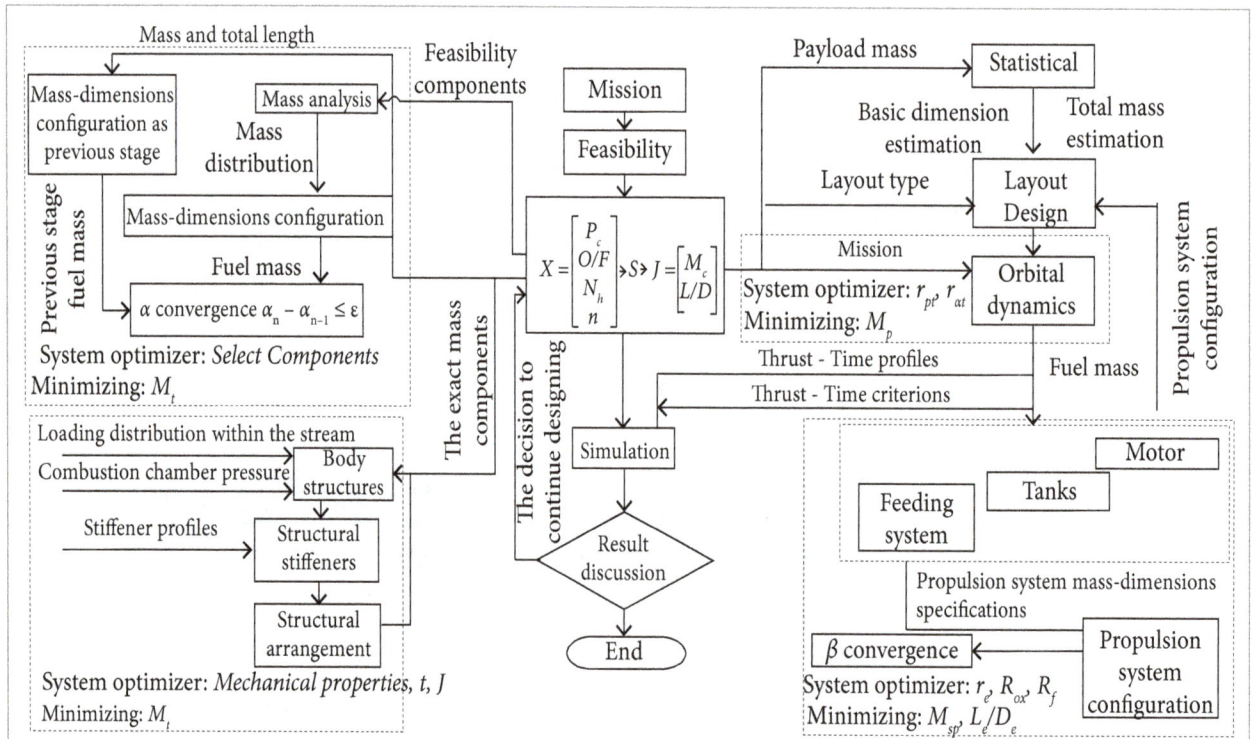

Figure 4. MSDO algorithm.

However, the most important outputs of the algorithms include the following:

- Multidisciplinary optimization system.
- Adaptation of all design parameters.
- Meeting all the requirements and limitations.

DESIGN METHODOLOGY OF SUBSYSTEMS

Statistical Design

Preliminary estimation of systematic parameters in upper stage design process is of utmost importance due to the following reasons: creating a basic configuration for upper stage and faster convergence and more optimal design.

In statistical design via obtained data, the created population and needed graphs were extracted to create initial input (Mirshams and Khaladjzadeh 2010). For instance, 2 sample graphs are given in Figs. 5 and 6. The payload mass is the first and the most important input in designing an upper stage.

$$M_k = 1.58 + 1.2 M_{pay} \qquad (R^2 = 0.96) \qquad (5)$$

where: M_k and M_{pay} are the final and payload mass.
The thrust to weight relative to burn time is represented by:

$$n = 19.38 - 0.035t + 2.6 \times 10^{-5} t^2 - 6.88 \times 10^{-9} t^3 \qquad (6)$$
$$(R^2 = 0.98)$$

Statistical equations are derived as follows: μ_p and μ_f payload mass ratio and dry mass ratio.

$$M_F = 0.026 M_{pay}^2 + 0.799 M_{pay} + 2.546 \qquad (7)$$

$$M_0 = -0.066 M_F^2 + 3.439 M_F + 3.004 \qquad (8)$$

$$M_P = 0.91 M_0^{0.917} \qquad (9)$$

$$\mu_p = 3.267 \mu_f^2 - 0.559 \mu_F + 0.273 \qquad (10)$$

$$n = 2880 t^{-0.94} \qquad (11)$$

$$T = 3.668 n^2 - 20.21 n + 107.7 \qquad (12)$$

$$\dot{m} = 0.525 n^2 - 0.937n + 14.45 \qquad (13)$$

$$T = 4422 (\dot{m} = \tfrac{M_P}{t}) - 0.497 \qquad (14)$$

$$L = 109.5 \, e^{-2.33D} \qquad (15)$$

$$LD^2 = 0.323 M_0^2 - 7.997 M_0 + 116.6 \qquad (16)$$

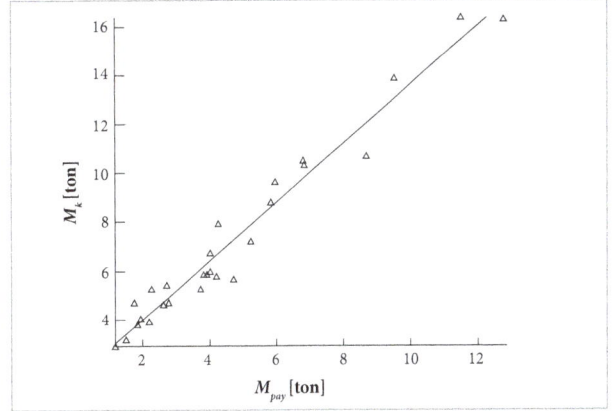

Figure 5. Dry mass *versus* payload mass.

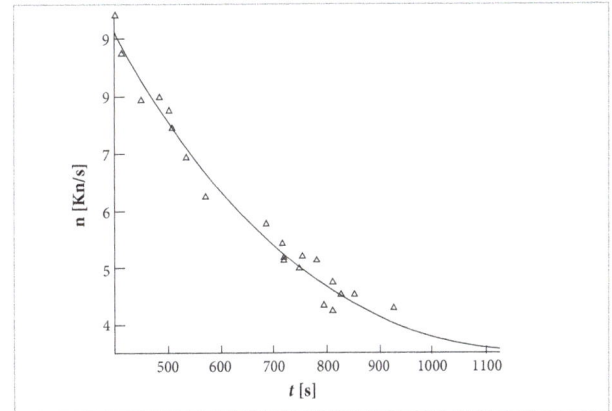

Figure 6. Thrust/weight *versus* burn time.

Trajectory Simulation and Design

In this study, trajectory design (Fig. 7) is done based on Hahman's approach as well as 2 references (Chobotv 1996; Curtis 2005).

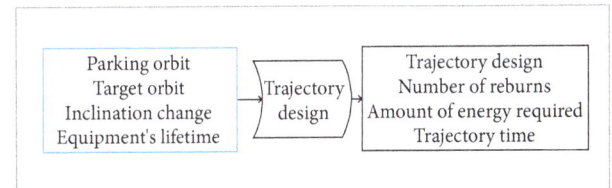

Figure 7. Trajectory design algorithm.

Mass Analysis

According to statistical studies, linear relationship between the dry mass of the rocket block with the portable mass of the same rocket block represents the similarity of mass ranges of the whole subsystems and their subcategories in each block with the same objectives. Thus, determining mass ranges of the main subcategories based on the main effective parameters is viable. In Table 4, all the objects and the way to achieve them are illustrated. Final mass of the upper stage is achieved via the table and considering all the mass parameters. The final mass of upper stage can be achieved with all mass parameters presented in Table 4.

Table 4. Distribution of the mass components.

Upper Stage		Estimated mass	Computational mass	Selective mass
Structure	Satellite installation stand	×		
	Body structure		×	
	Tank protector	×		
	Motor holder	×		
	Nacelle	×		
	Motor front and back flange		×	
	Separation equipment	×		
Feeding system	Helium		×	
	Helium tank		×	
	Other equipment	×		
Propulsion	Fuel tank		×	
	Oxidant tank		×	
	Connection tunes	×		
	Motor elements	×		
	Combustion chamber		×	
	Nozzle		×	
	Other components of motor	×		
Guidance and control hardware	Flight computer			×
	Guidance control block			×
	Inertia measurement block			×
	Valves	×		
	Accessories	×		
	Telemetric system	×		
Actuators	Electromechanical actuators			×
	Cables and electrical connections	×		
	Thrusters			×
	Braking motor			×
	Acceleration motor			×
Cases guidance	Central computer			×
	Sensors			×
	Gyro planes			×

Propulsion Subsystem Analysis

Different specifications of a space propulsion system with other propulsion systems of a rocket are summarized as (Friedman and Kenny 1965):

- Different outside conditions (space conditions).
- On-Off numbers (based on the trajectory design).
- Less thrust-to-weight ratio.
- The use of pressure feed system (high accuracy but low thrust) (Sutton and Biblarz 2001).

Figure 8 shows the propulsion analysis algorithm.

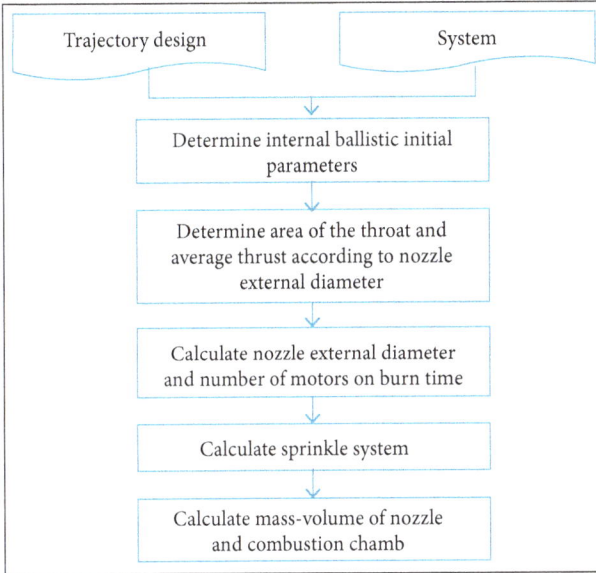

Figure 8. Propulsion sub-algorithm.

Propulsion System Convergence

In the first design loop, the amount of fuel mass and the final mass of upper stage in every burning is calculated through the following equations:

$$M_{f1} = M_0 e^{-\left(\Delta V_1 / I_{sp} g_0\right)} \tag{17}$$

$$M_{P1} = M_0 - M_{f1} \tag{18}$$

$$M_{fi} = \left(M_0 - \sum_{k=1}^{i-1} M_{pk}\right) e^{-\left(\Delta V_i / I_{sp} g_0\right)} \tag{19}$$

$$M_{Pi} = M_0 - \sum_{k=1}^{i-1} M_{pk} - M_{fi} \tag{20}$$

$$M_P = \sum_{k=1}^{i} M_{Pi} \tag{21}$$

After propulsion system design convergence, the amount of optimal fuel mass in every engine-on status is achieved. In the present study, the optimal propulsion system design is achieved through converging the structural convergence factor (β). Using upper stage mass in every design loop, all subsystems' propulsion design is achieved and then the new value for fuel mass and convergence factor is obtained.

$$M_{pj} = \frac{1-\beta}{\beta}(M_{shj} + M_{sej} + M_{sTj}) \tag{22}$$

where: j is the integration loop counter design; M_{sh}, M_{se}, and M_{sT} are dry mass of subsystems. The equation for convergence coefficient (β) is defined as follows:

$$\beta = \frac{M_s}{M_s + M_p} \tag{23}$$

The internal relations of the propulsion system are shown in Fig. 1.

Feeding System Analysis

Controlling pressure in fuel tanks is easily possible using the pressure system feed. Furthermore, the simplicity of adjusting pressure in pressure system feed determines its high reliability, thus the process of switching and flow control is easily possible. Output flow could be controlled with installing a heater or pressure control valves. Generally, the concurrent process of disembarkation of capsules containing compressed gas and filling propellant tanks could be shown via Eqs. 24 and 25 (Huzel et al. 1992).

High-pressure tanks (capsules):

$$I \begin{cases} \dfrac{dp}{dt} = \dfrac{(k-1)Z}{V}\left(\dfrac{dQ}{dt} - \dot{m}_d h_d - \dfrac{PV}{(k-1)Z^2}\dfrac{dZ}{dt}\right) \\[2mm] \dfrac{d\rho}{dt} = -\dfrac{\dot{m}_d}{V} \\[2mm] T = \dfrac{P}{R\rho} \end{cases} \tag{24}$$

Propellant tanks:

$$II \begin{cases} \dfrac{dp}{dt} = \dfrac{(k-1)}{V(t)}\left(\dfrac{dQ}{dt} - \dot{m}_i h_i - \dfrac{k}{(k-1)}P\dfrac{dV(t)}{dt}\right) \\[2mm] \dfrac{d\rho}{dt} = \dfrac{1}{V(t)}\left(\dot{m}_i - \rho\dfrac{dV(t)}{dt}\right) \\[2mm] T = \dfrac{P}{R\rho} \end{cases} \tag{25}$$

where: Q is heat transfer between the gas blowing and its environment; Z is the gas compressibility factor; V is the volume control; T is the temperature; P is the pressure; $\dot{m}_i$. $\dot{m}_d$ are blowing gas mass flow rate input and output volume control; h_i and h_d are blowing gas enthalpy entry and exit control volume.

The sub-algorithm of feeding system is shown in Fig. 9.

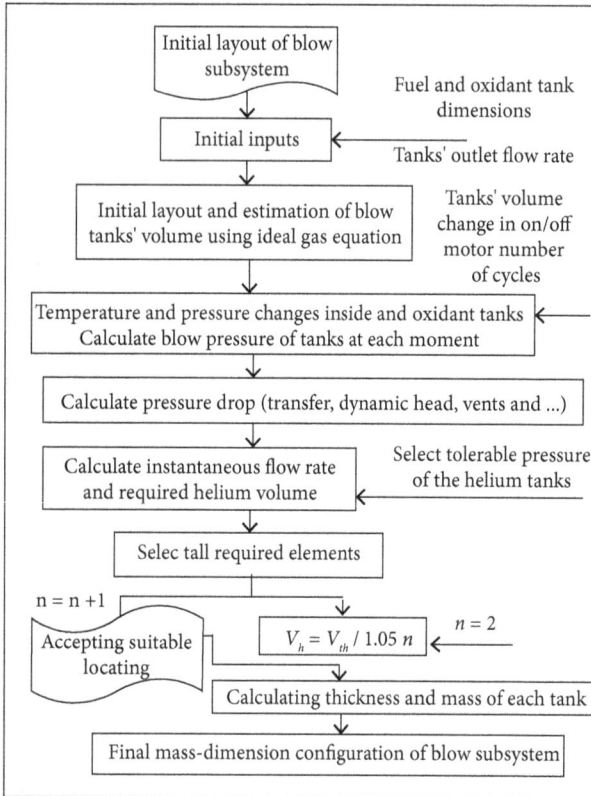

Figure 9. Blow system sub-algorithm.

Volume, thickness, and the mass of every helium tank is obtained from the following equations (Humble *et al.* 1995):

$$V_h = \frac{\gamma P_{po}}{P_{boo} - P_{boe}} V_{To} + \frac{\gamma P_{pf}}{P_{bfo} - P_{bfe}} V_{Tf} \quad (26)$$

$$t_G = \frac{P_{ho}}{2\sigma_{hw}} R_h \quad (27)$$

$$M_{Th} = \frac{3}{2} \frac{\rho_{hw}}{\sigma_{hw}} \frac{2\gamma V_b}{1 - \left(\frac{P_{Ge}}{P_{Go}}\right)} \quad (28)$$

where: V_h, t_G and M_{Th} are volume, thickness, and mass of blowing tanks; σ_{hw} and ρ_{hw} are the mechanical properties.

According to blow subsystem sub-algorithm, numbers of blow tanks are selected considering configuration and layout. The final mass of feeding system is calculated by the following equation:

$$m_{helium} = m_h + m_{line} + m_{he} + m_{valve} + m_{heater} \quad (29)$$

Tanks Analysis

The shape of the tank is a function of weight, leakage rate, tank volume, and locating restrictions. Spherical tanks have the best empty weight-to-loaded weight ratio (Hutchinson and Olds 2004). Tank design sub-algorithm is shown in Fig. 10. Other elements such as control valves are selected according to input pressure and flow rate.

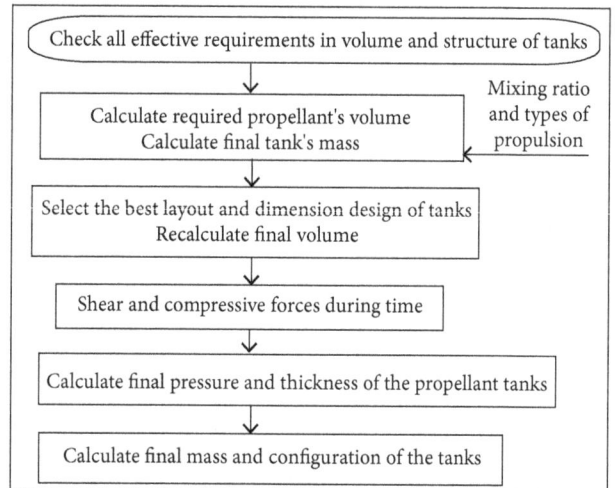

Figure 10. Tank design sub-algorithm.

Structure Analysis

In structural analysis, providing stability and structural strength to deal with all external pressures is the main goal. In order to determine the mass of the structure, at first, the loads on each section of the structure shall be determined via different stages of preparation to the end of the flight. Loads on each section mean axial force, shear force and bending torque which is applied under external loads during structure mission. Critical load for each section is occurred based on existing experiences in any of the selected above stages. Loading sub-algorithm and the thickness of the body structure is shown in Fig. 11.

After calculating longitudinal and lateral load flow, equivalent stresses are obtained and exerted to the structure which determines the thickness of the body (Ardema *et al.* 1996; Crawford and Burns 1963).

$$N_x = N_{xbend} + N_{xaxial} + N_{xullage} + N_{xhead} \tag{30}$$

$$\sigma_n = \frac{1}{\pi D t}\left(N + \frac{4M}{D}\right) \tag{31}$$

$$N_{equ}(x) = N(x) + \frac{4M(x)}{D(x)} \tag{32}$$

$$t = \frac{N_{equ}}{\pi D \sigma_n} \tag{33}$$

where: N is the axial force; M is the bending moment.

According to the critical situation exerted on the upper stage structure, lateral stiffeners with low number could be used to strengthen body structure. Design algorithm for strengthening stiffeners is shown in Fig. 12.

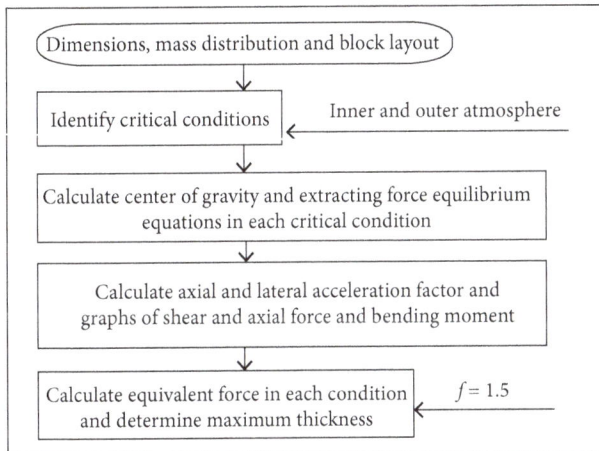

Figure 11. Body structure sub-algorithm.

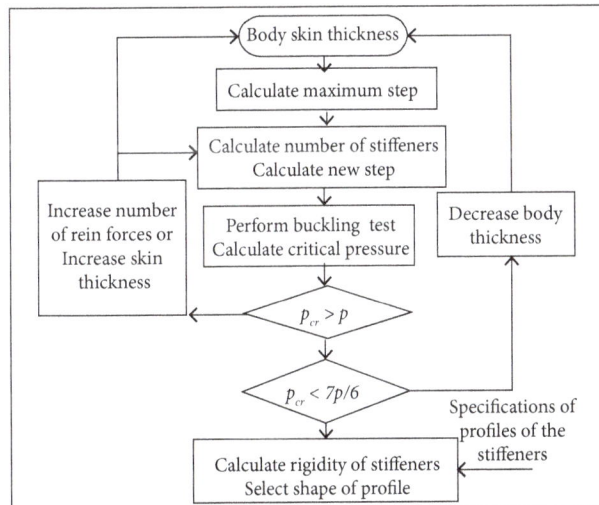

Figure 12. Stiffener structure sub-algorithm.

Stiffener rigidity and then the shape of the stiffener profile can be selected through the following equation:

$$J = \frac{(N+1)^{4/3} - 1}{(N+1)} \frac{t^3}{12(1-\mu^2)} \tag{34}$$

where: J is the rigidity stiffener.

Dimension Design

Dimension design is achieved with 2 different assumptions which, in one, upper stage diameter was determined and, in the other, it was calculated in the output. The calculation of the length and diameter can be achieved through the following equation:

$$L_t = L_f + L_o + L_{of} + L_e + L_s \tag{35}$$

$$D_t = D_p + 2(t_s + t_t + t_r) \tag{36}$$

RESULTS
SAMPLE SOLUTION

A sample solution for upper stage design is presented in this section to generally introduce the design method:

1. Mission definition
    *   Payload mass is 1.5 tones with parking orbit which is 200 KMs and destination orbit is 36000 KMs with inclination of 45 degrees change.
    *   Other design inputs are based on requirements and limitations.
2. Results of supplying design initial inputs are shown in Table 5.

Table 5. Initial input.

Input variable	Values
β	0.162
T (kn)	48.36
μ_p	0.75
M_p (ton)	10.23
M_F (ton)	3.47
M_0 (ton)	13.70

3. Trajectory design:
    *   Hohmann's transfer (3 times Re burn).
    *   Determining transfer orbit (according to energy limitation and optimized mode).

4.   Determining material and construction inputs of the subsystem:
   •   Used mechanical properties, environmental conditions, used fluid properties.
   •   Determining fuel and oxidizer (hydrogen/oxygen).
5.   Initial feasibility based on ability to transport about 13 tones to the parking orbit through stages of launcher rocket.
6.   System design:
   •   Determining configuration parameters according to Table 4.
   •   Selecting initial mass factor.
7.   Propulsion design parameters (Table 6).

Table 6. Propulsion engine parameters.

Parameter name	Input variable	Values
Thrust factor	C_f	1.68
Nozzle expansion ratio	ε	37
Specific impulse	Isp	326
Burning time	t_b (s)	680
Thrust	T (kn)	60
Propulsion flow rate	$\dot{m}$ (kg/s)	20.52
Motor mass	m_m (m)	497
Motor diameter	d_m (m)	0.852
Motor length	L_m (m)	2.04
Nozzle length	L_n (m)	1.14
Nozzle outlet diameter	d_e (m)	0.78

8.   Feeding system design (Table 7):

Table 7. Feeding system parameters.

Parameter name	Input variable	Values
Blowing system mass	m_h (kg)	66
Cases blowing mass	M_{Th} (kg)	34.5
Helium mass	m_{he} (kg)	20.6
Thickness of helium tanks	t_h (mm)	4.45
Radius of helium tanks	r_h (mm)	257
Number of helium tank	N_h	6

a. Selecting pressure-feed system.
b. Calculating pressure drop in fuel flow which is 3.2 bars and for oxidant is 4 bars.
c. Fuel tank pressure 11.2 bars and oxidant 12 bars.

9.   Design of tanks: mass and volume of tanks (Table 8).

Table 8. Tank parameters.

Parameter name	Values
Oxidant volume	7.050 m²
Oxidant mass	9.669 ton
Fuel volume	4.2827 m²
Fuel mass	3.581 ton
Oxidant tank radius	1.1895 m
Fuel tank radius	1.1895 m
Oxidant tank pressure	12 bar
Fuel tank pressure	11.2 bar
Fuel tank thickness	0.72257 m
Oxidant tank thickness	3.801 mm
Fuel tank thickness	3.5485 mm

10.   Body structure: determining body thickness and stiffener which requires determining critical modes properties of previous steps, fairing and types of stiffener.
11.   Mass distribution:
   •   Determining mass distribution (Table 9).
   •   Upper stage primary masses.

Table 9. Other components mass distribution.

Parameter name	Values [kg]
Control block	245.88
Actuators	94.56
Cables	113.48
Body	189.1397
Telemetry	151
Stand	150
Disposal system	60
Engine maintenance	28.28
Control segments	18.91
Long tube	31
Flanges	28.28
Separation	75
Gyro planes and IMU	35

12.   Upper stage mass and dimensions:
   •   Upper stage dimensions (Table 10).
   •   Upper stage primary masses (Table 11).

Table 10. Diameter and total length.

Parameter name	Values
Dimensional ratio	2.7812
Diameter (m)	2.4921
Total length (m)	6.9314

Table 11. Mass parameters.

Parameter name	Values
Fuel mass	13.249
Final mass	3.391
Structure mass	1.891
Total mass	16.64

DIFFERENT VARIANTS AND VALIDATION

Technology for constructing a launch vehicle in countries varies from each other, however, the statistical graphs indicate the approximate similarity of systematic parameters for upper-stage systems. For instance, Fig. 13 shows the relationship of payload mass and dry mass in the upper-stage system. The real samples are circular and the samples derived from this paper for UDMH/N204 fuel are square-shaped. The comparative curve of burn time due to thrust/weight is presented in Fig. 14. Convergence factors (α and β) are illustrated in Figs. 15 and 16.

The close similarity of these graphs is the main reason to validate the results derived from the study. At this point another validation was compared with Cent.D-5 upper-stage of Atlas V (401) which is presented in Table 12. Problem inputs: M_{pay} = 4.75 ton; D = 3.05m and I_{sp} = 4378 N*s/kg.

In Table 13, the errors in the statistical design methods and MSDO compared to the systematic data of Cent. D-5 upper stage system. As shown in Table 5, errors of fuel mass, dry mass and total mass in MSDO design are decreased by 9 to 16% compared with Cent. D-5. Furthermore, the accuracy of primary data derived from the statistical design is obvious in this table.

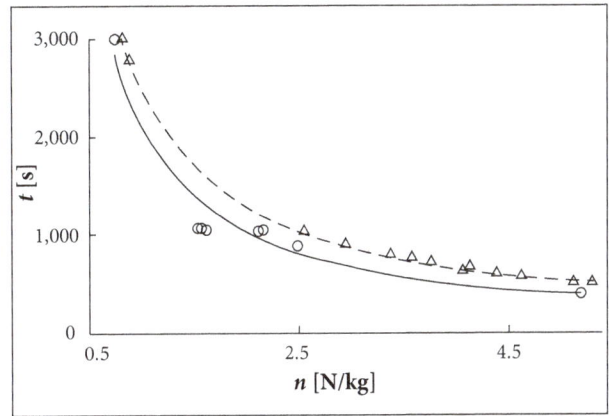

Figure 14. Comparative curve of burn time due to T/W.

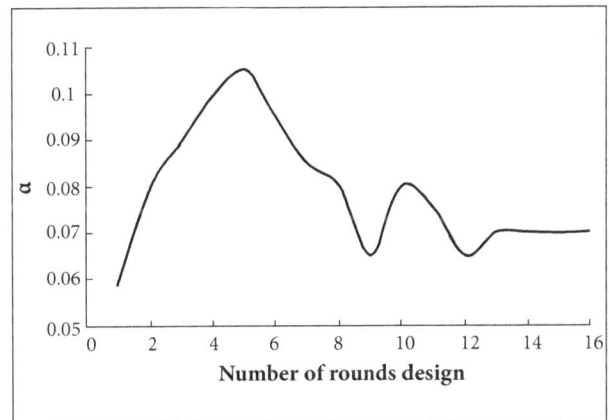

Figure 15. Convergence factor α.

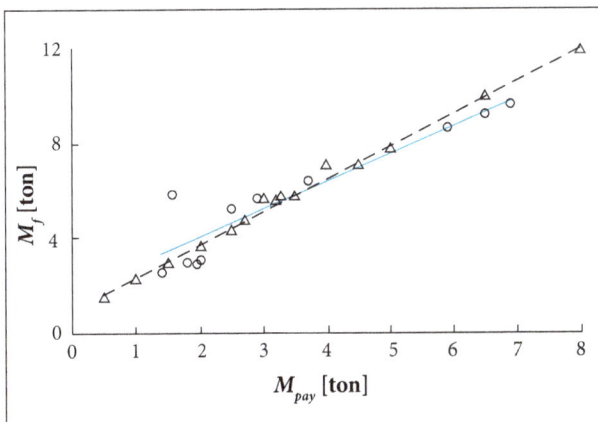

Figure 13. Final mass *versus* payload mass.

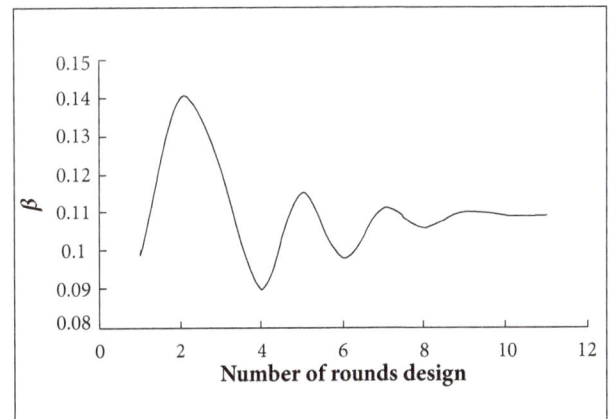

Figure 16. Convergence factor β.

Table 12. Comparison of validation to Cent D-5 upper-stage.

Output data	Mp (ton)	M_o (ton)	μ_t	μ_p	β	n (N/kg)
System parameters Cent. D-5	19.65	26.78	0.266	0.18	0.10	3.64
Statistical design	16.8	23.66	0.29	0.2	0.11	3.88
MSDO	18.93	26.01	0.27	0.182	0.109	3.61

Table 13. Methods errors compared to the systematic data of Cent. D-5 upper stage system

Design type method	Propellant weight [%]	Dry weight [%]	Lift off weight [%]
Statistical	11.6	11.3	14.5
MSDO	0.2	2.1	3.66
0.27	0.182	0.109	3.61

CONCLUSION

Design processes of all elements are done simultaneously in this paper. In each design loop, calculations become more accurate and results of each section, more suitable according to other systems. In this design method, a mistake can be found by system (because it affects all elements) and it would be easier to fix it. The main results are:

- Proportional with technology and ability for construction.
- Logical relation of all segments and subsystems.
- Developing a national design and method.

- Presenting a sample upper stage according to different inputs.
- Logical convergence of parameters to logical values.
- Ability of performing general feasibility.
- Extensibility to accurate initial design.
- Ability of networking the design under supervision of the system.
- Designing systematic sample of upper stage.
- General guideline based on implementation of detail design (limitations of methods and range of parameters).
- All the important design points are specified in order to divide the algorithm for performing different projects.

AUTHOR'S CONTRIBUTION

Zakeri M has provided the article's text; Nosratollahi M and Novinzade A carried out the guidance and control of the paper; the final set was conducted by Zakeri M and Nosratollahi M; the idea was developed by all authors.

REFERENCES

Adami A, Mortazavi M, Nosratollahi M (2015) A new approach in multidisciplinary design optimization of upper-stages using combined framework. J Acta Astronautica 114:174-183. doi: 10.1016/j.actaastro.2015.04.011

Aldheeb MA, Kafafy R, Idres M, Omar HM, Abido MA (2012) Design optimization of micro air launch vehicle using differential evolution. J Aerosp Technol Manag 4(2):185-196. doi: 10.5028/jatm.2012.04020112

Balesdent M, Bérend N, Dépincé P, Chriette A (2011) A survey of multidisciplinary design optimization. Structural and Multidisciplinary Optimization 45(5):619-642. doi: 10.1007/s00158-011-0701-4

Brown NF, Olds JR (2006) Evaluation of multidisciplinary optimization techniques applied to a reusable launch vehicle. J Spacecraft Rockets 43(6):1289-1300. doi: 10.2514/1.16577

Casalino L, Pastrone D (2010) Optimal design of hybrid rocket motors for launchers upper stages. J Propul Power 26(3):421-427. doi: 10.2514/1.41856

Chambers MC, Ardema MD, Patron AP, Hahn AS, Miura H, Moore MD (1996) Analytical fuselage and wing weight estimation of transport aircraft. NASA TM-110392.

Chobotv VA (1996) Orbital mechanics. Reston: American Institute of Aeronautics and Astronautics.

Crawford RF, Burns AB (1963) Minimum weight potentials for stiffened plates and shells. AIAA J 1(4):879-886. doi: 10.2514/3.1658

Curtis HD (2005) Orbital mechanics for engineering students. Oxford: Elsevier Butterworth-Heinemann.

Diedrich A, Hileman J, Tan D, Willcox K, Spakovszky Z (2006) Multidisciplinary design and optimization of the silent aircraft. Proceedings of the 44th AIAA Aerospace Sciences Meeting and

Exhibit; Reno, USA.

Fleeman EL (2001) Technologies for future precision strike missile systems - missile design technology. Paper presented at: RTO SCI Lecture Series on Technologies for Future Precision Strike Missile Systems. Published in RTO-EN-018; Madrid, Spain.

Friedman PA, Kenny RJ (1965) Chemical pressurization of hypergolic liquid propellants. *J Spacecraft Rockets* 2(5):746-753. doi: 10.2514/3.28273

Humble RW, Henry GN, Larson WJ (1995) Space propulsion analysis and design. New York: McGraw-Hill.

Hutchinson VL, Olds JR (2004) Estimation of launch vehicle propellant tank structural weight using simplified beam approximation. Proceedings of the 40th AIAA/ASME/SAE/ASEE Joint Propulsion Conference and Exhibit; Florida, USA.

Huzel DK, Huang DH, Arbit H (1992) Modern engineering for design of liquid-propellant rocket engines. Washington: American Institute of Aeronautics and Astronautics.

Mirshams M, Khaladjzadeh L (2010) Drivation of system level characteristics of a manned spacecraft by applying statistics models. Journal of Space Science and Technology 3(1-2):25-36. In Persian.

Motlagh JA, Novinzadeh AB, Zakeri M (2013) New approach in designing solid upper stage for interplanetary missions using finite burn assumption. IEEE Aero Electron Syst Mag 28(10):36-43. doi:

10.1109/MAES.2013.6642830

Olds JR (1993) Multidisciplinary design techniques applied to conceptual aerospace vehicle design (PhD thesis). Raleigh: North Carolina State University.

Peoples R, Schuman T (2003) A joint performance and financial approach to aircraft design optimization; [accessed 2016 Sept 27]. https://dspace.mit.edu/bitstream/handle/1721.1/68163/16-888-spring-2004/contents/projects/6peoples_schuman.pdf

Riddle E (1998) Use of optimization methods in small satellite systems analysis. Paper presented at: 12th AIAA/USU Conference on Small Satellites; Logan, USA.

Sutton GP, Biblarz O (2001) Rocket propulsion elements. Lewiston: John Wiley & Sons.

Tsuchiya T, Mori T (2002) Multidisciplinary design optimization to future space transportation vehicles. AIAA-2002-5171.

Villanueva FM, Linshu H, Dajun X (2013) Kick solid rocket motor multidisciplinary design optimization using genetic algorithm. J Aerosp Technol Manag 5(3):293-304. doi: 10.5028/JATM.v5i3.225

Wronski J, Gray JM (2004) Multi-disciplinary system design optimization of the F-350 rear suspension; [accessed 2016 Sept 27]. http://core.csu.edu.cn/NR/rdonlyres/Aeronautics-and-Astronautics/16-888Spring-2004/CAC1955D-2E22-4307-A7A8-905105DD9900/0/wronski_gray.pdf

11

The Use of an Atmospheric Model to Simulate the Rocket Exhaust Effluents Transport and Dispersion for the Centro de Lançamento de Alcântara

Daniel Schuch[1], Gilberto Fisch[2]

ABSTRACT: This paper introduces a new approach to represent the rocket exhaust effluents into an atmospheric dispersion model considering the trajectory and variable burning rates of a Satellite Vehicle Launcher, taking into account the buoyancy of the exhausted gases. It presents a simulation for a Satellite Vehicle Launcher flight at 12:00Z in a typical day of the dry season (Sept 17, 2008) at the Centro de Lançamento de Alcântara using the Weather Research and Forecasting Model coupled with a modified chemistry module to take into account the gases HCl, CO, CO_2, and particulate matter emitted from the rocket engine. The results show that the HCl levels are dangerous in the first hour after the launching into the Launch Preparation Area and at the Technical Meteorological Center region; the CO levels are critical for the first 10 min after the launching, representing a high risk for human activities at the proximities of the launching pad.

KEYWORDS: Satellite Vehicle Launcher, Mesoscale model, Atmospheric dispersion model, HCl.

INTRODUCTION

The Centro de Lançamento de Alcântara (CLA) is the Brazilian access to the space, located at the north part of the northeastern region of Brazil. It has some advantages due to its geographical position close to the Equator, which allows rocket launchings that consume less propellant for geostationary satellite missions. Other advantages are associated with its proximity of São Luís (capital of Maranhão State) as well as its low population density, so the health risks of contamination by gases sent out from launchings are reduced. Rockets such as the Veículo Lançador de Satélites (VLS) are launched from this Range Center.

During the first few seconds following the ignition of the engine, the VLS releases a large cloud of hot, buoyant exhaust products near the ground level which rise and entrain into atmosphere until reach an approximate equilibrium with the ambient conditions. This cloud is composed by the products of the combustion of perchlorate and aluminum: hydrogen chloride (HCl), water (H_2O), carbon monoxide (CO), carbon dioxide (CO_2), and particulate material composed by aluminum oxide (Al_2O_3) used into the grain composition of the solid propellant (Denison *et al.* 1994).

All the Space Centers around the world have adopted some models in order to predict these gases dispersions. For instance, the East US Space Ranger Center (like NASA JFK/U. S. Cape Canaveral) uses an operational model known as Rocket Exhaust Effluent Diffusion Model (REEDM) and it has been used to assess the environmental impact of aerospace activities (Bjorklund

1.Departamento de Ciência e Tecnologia Aeroespacial – Instituto Tecnológico de Aeronáutica – Programa de Pós-Graduação em Ciências e Tecnologias Espaciais – São José dos Campos/SP – Brazil. **2.**Departamento de Ciência e Tecnologia Aeroespacial – Instituto de Aeronáutica e Espaço – Divisão de Ciências Atmosféricas – São José dos Campos/SP – Brazil.

Author for correspondence: Daniel Schuch | Departamento de Ciência e Tecnologia Aeroespacial – Instituto Tecnológico de Aeronáutica – Programa de Pós-Graduação em Ciências e Tecnologias Espaciais | Praça Marechal Eduardo Gomes, 50 – Vila das Acácias | CEP: 12.228-901 – São José dos Campos/SP – Brazil | Email: underschuch@gmail.com

et al. 1982). This dispersion model is based on Gaussian model concepts: the exhaust material (mixture amongst CO, CO_2, HCl, and Al_2O_3) is assumed to be uniformly and vertically distributed and to have a bivariate Gaussian distribution in the plane of the horizon at the point of cloud stabilization, which is determined by the cloud rise theory.

The model used at the European Spaceport of Kourou (French Guyana) is the SARRIM Software (Cencetti *et al.* 2011), which considers the emissions divided into "puffs" from the launching pad up to the stabilization height, dealing with the local and large scale impact assessment for propellant and hypergolic rocket releases. It is an operational and fast running tool taking into account atmospheric thermal stratifications inside the boundary layer using *in situ* data like radiosondes.

The India Space Center (Satish Dhawan Space Center SDCS SHAR) has coupled a Hybrid Single-Particle Lagrangian Integrated Trajectory (HYSPLIT) model with an atmospheric mesoscale one (in this case, it was used the mesoscale meteorological model-MM5) to predict the dispersion of exhaust pollutant in the form of vapor and ground level concentrations (Rajasekhar *et al.* 2011).

The Brazilian community is also addressing this problem for CLA since 2010 and it was developed the Modelo Simulador da Dispersão de Efluentes de Foguetes (MSDEF), which represents the solution for time-dependent advection-diffusion equation applying the Laplace transform considering the Atmospheric Boundary Layer as a multilayer system. This solution allows a time evolution description of the concentration field emitted from a source during a release lasting time; it takes into account deposition velocity, first-order chemical reaction (decay), gravitational settling, precipitation scavenging, and plume rise effect. A detail description of this model can be found in Moreira *et al.* (2011). In Nascimento *et al.* (2014), the authors coupled this model to the Weather Research and Forecasting Model (WRF), for the meteorological forecast, and to the Community Multi-scale Air Quality model (CMAQ), for the chemistry. Moreover, Iriart and Fisch (2016) used the WRF coupled with this chemistry module (Chem). Both studies addressed the CLA dispersion problem for air quality models.

In this study we propose a new representation of a rocket emission, taking into account the plume rise effect, trajectory, and variable emissions rates (in time and space), into a meteorological/chemical model in order to achieve a better vertical distribution of the emissions and then predict the transport, dispersion, and atmospheric reaction of the gas exhausted. A simulation using data of the Brazilian VLS was assessed using the WRF

model (version 3.7.1) with a modified Regional Atmospheric Chemistry Mechanism (which includes HCl) for the CLA region.

METHODOLOGY

The WRF model is a numerical weather prediction system, considered the state-of-the-art, designed for both atmospheric research and operational forecasting needs, being applied to a wide range of meteorological problems with scales from tens of meters to thousands of kilometers. The coupling of the meteorology and chemistry (into the WRF-Chem coupled) is calculated on-line (without loss of information), and the meteorological process of transport, radiation, and reactions is fully coupled (interacting with each other) and solved simultaneously without any type of interpolation (Grell *et al.* 2005; Skamarock *et al.* 2008).

For the simulations, 3 nested domains centered into the Setor de Preparação de Lançamento (SPL), where the rocket is launched, were chosen. The outer domain (d1) is a 100×100 grid points with 9 km of horizontal resolution; the middle domain (d2) has 70×70 points and 3 km of resolution; the inner domain (d3) has 40×40 points with 1 km of horizontal resolution. All domains have 43 vertical levels, from surface up to 30 km, distributed mainly close to the surface. Figure 1a shows an image of the region and the 3 domains.

The static data (topography, land mask, vegetation, etc.) used was provided by the United States Geological Survey (available from: http://www.mmm.ucar.edu/wrf/src/WPS_files/geog.tar.gz) with spatial resolution of 30" (this data is usually used to weather forecasts). Figure 1b shows the topography (lines) and landmask (colors) for the inner domain (d3). Also, it was marked the position of some buildings that are part of the CLA's structure: SPL and the Setor de Meteorologia (SMT), as well as the only habited areas nearby: the city of Alcântara (CLA) and the Tapireí village (VTA).

The meteorological variables data were extracted from analysis of the Global Forecasting System (GFS) with spatial resolution of 0.5° and available every 3 h. This option was chosen as this meteorological inputs are operational from the National Centers for Environmental Prediction (NCEP). The chemical species used was obtained from the Model for Ozone and Related chemical Tracers-4 (Emmons *et al.* 2010), and the initial and boundary states were modified with the MOZBC pre-processor (Pfister *et al.* 2011).

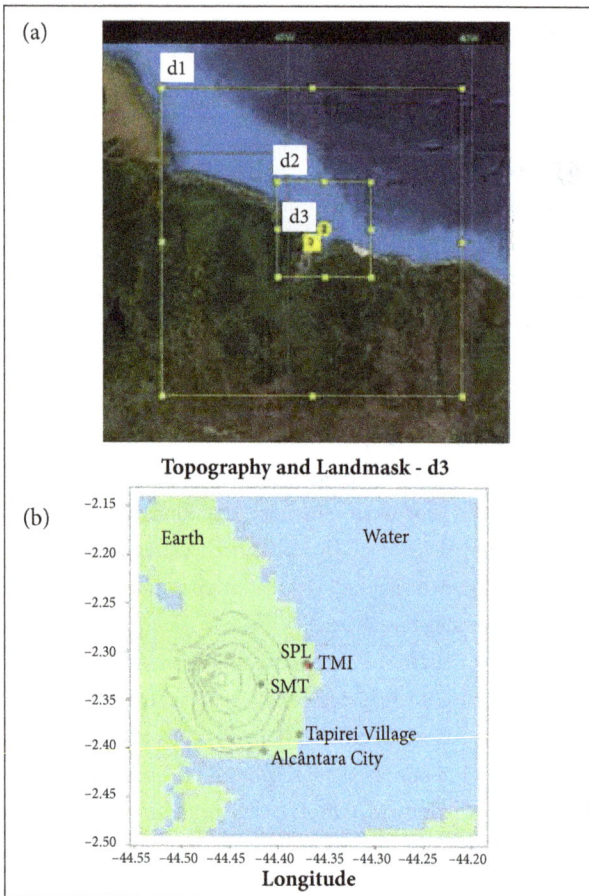

Figure 1. (a) Domains of simulation; (b) Topography (lines) and landmask (green for continent and blue for water) of inner domain.

For the WRF configuration, it was used the following set of parametrizations:

- Microphysics: WRF Single-Moment 3-class Scheme, simple and efficient scheme which contains ice process adapted to large scale.
- Long wave radiation: Rapid Radiative Transfer Model (RRTM), a scheme which utilizes tables of radiation efficiency.
- Short wave radiation: Dudhia scheme, simple scheme of integration which allows the absorption of radiation in clear sky from the clouds and scattering by atmosphere.
- Superficial layer: MYNN surface layer, Nakanishi and Niimo scheme.
- Surface: Noah Land Surface Model, scheme of soil temperature and humidity with 4 layers.
- Planetary Boundary Layer (PBL): Mellor Yamada Nakanishi and Nimo level 2.5, scheme with prediction of turbulent kinetic energy to the model sub-grid.

- Cumulus: Grell 3D, enhanced scheme of Grell-Devenyi which can be used for high spatial resolutions (turned off in the d3 domain).

Due to the necessity of representing all the emitted species by the combustion of the rocket solid propellant into the WRF model, especially the species HCl, a new chemistry mechanism based on the RACM (Stockwell *et al.* 1997) was created. HCl and more 5 chlorinated species were added (Cl, Cl_2, HOCl, ClO, and formyl fluoride), as well as 3 photoreactions, 5 inorganic reactions, and 11 organic reactions from the CB05 mechanism (Yarwood *et al.* 2005) inside the RACM.

Since version 2.2, the WRF model was released with the kinetic preprocessor (KPP) built in the source code. It was designed as a general analysis tool to facilitate the numerical solution of chemical reaction network problems. The KPP subroutines automatically generate FORTRAN code that computes the time-evolution of chemical species, starting with a specification of the chemical mechanism. KPP further allows a rich selection of numerical integration schemes and provides a framework for evaluation of new integrators and chemical mechanisms (Damian *et al.* 2002).

The interaction of the KPP and WRF is made by a subprogram named WRF-KPP-Coupler (WKC). Once the KPP option is enabled, the WRF's compilation script compiles and executes the WKC, and the KPP generates all the code to compile WRF, including the new mechanism. Even though the WKC generates the code to be integrated into the model, the mechanism needs to be included to the file Registry.chem in order to be recognized by the model. Consequently, the Chem module must be modified in order to include the new mechanism in the processes of initialization, calculation of velocities of deposition, different emission options, and optical proprieties.

REPRESENTATION OF SATELLITE VEHICLE LAUNCHER EMISSIONS

In air quality (AQ) models, the way emissions are represented is the most critical input parameter and has greater impact into the final concentration of the contaminant. The inclusion of a source like the VLS into WRF needs some considerations: the rocket like the fuel expenditure, vertical trajectory, and the type of propellant used should be taken into account as well as the limitations of the model for time and space scales.

The total mass of gases released into atmosphere in a VLS launch is given by:

$$M = \int_{t_0}^{t_N} \frac{\partial M}{\partial t} dt \tag{1}$$

where: $\partial M/\partial t$ is the fuel expenditure rate (g/s), which varies with time along the trajectory of the rocket; t_0 is the ignite time; and t_N is the time when the fuel is completely burned.

As the shortest interval for data input of WRF is 1 min, the emissions were split in min by min as

$$M = M_1 + M_2 + ... + M_n + ... + M_N = \tag{2}$$

$$= \int_{t_0}^{t_1} \frac{\partial M}{\partial t} dt + \int_{t_1}^{t_2} \frac{\partial M}{\partial t} dt + ... + \int_{t_{n-1}}^{t_n} \frac{\partial M}{\partial t} dt + ... + \int_{t_{N-1}}^{t_N} \frac{\partial M}{\partial t} dt$$

where: $t_n = 60n$ (s), and each term represents an emission file for the model.

For the vertical domain, the WRF model was discretized into sigma levels, which is a normalized coordinate (assuming unitary value at surface and 0 on the top of the domain) following the topography (Fig. 2).

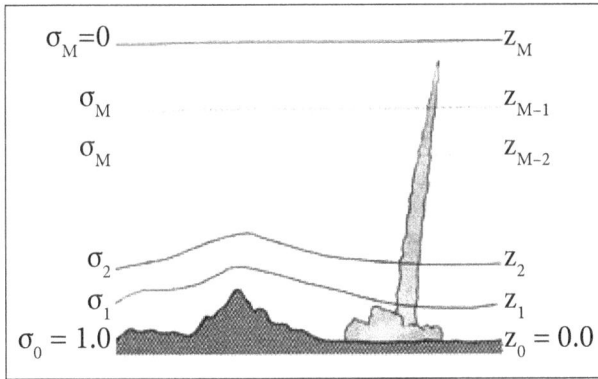

Figure 2. Layers and the rocket trajectory.

Whereas the rocket it is not a fixed source, the distribution of emission should follow the rocket trajectory and is allocated into M model layers as follows:

$$M_n = \sum_{m=1}^{M} M_{nm} = M_{n1} + M_{n2} + ... + M_{nm} + ... + M_{nM} \tag{3}$$

where:

$$M_{nm} = \int_{t_{n-1}}^{t_n} H\left((z_{m-1} - z)(z_m - z)\right) \frac{\partial M(t,z)}{\partial t} dt \tag{4}$$

H is the Heaviside function (whose value is 0 for negative and 1 for positive argument); z_m is the height of the top of the layer m; z_{m-1} is the top of the layer below the surface (z_0); $\partial M/\partial t$ is function of height and time.

For VLS emission scenarios, the exhaust from the rocket combustion is at several thousand Kelvin degrees and highly buoyant. The high temperature of these exhaust emissions causes the plume to be less dense than the surrounding atmosphere, and buoyancy forces acting on the cloud can cause it to lift off the ground and accelerate vertically. As the buoyant cloud rises, it entrains ambient air and grows in size while also cooling. In this initial cloud rise phase, the growth of the cloud volume is due primarily to internal velocity gradients and mixing induced by large temperature gradients within the cloud itself. Even though the cloud is entraining air and cooling due to the mixing hot combustion gases with cooler ambient air, the net thermal buoyancy in the cloud is conserved, and the cloud will continue to rise until it either reaches a stable layer in the atmosphere or the cloud vertical velocity becomes slow enough to be damped by viscous forces (Nyman 2009).

Considering that the WRF model does not have a general scheme for plume rise, we have assumed that the plume is released into a height matching the rocket trajectory plus a plume rise (Δz) height. For the determination of this height rise, the following parametrizations were adopted, in which the height is calculated directly by a model based on Briggs (1975).

For an instantaneous cloud rise scheme, this rise is defined as

$$\Delta z_i = \left(\frac{8F_i}{\gamma^3 s}\right)^{1/4} \tag{5}$$

where: F_i = buoyancy term = $3gq/4\pi\rho C_p T_a$; T_a (m^{4}s^2) is the ambient temperature (K); ρ is the air density (kg/m^3); C_p is the specific heat of exhaust cloud gases = 1.7755 (cal/kgK); q = initial heat of the plume = $H (\partial M/\partial t) \delta t$ (cal); H is the effective fuel heat content (cal/g); γ is the air entrainment coefficient = 0.64 (dimensionless); s = atmosferic stability parameter $g/\theta_0(\Delta\theta_0/\Delta z)$(s^{-1}); g is the gravitational acceleration constant = 9.81 (m/s^2); θ_0 = potencial temperature of ambient air = $T_a(p_0/p)^{R/C_p}$ (K).

The time for the ground cloud reaches a height z_k in a stable atmosphere given by

$$t_i = s^{-1/2} \cos^{-1}\left(1 - \left(\frac{s\gamma^3 z_k^4}{4F_i}\right)\right) \tag{6}$$

where: t_i is constrained to be less than the cloud stabilization time:

$$t^* = \frac{\pi}{\sqrt{s}} \quad (7)$$

The height of the rise in a continuous plume in a stable atmosphere is defined as

$$\Delta z_c = \left(\frac{6F_c}{u\gamma^2 \sqrt{s}} \right)^{1/3} \quad (8)$$

where: F_c = buoyancy term = $gq/\pi\rho C_p T_a$ (m⁴s²) ; γ is the air entrainment coefficient = 0.5 (dimensionless).

And the time for a continuous plume to reach the height z_k in a stable atmosphere is given by

$$t_c = s^{-1/2} \cos^{-1}\left(1 - \left(\frac{su\gamma^2 z_k^3}{3F_c}\right)\right) \quad (9)$$

where: t_c is constrained to be less than t^*, described by Eq. 7, and u is the wind velocity (Bjorklund *et al.* 1982).

The following equation was based on a solution of the Newton's second law and solved iteratively to predict the motion of a buoyant cloud in the atmosphere, resulting in cloud stabilization height:

$$\Delta z_n(t) = \left[\frac{3F_m}{u\gamma^2\sqrt{s}}\sin(t\sqrt{s}) + \frac{3F_c}{u\gamma^2\sqrt{s}}\left(1-\cos(t\sqrt{s})\right) + \left(\frac{r_0}{\gamma}\right)^3 \right]^{1/3} - \frac{r_0}{\gamma} \quad (10)$$

where: F_m is the initial vertical momentum = $r_0 w_0 u$ (m⁴s²); r_0 is the initial plume cross-sectional radius = 3.5 (m); w_0 is the initial vertical velocity (m/s); u is the mean ambient wind speed (m/s); γ is the air entrainment coefficient = 0.33 (dimensionless); ρ is the density of exhaust gases = 0.109 (kg/m³).

A critical parameter in the cloud rise equation is the rate of ambient air entrainment that is defined by γ. Cloud growth as a function of altitude is assumed to be linearly proportional, and the air entrainment coefficient have been compared from the literature (observations and measurements of Titan IV rocket ground clouds), and an empirical cloud rise air entrainment coefficient has been derived from the test data (Nyman 2009). It should be noticed that there is no data of this nature for VLS.

Considering the different formulations from Eqs. 5 to 10, Briggs (1975) suggests that the value of the rise height

increment to be used is the smallest between Δz_i, Δz_c, and Δz_n. This suggestion is the most prudent due to the fact that, as the rise effect is larger, the concentration values obtained at surface level are smaller, reducing the risk of underestimating the value of these concentrations.

Once the plume rise height is calculated, the emission into the WRF model can be written as

$$E_n(i,j,m) = \int_{t_{n-1}}^{t_n} H\left((z_{m-1}-z)(z_m-z)\right)\frac{\partial M(t,z+\Delta z(t^*))}{\partial t} dt \quad (11)$$

when $i = i_{TMI}$ and $j = j_{TMI}$ or

$$E_n(i,j,m) = 0 \quad (12)$$

where: $i \neq i_{TMI}$ or $j \neq j_{TMI}$. The indexes i and j are horizontal grid coordinates of the model.

The rocket is launched from the point (i_{TMI}, j_{TMI}) of the WRF inner domain of simulation (d3). Note that Eqs. 11 and 12 disregard the horizontal component of the rocket displacement, but it is a fine assumption even into the finest scale models.

Table 1 shows the composition of the exhaust gases of the VLS, the mass percentage gas, and the variable into WRF. HCl, CO, and CO_2 are pollutants into gaseous form, and Al_2O_3 is divided into 2 sizes of particulate matter of 2.5 mm (pm 2.5) and 10 mm (pm 10).

Table 1. Composition of the exhaust gases of VLS.

Species		%	Variable
Alumina	Al_2O_3	28.4	pm 2.5 + pm 10
Carbon monoxide	CO	28.7	CO
Hydrogen chloride	HCl	21.4	HCl
Nitrogen	N_2	8.3	-
Water vapor	H_2O	6.8	-
Carbon dioxide	CO_2	3.6	CO_2
Hydrogen	H_2	2.8	-

The emissions were generated with a pre-processor named ANTHRO_EMISS, and these files were edited into R software, a language for statistical computing (R Core Team 2015). All emissions were set to 0, and just the rocket emissions are presented in the simulations at the time of launching, which was 12:00Z, using the meteorological output from WRF, the VLS data, and the equations described above.

RESULTS AND DISCUSSION

For the simulations, we have chosen a typical clear sky day from the dry season (Sept 17, 2008), where the model started at 00:00Z and the release of VLS gases occurred at 12:00Z, considering 12 h for the model spin-up. With a radiosonde released at 11:32Z, the mean wind within the planetary boundary layer (about 600 m) is about 13 m/s and the atmospheric stability is unstable, favoring the turbulence. These are characteristics of a suitable day for a rocket launching (wind speed below 10 m/s at surface level, lower wind speed up to 5 km, no rain, etc.) and good conditions for dispersion (presence of an atmosphere unstable) and transport (strong influences of the trade winds) of the rocket effluents.

Figure 3 shows the vertical distribution of the VLS emissions using Eq. 12 for the VLS data and the distribution used by Nascimento *et al.* (2014) for a hypothetical launching of a Titan-IV rocket. It can be observed that the distribution of emissions for the VLS is quite different from Titan-IV: it is lower close to the surface and higher at 1 – 2 km.

Figure 4 shows a time sequence of arrow plots of the horizontal component of the wind at the surface level, whose direction is predominantly from east. The wind speed is above 7 m/s in the ocean region and becomes weaker

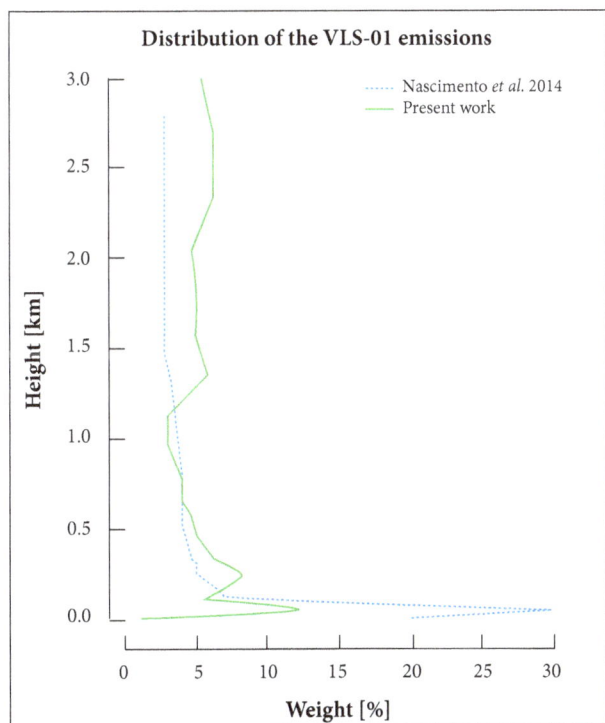

Figure 3. Emissions for the VLS and Titan-IV.

inland the continent (about 6 m/s in the SPL). The shaded areas represent the HCl concentration at the first layer of the model (approximately 40 m) for the initial 30 min after the launching of VLS. These plots have a log scale for the concentration for a better visualization. In this figure we can observe the exhausted cloud behaviors after the rocket launching: initially it presents a maximum concentration at the launch pad (SPL and TMI) and it is advected and dispersed with time. This plume reached the SMT with a still higher concentration (around 176 ppmv) within about 10 min. The other locations (CLA and VTA) were not reached by this plume. HCl is a colorless gas with an irritating pungent odor perceivable at 0.8 ppmv (Lide 2003). Table 2 presents the hydrogen chloride exposure limits (Braxter *et al.* 2000).

Table 2. Health effects of respiratory exposure to HCl concentration.

HCl exposure limits (ppmv)	Health effects
5 <	Coughing
35	Throat irritation occurs after only a short time
35 <	Severe breathing difficulties and skin inflammation or burns
10 – 50	Maximum level that can be sustained for several hours
100 <	Swelling of the lungs and often throat spasm
50 – 1,000	Maximum possible exposure: 1 h
1,000 – 2,000	Very dangerous even for a very short exposure

Source: extracted from Braxter *et al.* (2000).

Figure 4b presents the same information of Fig. 4a for CO concentrations and it is different from Iriart and Fisch (2016) by 2 reasons: the amount of gases exhausted is associated with a real value for VLS launching and this material is released during the flights trajectory (vertical dependence). For this variable we can notice the larger areas with higher concentrations that persist beyond the plume passage. However, this variable is not so toxic as HCl. The CO is a colorless and odorless gas that is slightly less dense than air. It is toxic to humans (and another hemoglobic animals) when observed in concentrations above about 35 ppmv. In the atmosphere, it is spatially variable and short-lived, having a role in the formation of ground-level ozone. The recommendation of the World Health Organization (1999) is that the exposure times should not exceed those shown in Table 3.

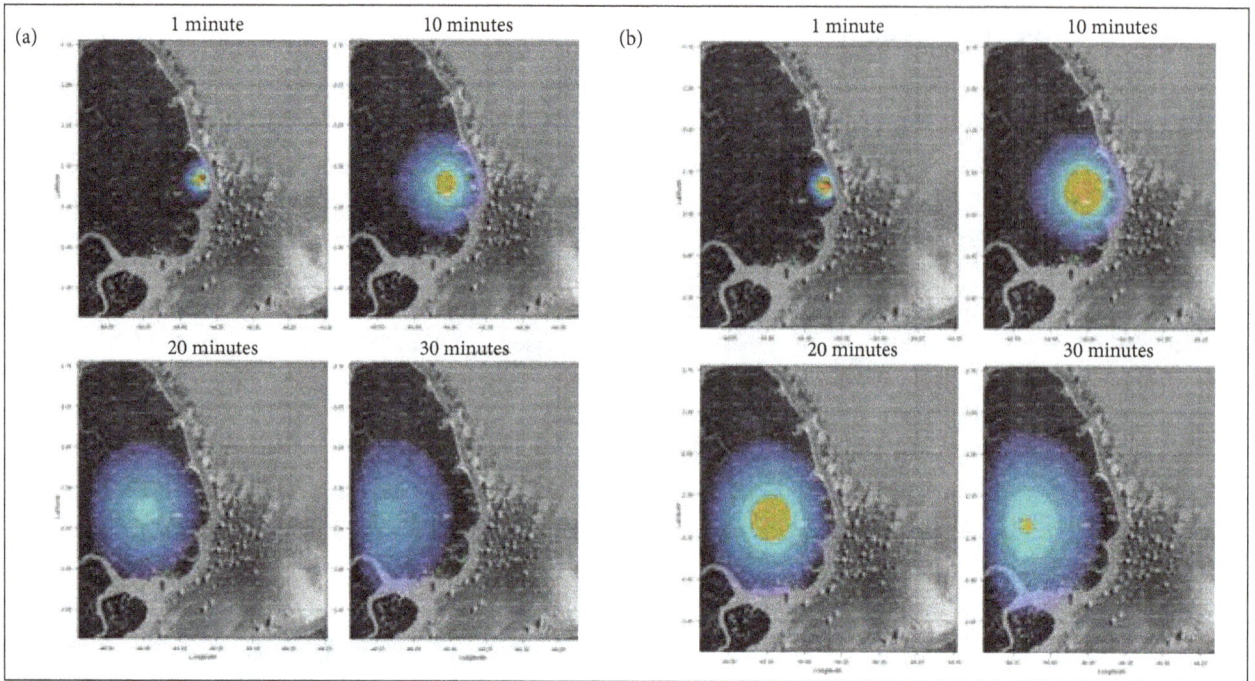

Figure 4. Surface concentrations after launching of (a) HCl and (b) CO.

Table 3. Exposure limits for CO.

CO exposure limits (ppmv)	Maximum exposition time
9	8 h
26	1 h
52	30 min
87	15 min
1,950	Rapidly fatal

Source: extracted from Winter and Miller (1976).

Figure 5a shows a time series of the HCl concentrations (presented in a logarithm scale) for up to 2 h after the launching for the locations SPL, SMT, CLA, and VTA. The higher levels are between 1,540 ppmv (SPL) and 176 ppmv (SMT) and, according to Table 2, the maximum exposure time is only 1 h; the inhalation of the gas can cause swelling of the lungs, throat spasm, and irritation. In the surrounding area, it presents concentration levels between 0.11 ppmv (VTA) and 0.53 ppmv (CLA), which are safety values of HCl, with effects imperceptible to the majority of the population.

Figure 5b presents the CO concentrations in function of time. The higher levels are in the range between 2,735 ppmv at SPL and 176 ppmv at SMT. From Table 2, a maximum exposure time is between 30 min and 2 h, and this may cause headache, increased heart rate, dizziness, nausea, and even

death. In the surrounding area, it presents concentration levels below 2 ppmv (CLA and VTA), which are safety for the mankind.

Figure 5c shows the concentration of the CO_2 with a peak of 598 ppmv at SPL. It is considered the minimal value for an effect on health by CO_2 inhalation of 15,000 ppmv during 1 month of exposure. In practical sense these values of CO_2 concentration have no effect for the humans.

Figure 6 shows the particulate material with 10- and 2.5-mm concentrations composed by Al_2O_3 in function of time. The levels for the CLA and the VTA are below 1 mg/m^3.

The World Health Organization (2000) guidelines do not recommend the use of levels of pm 10 and pm 2.5 as long as there are no sufficient data to enable the derivation of specific values at present. Nevertheless, the large body of information on studies relating day-to-day variations in particulate matter to day-to-day variations in health provides quantitative estimates of the effects of particulate matter that are generally consistent. The available information does not allow a judgment to be made on concentrations below, in which no effects would be expected.

On the other hand, the Conselho Nacional do Meio Ambiente (Brasil 1990) presents some critical levels for monitoring purposes of the air quality in Brazilian territory for pm 10 and pm 2.5: attention case for 250 mg/m^3, alert for

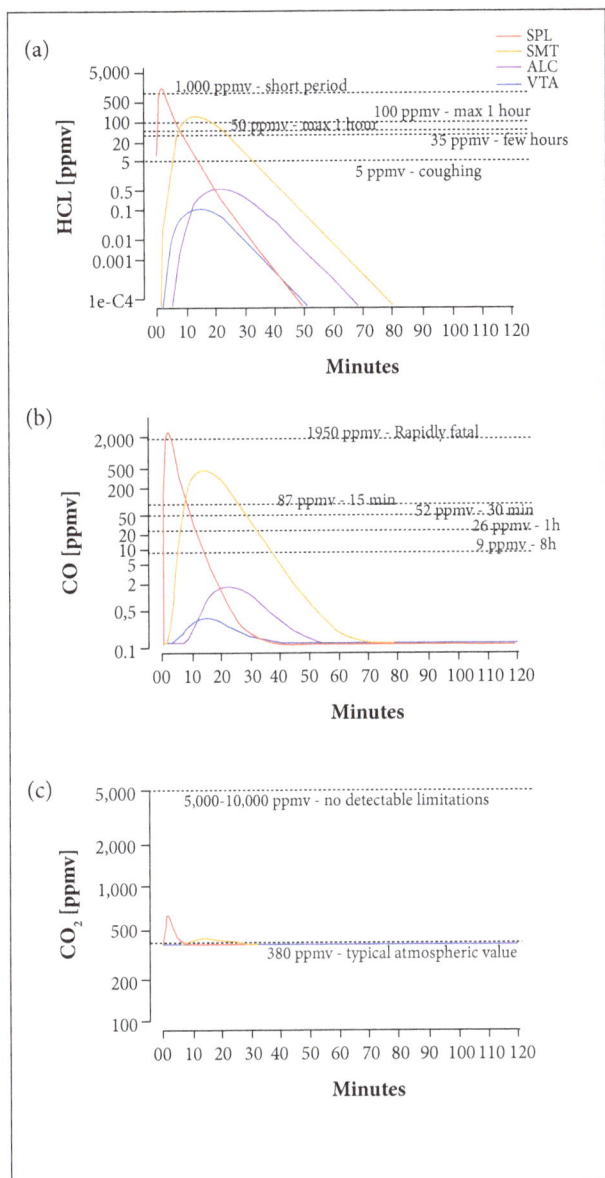

Figure 5. Time series of simulated and the levels of health effects for (a) HCl; (b) CO and (c) CO_2.

420 mg/m^3, and emergency for values beyond 500 mg/m^3. These values are for both pm 10 and pm 2.5, being calculated as an average during a time interval over 24 h. The maximum values for the pm 10 and pm 2.5 are 30.3 and 273.5 mg/m^3, respectively. The peak concentration of pm 10 is below the regular level adopted (50 mg/m^3), considered a secondary air quality standard or, in other words, a level in which it provides the minimum adverse effect on human health. However, the pm 2.5 concentration peaks are next to the attention level, but the mean value for the first hour is approximately 14 mg/m^3.

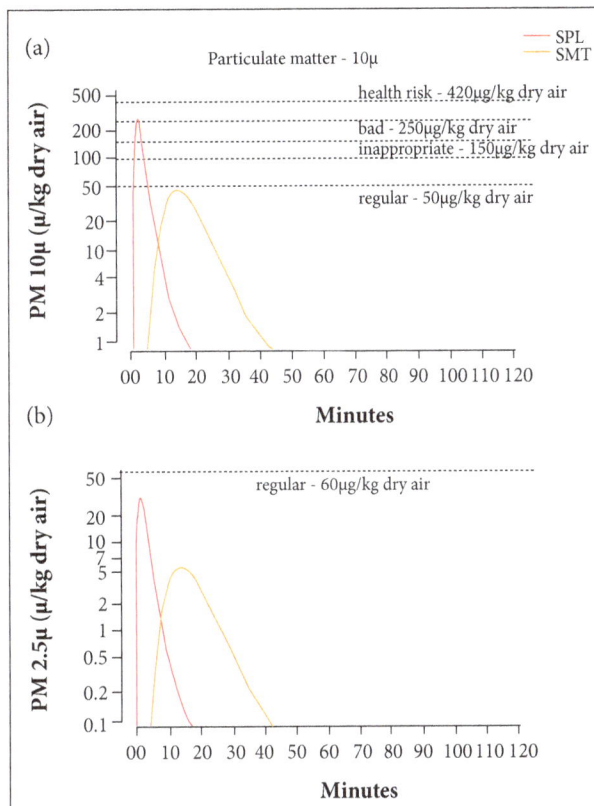

Figure 6. Time series of simulated and the levels of health effects for (a) pm 10 and (b) pm 2.5.

CONCLUSION REMARKS

In this paper we present a way to include the major atmospheric pollutants emissions from the VLS into a weather model (WRF-Chem) taking into account the vertical distribution of emission and the effect of buoyancy of the hot cloud formed by the effluents. This implementation represents an improvement to the study of Iriart and Fisch (2016), in which the emissions were made only at the surface level (without accounting the effect of buoyancy, trajectory, and variable emissions rates), not considering HCl; this features allowed us to use data from the Brazilian rocket VLS (instead of Titan-IV data) and analyze more realistic concentrations at the closer sites of the VLS launcher pad as the plume behavior for the CLA region.

The results show that the HCl levels are dangerous in the first hour after the launching at the SPL and SMT regions; the CO levels are more critical for the first 10 min after the launching, representing a high risk at the proximities of the SPL and attention state on the SMT. The concentrations of CO_2, pm 2.5, and pm 10 showed secure levels even for the proximities of the SPL; finally, the exhaust cloud does not reached the CLA or the VTA, mainly due to the wind direction.

After 40 min of the VLS launching the clouds were dispersed and left the inner domain, which is the region of interest. This means that the levels of each pollutant are below the minimum concentration for a detectable effect on the human health.

As the transport and dispersion of the rocket effluents depend tightly on the atmosphere state, the simulations are not a general solution for the problem, but represent a very good prediction scenario. Stronger or weaker winds (or different wind directions) can make the transport more or less efficient (as well as allow the plume to reach different locations), and different stability conditions (stable or unstable) will affect the dispersion by reducing the turbulent mixing, resulting in higher (or lower) concentrations. Other simulations for different atmospheric stability and wind directions will be made for more general assumptions and case studies.

In a future study, we will focus our attention on the chemical mechanism, where the afterburn and HCl atmospheric reactions must be included into the chemical module of WRF for a more realistic representation of the atmospheric chemical influence of the effluents of rocket exhausts.

ACKNOWLEDGEMENTS

The authors thank the Coordenação de Aperfeiçoamento de Pessoal de Nível Superior (CAPES), who funded part of this research through the project PRO-ESTRATEGIA (number 2240/2012), as well as provided the scholarship grant PQ (number 308011/2014). At last, but not least, a special thanks to Dr. Stacy Walters, from The University Corporation for Atmospheric Research (UCAR), who contributed for this study with a very important support.

AUTHOR'S CONTRIBUTIONS

Schuch D performed the experiments and prepared the figures. Schuch D and Fisch G discussed the results and commented on the manuscript.

REFERENCES

Bjorklund JR, Dumbauld JK, Cheney CS, Geary HV (1982) User's manual for the REEDM (Rocket Exhaust Effluent Diffusion Model) computer program. NASA Contractor Report 3646. Huntsville: NASA George C. Marshall Space Flight Center.

Brasil. Ministério do Meio Ambiente (1990) Resolução CONAMA n° 3, de 28 de junho de 1990; [accessed 2017 Jan 19]. http://www.mma.gov.br/port/conama/legiabre.cfm?codlegi=100

Braxter PJ, Adams PH, Cockcroft A, Harrington JM (2000) Hunter's diseases of occupations. 9th ed. London: Arnold; New York: Oxford University Press.

Briggs GA (1975) Plume rise predictions. Lectures on Air Pollution and Environmental Impact Analysis. Amer Meteor Soc (72-73):59-111. doi: 10.1007/978-1-935704-23-2_3

Cencetti M, Veilleur V, Albergel A, Olry C (2011) SARRIM: A tool to follow the rocket releases used by the CNES Environment and Safety Division on the European Spaceport of Kourou (French Guyana). Int J Environ Pollut 44(1-4):87-95. doi: 10.1504/IJEP.2011.038406

Damian V, Sandu A, Damian M, Potra F, Carmichael GR (2002) The kinetic preprocessor KPP – a software environment for solving chemical kinetics. Comput Chem Eng 26(11):1567-1579. doi: 10.1016/S0098-1354(02)00128-X

Denison MR, Lamb JJ, Bjorndahl WD, Wong EY, Lohn PD (1994) Solid rocket exhaust in the stratosphere: plume diffusion and chemical reactions. J Spacecraft Rockets 31(3):435-442. doi: 10.2514/3.26457

Emmons LK, Walters S, Hess PG, Lamarque JF, Pfister GG, Fillmore D, Granier C, Guenther A, Kinnison D, Laepple T, Orlando J, Tie X, Tyndall G, Wiedinmyer C, Baughcum SL, Kloster S (2010) Description and evaluation of the Model for Ozone and Related chemical Tracers, version 4 (MOZART-4). Geosci Model Dev 3:43-67. doi: 10.5194/gmd-3-43-2010

Grell GA, Peckham SE, Schmitz R, McKeen SA, Frost G, Skamarock WC, Eder B (2005) Fully coupled "online" chemistry within the WRF model. Atmos Environ 39(37):6957-6975. doi: 10.1016/j.atmosenv.2005.04.027

Iriart PG, Fisch G (2016) Uso do modelo WRF-Chem para a simulação da dispersão de gases no Centro de Lançamento de Alcântara. Rev Bras Meteorol 31(4):610-625. doi: 10.1590/0102-7786312314b20150105

Lide DR (2003) CRC handbook of chemistry and physics. 84th ed. Boca Raton: CRC Press.

Moreira DM, Trindade LB, Fisch G, Moraes MR, Dorado RM, Guedes RL (2011) A multilayer model to simulate rocket exhaust clouds. J Aerosp Technol Manag 3(1):41-52. doi: 10.5028/jatm.2011.03010311

Nascimento EGS, Moreira DM, Fisch G, Albuquerque TTA (2014) Simulation of rocket exhaust clouds at the Centro de Lançamento de Alcântara using the WRF-CMAQ modeling system. J Aerosp Technol Manag 6(2):119-128. doi: 10.5028/jatm.v6i2.277

Nyman RL (2009) NASA Report: Evaluation of Taurus II Static Test Firing and Normal Launch Rocket Plume Emissions.

Pfister GG, Parrish DD, Worden H, Emmons EK, Edwards DP, Wiedinmyer C, Diskin GS, Huey G, Oltmans SJ, Thouret V, Weinheimer A, Wisthaler A (2011) Characterizing summertime chemical boundary conditions for airmasses entering the US West Coast. Atmos Chem Phys 11:1769-1790. doi: 10.5194/acp-11-1769-2011

R Core Team (2015) R: A language and environment for statistical computing; [accessed 2016 Apr 28]. https://www.R-project.org/

Rajasekhar M, Kumar MD, Subbananthan T, Srivastava V, Apparao B, Rao VS, Prasad M (2011) Exhaust dispersion analysis from large solid propellant rocket motor firing using HYSPLIT model over Satish Dhawan Space Centre (SDSC SHAR). Proceedings of the Indo-US Conference-cum-Workshop on "Air Quality and Climate Research"; Hyderabad, India.

Skamarock WC, Klemp JB, Dudhia J, Gill DO, Baker DM, Duda MG, Huang XY, Wang W, Powers JG (2008) A description of the Advanced Research WRF version 3. Technical Note. Boulder: National Center for Atmospheric Research.

Stockwell WR, Kirchner F, Kuhn M, Seefeld S (1997) A new mechanism for regional atmospheric chemistry modeling. J Geophys Res 102(D22):25847-25879. doi: 10.1029/97JD00849

Winter PM, Miller JN (1976) Carbon monoxide poisoning. J Am Med Assoc 236(13):1502-1504. doi: 10.1001/jama.1976.03270140054029

World Health Organization (2000) Air quality guidelines for Europe. WHO regional publications: European series, n. 91; [accessed 2016 Apr 20]. http://www.euro.who.int/__data/assets/pdf_file/0005/74732/E71922.pdf

World Health Organization (1999) Environmental Health Criteria 213: carbon monoxide. 2nd ed.; [accessed 2015 Oct 28]. http://www.who.int/ipcs/publications/ehc/ehc_213/en/

Yarwood G, Rao S, Yocke M, Whitten G (2005) Updates to the carbon bond chemical mechanism: CB05. Final report to the U.S. EPA, RT-0400675; [accessed 2016 May 19]. http://www.camx.com/files/cb05_final_report_120805.aspx

12

Reduced Order Modeling of Composite Laminates Through Solid-Shell Coupling

Gigliola Salerno[1], Stefano Mariani[2], Alberto Corigliano[2]

ABSTRACT: Composite laminates display a complex mechanical behavior due to their microstructure, with a through-thickness variation of the displacement and stress fields that depends on the fiber orientation in each layer. Aiming to develop reduced-order numerical models mimicking the real response of composite structures, we investigated the capability and accuracy of finite element analyses coupling layered shell and solid kinematics. This study represents the first step of a work with the goal of accurately matching stress evolution in regions close to possible impact locations, where delamination is expected to take place, with reduced computational costs. Close to such locations, a 3-D modeling is adopted, whereas in the remainder of the structure, a less computationally demanding shell modeling is chosen. To test the coupled approach, results of numerical simulations are presented for a quasi-statically loaded cross-ply orthotropic plate, either simply supported or fully clamped along its boundary.

KEYWORDS: Composite structures, Reduced-order modeling, Solid-shell coupling.

INTRODUCTION

The numerical analysis of complex structures often requires very detailed space discretizations, especially in regions where high-field gradients are expected or where an enhanced solution accuracy is required for reliability issues. To speedup the analysis, reduced-order models can be formulated from a purely mathematical standpoint; alternatively, a wise coupling of refined and coarse-grained discretizations can be adopted. According to the notation proposed in Sellitto *et al.* (2011), a coupling of local (namely, accurate) and global (namely, time efficient) models can be envisaged.

The aforementioned problem can be relevant also in the analysis of composite laminates, which typically display a small thickness in comparison to their in-plane dimensions. Accordingly, a plate- or shell-like kinematics proves efficient to study problems mainly ruled by bending, if the structural response is of concern. If instead one focuses on decohesion mechanisms like delamination (Allix and Ladevèze 1992; Corigliano 1993; Abrate 1998; Schoeppner and Abrate 2000), a so-called meso-mechanical approach will be necessary. By allowing for the small thickness of the resin-enriched regions, simplified computational models resting on a lumping of the interlaminar decohesion onto 0-thickness surfaces have been developed in Corigliano and Mariani (2002), Corigliano *et al.* (2003), Mariani and Corigliano (2005), and Corigliano *et al.* (2006).

To model the decohesion processes, the through-thickness variation of the displacement field in displacement-based finite element (FE) simulations is required, through 3-D space discretizations. A fully solid discretization of the structural

1.Centro Universitário FEI – Departamento de Engenharia de Materiais – São Bernardo do Campo/SP – Brazil. **2.**Politecnico di Milano – Dipartimento di Ingegneria Civile e Ambientale – Milano/Milano – Italy.

Author for correspondence: Gigliola Salerno | Centro Universitário FEI – Departamento de Engenharia de Materiais | Avenida Humberto de Alencar Castelo Branco, 3.972 | CEP: 09.850-901 – São Bernardo do Campo/SP – Brazil | Email: gsalerno@fei.edu.br

component would result in a heavy computational burden for real-life case studies. The other way around, a shell-like 2-D modeling proves sufficient for regions not exposed to delamination events. Here a hybrid approach is adopted, able to ensure accuracy with limited computational costs by coupling a solid kinematics to a shell kinematics away from the delaminating zones.

Dealing with ductile fracture processes in homogeneous metals, this approach was already adopted in Corigliano *et al.* (1999) to model the through-thickness propagation of a crack in pressurized pipelines, in Stringfellow and Paetsch (2009) to model the collision-induced failure of structural parts of a cab car, in Kim (2003) for the fatigue assessment of welded joints, and in Gong *et al.* (2016) to study the effect of the residual stress induced by welding on the buckling of storage tanks; in all the cases, stress triaxiality in the process regions plays a prominent role, and a shell-like model, basically missing the out-of-plane constraint on the stress field, does not necessarily prove accurate. In Reinoso *et al.* (2012), the envisioned approach was adopted to model the damage evolution in the skin-stringer joint of a complex composite sample for aeronautical applications (Li et al. 2013). In most of these studies, the commercial FE code Abaqus was adopted, and its solid-to-shell coupling feature was exploited to match the displacement fields at the boundaries of the facing solid and shell domains as well as to prevent unphysical kinking. This coupling is locally enforced by Abaqus in a strong form (Dassault Systems 2010); an alternative, variational enforcement was instead proposed in Blanco *et al.* (2008).

Focusing on coupled modeling for thin plates undergoing bending deformations induced by a distributed load, in this paper we provide an assessment of the hybrid approach in terms of plate deflection, elastic energy stored in the laminate, through-thickness variation of the stress fields, and speedup with respect reference solid models. Results are compared not only to those of fully solid 3-D simulations, but also to the outcomes (if available) of the Pagano approach (Pagano 1969, 1970) and of the first-order shear deformation theory (FSDT), (Reddy 2002; Wang *et al.* 2000).

In what follows, we highlight some computational features of the coupled procedure in the section "Solid-Shell Coupling and Computational Issues". In the subsequent section, results are discussed for a $[90/0]_{6s}$ composite plate, either simply supported or fully clamped along its boundary. Finally, some concluding remarks are gathered.

SOLID-SHELL COUPLING AND COMPUTATIONAL ISSUES

Let us consider a flat rectangular plate, as depicted in Fig. 1. The plate is either simply-supported or fully-clamped along its boundary and is quasi-statically loaded over its top surface by a uniform transversely-distributed load q_0, which is positive if pointing outward (the solution can be obviously generalized to the non-uniform load case). An orthonormal reference frame is introduced, with x and y axes located on the mid-plane of the plate and z axis pointing in the same direction of the load.

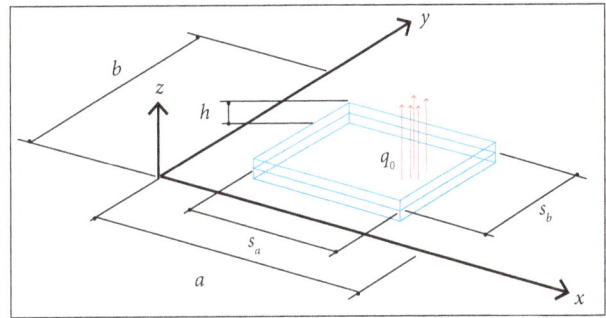

Figure 1. Plate bending problem, coupled solid-shell modeling: geometry and notation.

The central portion of the plate, whose side lengths are s_a and s_b, is modeled with a full 3-D kinematics (Fig. 1); the remainder of the plate is instead modeled with a 2-D shell-like kinematics. Accordingly, the former region is discretized with 3-D brick elements, whereas the latter region is discretized with 2-D shell elements. Numerical simulations here collected have been run using the commercial FE code Abaqus (Dassault Systems 2010).

In the remained of this section, we discuss 2 topics that affect the accuracy of the solutions provided by the hybrid 2- and 3-D model: the elements' kinematics as well as the coupling between the solid and shell structural regions.

As far as the shell kinematics is concerned, restricting the analysis to the small displacement regime, the 4-node (reduced-integration) S4R element has been adopted. The through-thickness variation of the strain and stress fields featured by such element is compliant with the FSDT; this kinematics implies that segments normal to the mid-plane of the plate retain their straight shape when deformed. Even if shear correction factors are adjusted in the case of anisotropic materials, and an inhomogeneous material response can be allowed for along the thickness direction, the transverse shear deformations are assumed constant throughout the whole thickness by this element kinematics. This assumption represents an approximation to the real composite behavior, which is instead characterized by the so-called zigzagging of the stress and strain fields (Bogdanovich and Pastore 1996; Reddy 2002; Carrera 2003). The adopted reduced-integration is developed based on an assumed strain approach via Hu-Washizu variational principle plus stabilization to avoid 0-energy deformation patterns and shear locking in case of thin plates (Bathe and Dvorkin 1984; Simo *et al.* 1989; Dassault Systems 2010).

In the solid region, linear brick elements have been adopted. Due to the considered bending-dominated deformation of the plate, the C3D8I elements featuring incompatible modes have been selected to avoid, or reduce as much as possible, parasitic shear and volumetric locking effects (Simo and Rifai 1990).

The coupling between solid and shell regions of the plate has been obtained through a surface-based interaction. This technique allows matching the displacements of nodes on the border of the solid region to the displacements and rotations of nodes on the border of the shell region, so as not only displacement jumps but also rotation jumps (*i.e.* kinks) are locally prevented (Fig. 2). Because of the S4R shell kinematics mentioned above, which cannot model the zigzagging of the fields along the through-thickness direction, this coupling introduces a local perturbation in the numerical solution. To assess the effects of such perturbation on the modeled structural response, in the next section the size ratios $\alpha = s_a/a$ and $\beta = s_b/b$ (Fig. 1) are varied in the range of $0.25 \leq \alpha, \beta \leq 0.75$.

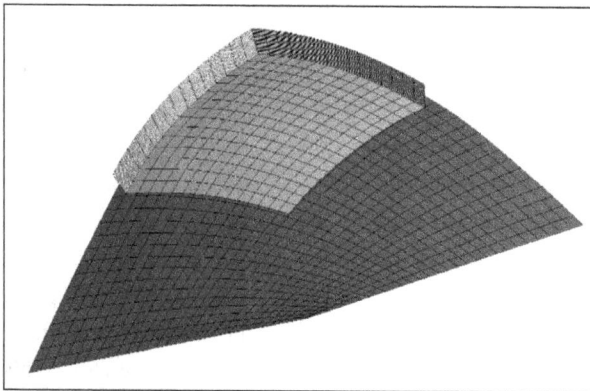

Figure 2. Example of the deformed configuration of a bent composite plate modeled with the hybrid solid-shell approach (the plate deflection has been magnified to highlight the quality of coupling along the border between the solid and shell regions). Solid and shell elements are respectively depicted using light and dark grey colors.

RESULTS AND DISCUSSION: $[90/0]_{6s}$ LAMINATE

A symmetric cross-ply $[90/0]_{6s}$ laminate is considered; the side lengths of the composite plate are $a = b = 127$ mm, while the thickness is $h = 3.36$ mm. A uniform transversely-distributed load $q_0 = 2$ MPa is adopted. The elastic properties of each transversely isotropic layer, in a local x_1, x_2, x_3 reference frame, with axis x_1 aligned with the fiber longitudinal direction, are: $E_{11} = 144{,}000$ MPa, $E_{22} = E_{33} = 9{,}690$ MPa, $G_{12} = 55{,}385$ MPa, $G_{13} = G_{23} = 5{,}760$ MPa, and $v_{13} = v_{23} = 0.3$ (Salerno 2009). In what precedes, E is a Young's modulus, G represents a shear modulus, and v is a Poisson's ratio.

A preliminary investigation of the effects of meshing on the results has been carried out; what turned out is that the stacking sequence, and the relevant zigzagging of the in-plane strain and stress fields require a meshing with more than 1 element across the thickness of each layer to approach the FSDT solution (Salerno 2009). Anyhow, results are here presented in the case of 1 element discretizing each layer in the through-thickness direction to assess the accuracy of the numerical solutions when minimal meshing and, therefore, minimal computing time are enforced. To assess the relevant accuracy, the outcomes of the simulations will be compared with those obtained with a so-called overkill model featuring 4 elements to discretize across the thickness each layer of the composite panel; such model has been checked to provide mesh-independent results, hence no benefits would be obtained by further refining the space discretization in the through-thickness direction and over the mid-plane of the plate.

As far as the simply supported plate case is concerned, results are reported in Fig. 3 in terms of the through-thickness variation of the stress fields for $\alpha = \beta = 0.5$; while stress components σ_{xx} and σ_{yy} are provided at the plate center, the shear component σ_{xy} is given as measured at the plate corner due to the symmetry in the solution (for cross-ply laminates) leading to a vanishing σ_{xy} close to the center. While the fully solid discretization provides results almost perfectly matching the FSDT ones, at least at the plate center, the solid-shell coupling introduces a perturbation, leading to reduced stress amplitudes close to the top and bottom surfaces of the plate. Since σ_{xy} is given at the plate corner, it must be noted that with the coupled solid-shell kinematics only 2 points are shown for this field in Fig. 3, as provided by the FE code for shell elements.

Results in terms of central plate deflection and overall stored strain energy are shown in Fig. 4 to understand the effect of the modeling parameters α and β. The strain energy stored in the whole plate is here adopted to implicitly assess the in-plane size of the region affected by the distortion in the stress and strain fields due to the coupling of the solid and shell kinematics, keeping in mind that the zigzagging of the fields is not enforced in the computational model. In the investigated intervals, it is shown that the impact of α and β is marginal: the maximum discrepancy with respect to the reference overkill simulation (whose results are represented by the dashed lines in the graphs) amounts to 2.6% as for the central deflection and 7.6% as for the overall strain energy.

If the laminate is instead assumed to be fully clamped along its boundary, as before a comparison, in terms of the

through-thickness variation of the stress field, it is reported in Fig. 5 between the fully-solid and the coupled solid-shell (for $\alpha = \beta = 0.5$) models featuring the minimal discretization in the out-of-plane direction, with a single element for each lamina. In this example, the coupled model provides a slight overestimation of the stress level at the center of the plate, amounting to around 1.8% in terms of the extreme values

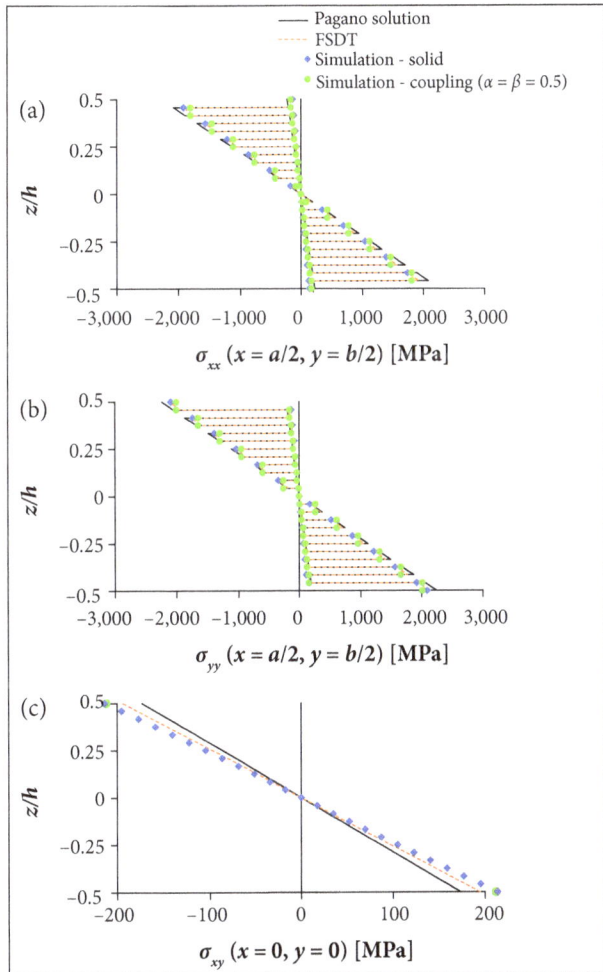

Figure 3. Simply supported laminate. Comparison among Pagano and FSDT solutions as well as results of the simulations in terms of: (a) σ_{xx} and (b) σ_{yy} at the plate center; (c) σ_{xy} at the plate corner.

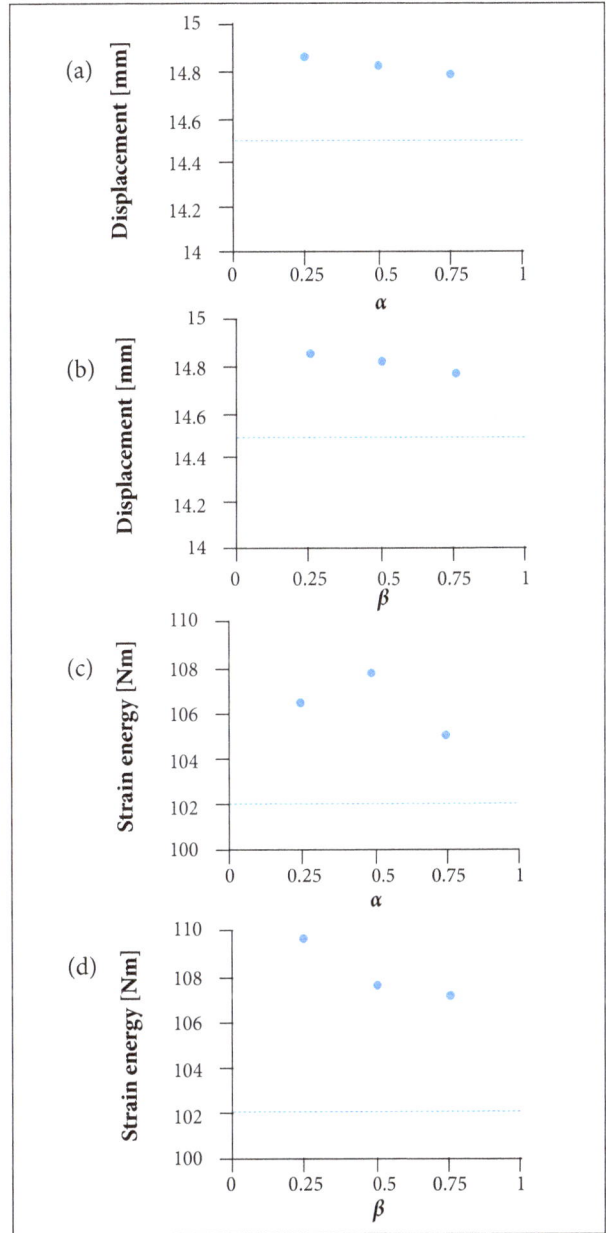

Figure 4. Simply-supported laminate. Effect of coupling coefficients α (with $\beta = 0.5$) (a and c) and β (with $\alpha = 0.5$) (b and d) on: (a and b) central plate deflection; (c and d) stored strain energy. Dashed lines in the graphs represent the results of the reference overkill simulation.

at the top and bottom surfaces of the plate as well as an underestimation at the plate corner, amounting to about 14.6% at the same top and bottom surfaces. As in Fig. 3, in these plots the stress components σ_{xx} and σ_{yy} are provided for all the nodes across the plate thickness in the region modeled with solid elements; the stress component σ_{xy} at the plate corner is instead provided for the coupled solid-shell model only at the top and bottom plate surfaces, as furnished by the FE code.

Figure 6 shows the results in terms of central deflection and overall stored strain energy, compared to the reference overkill model (whose results are again represented by the dashed lines in the graphs). Even with a fully-clamped boundary, the α and β ratios do not affect much the solution; at most, a discrepancy of 5.4% with respect to the fine mesh solution is reported.

Figures 4 and 6 show that parameters α and β might have a different effect on the accuracy of the solution, especially in terms of the elastic energy stored in the whole plate. This can be basically linked to the transversely isotropic behavior of each lamina, to the panel lay-up and so to the mechanical interaction among laminae when the plate is bent; since α and β move the boundary between the solid and shell regions (the latter is not

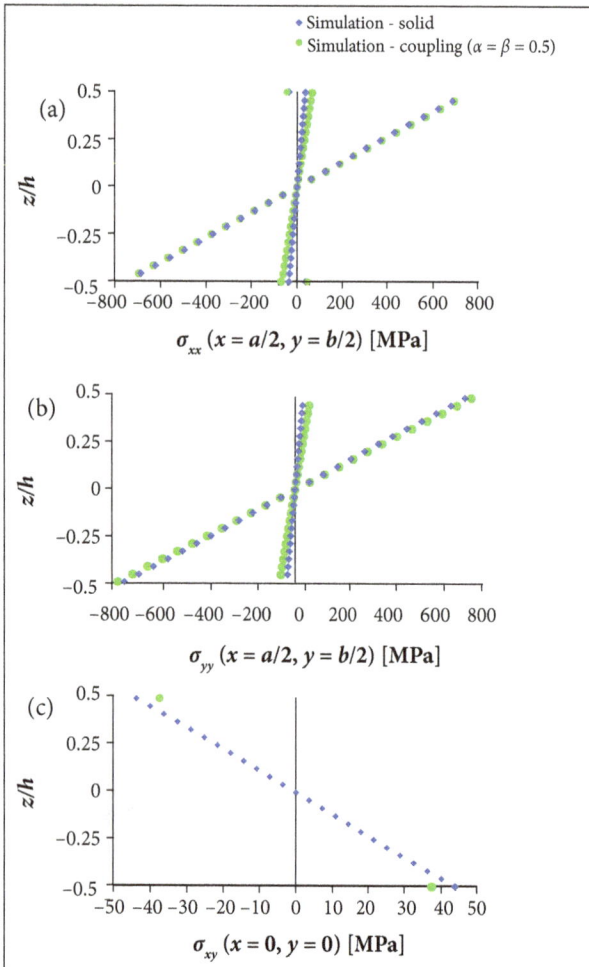

Figure 5. Clamped laminate. Comparison between results of the simulations, either allowing for solid-shell coupling or adopting a uniform 3-D space discretization in terms of: (a) σ_{xx} and (b) σ_{yy} at the plate center; (c) σ_{xy} at the plate corner.

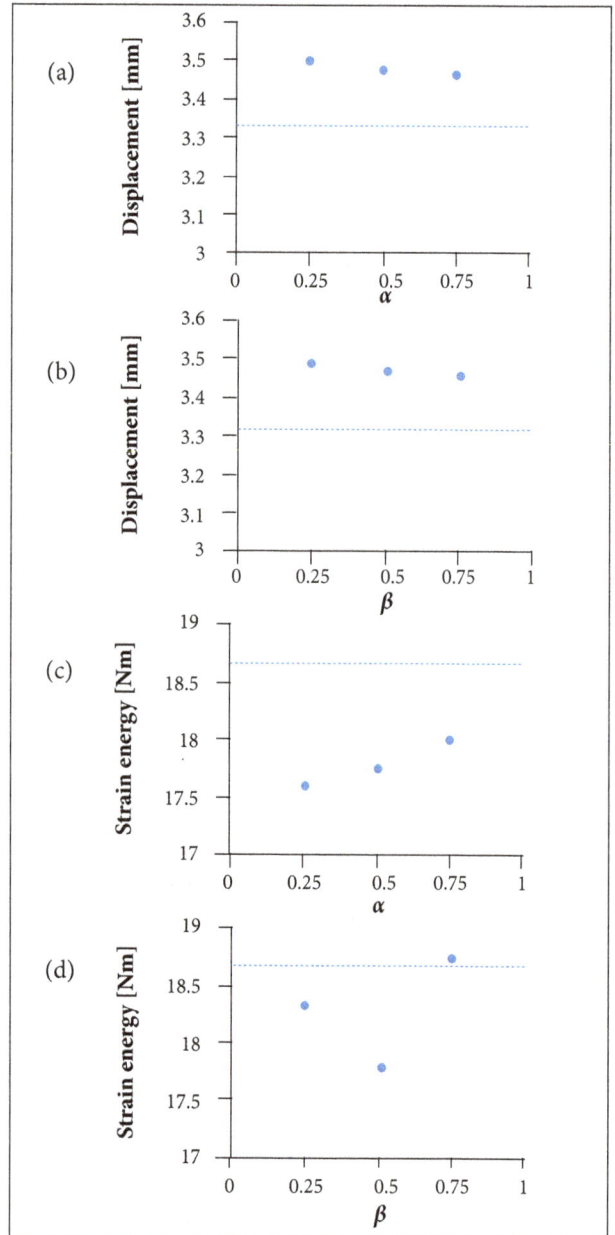

Figure 6. Clamped laminate. Effect of coupling coefficients α (with β = 0.5) (a and c) and β (with α = 0.5) (b and d) on: (a and b) central plate deflection; (c and d) stored strain energy. Dashed lines in the graphs represent the results of the reference overkill simulation.

able in the present model to capture the actual zigzagging of the stress and strain fields) along the 2 orthogonal in-plane directions, they can have a different impact on the solution, as shown here.

To understand the computational gain achieved through the hybrid kinematics, which looks necessary to motivate the adoption of a complicated modeling of the composite plates, outcomes are gathered in Fig. 7 in terms of the speedup as a function of the inverse of the relative in-plane size of the solid region, *i.e.* , for both the considered boundary conditions. The speedup is measured as the ratio between the CPU time of the fully-solid analysis and the CPU time of the hybrid one. The same in-plane discretization has been adopted in the 2 aforementioned solutions; in the hybrid analyses, the same through-thickness discretization of the fully-solid simulation has been adopted at the central plate portion. As a term of assessment, the dashed line in the plot shows how the speedup would scale if linearly depending on $(\alpha\beta)^{-1}$; as the computing time depends also on the number of layers in the composite, having used 1 element across each lamina in the considered minimal meshing strategy, such linear dependence is here reported only as a qualitative, although rather good, estimation of the computational gain attained with the hybrid kinematics. These results have been obtained by running Abaqus on a laptop with Windows 7-64 bit as operating system and an Intel Core I7-2620 M @ 2.70 GHz as CPU.

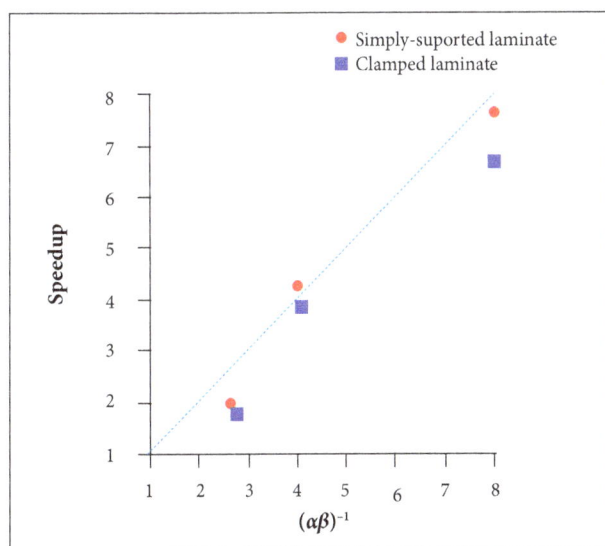

Figure 7. Effect of the relative in-plane size $(\alpha\beta)^{-1}$ of the solid region on the speedup.

CONCLUSIONS

In this paper, a numerical investigation to assess the accuracy and efficiency of coupled solid-shell modeling for composite laminates has been provided.

Quasi-static or low-velocity impact loadings may lead to delamination processes, located along the interlaminar regions; hence, standard shell elements cannot be used in finite element modeling, as they do not account for jumps in the displacement field across the thickness. To capture the post-impact response of composite structures for aeronautical applications, a full 3-D kinematics for the plate region around the impact location looks necessary; in the remainder of the structure, where delamination is not expected to occur, a shell kinematics can instead prove sufficient.

In case of thin laminated plates, we have accordingly approached the problem by coupling solid and shell elements. For a $[90/0]_{6s}$ composite plate under a uniform distributed load, it has been shown that the through-thickness variation of the stress field is matched by the hybrid modeling with a rather good level of accuracy: a discrepancy of at most 10 – 15%, with respect to reference numerical solutions, has been reported. It has been also shown that the speedup (computed as the ratio between the CPU time required to run the 3-D solid analysis and the CPU time required to run the hybrid model) basically scales with the ratio between the volume of the whole composite and the volume of the 3-D modeled region. In case of real-life composite structures, such approach is therefore expected to provide an excellent reduction of the computational costs.

Maximum computational gain can therefore be attained if the 3-D region is assumed large enough to enclose all the possible delaminating regions. Since the extent of delamination cannot be known *a priori*, being dependent on the loading conditions, a mesh updating procedure would be also designed so as the kinematics is locally switched from a 2-D shell one to a 3-D solid one every time a critical stress threshold is approached throughout the laminate thickness.

AUTHOR'S CONTRIBUTION

All the authors worked on this paper in all topics: idea, text and simulations. The authors contributed equally to it.

REFERENCES

Abrate S (1998) Impact on Composite Structures. Cambridge: Cambridge University Press.

Allix O, Ladevèze P (1992) Interlaminar interface modelling for the prediction of delamination. Compos Struct 22(4):235-242. doi: 10.1016/0263-8223(92)90060-P

Bathe KJ, Dvorkin EN (1984) A continuum mechanics-based four-node shell element for general non-linear analysis. Eng Computation 1(1):77-88. doi: 10.1108/eb023562

Blanco PJ, Feijoo RA, Urquiza SA (2008) A variational approach for coupling kinematically incompatible structural models. Comput Meth Appl Mech Eng 197(17-18):1577-1602. doi: 10.1016/j.cma.2007.12.001

Bogdanovich A, Pastore C (1996) Mechanics of textile and laminated composites with applications to structural analysis. London: Chapman & Hall.

Carrera E (2003) Historical review of Zig-Zag theories for multilayered plates and shells. Appl Mech Rev 56(3):287-308. doi: 10.1115/1.1557614

Corigliano A (1993) Formulation, identification and use of interface elements in the numerical analysis of composite delamination. Int J Solid Struct 30(20):2779-2811.

Corigliano A, Maier G, Mariani S (1999) Analysis of ductile fracture in damaged pipelines by a geometric parameter method. Eng Struct 21(10):924-936. doi: 10.1016/S0141-0296(98)00040-6

Corigliano A, Mariani S (2002) Identification of a constitutive model for the simulation of time-dependent interlaminar debonding processes in composites. Comput Meth Appl Mech Eng 191(17-18):1861-1894. doi: 10.1016/S0045-7825(01)00357-7

Corigliano A, Mariani S, Pandolfi A (2003) Numerical modeling of rate-dependent debonding processes in composites. Compos Struct 61(1-2):39-50. doi: 10.1016/S0263-8223(03)00030-8

Corigliano A, Mariani S, Pandolfi A (2006) Numerical analysis of rate-dependent dynamic composite delamination. Compos Sci Tech 66(6):766-775. doi: 10.1016/j.compscitech.2004.12.031

Dassault Systems (2010) Abaqus 6.10 Theory Manual. Vélizy-Villacoublay: Abaqus Software.

Gong J, Yu L, Wang F, Xuan F (2016) Effect of welding residual stress on the buckling behavior of storage tanks subjected to harmonic settlement. J Pressure Vessel Technol 139(1):011401-011401-9. doi: 10.1115/1.4033941

Kim D (2003) Welding simulation of ship structures using coupled shell and solid volume finite elements (Master's thesis). Bethlehem: Lehigh University.

Li D, Qing G, Liu Y (2013) A layerwise/solid-element method for the composite stiffened laminated cylindrical shell structures. Compos Struct 98:215-227. doi: 10.1016/j.compstruct.2012.11.013

Mariani S, Corigliano A (2005) Impact induced composite delamination: state and parameter identification via joint and dual extended Kalman filters. Comput Meth Appl Mech Eng 194(50-52):5242-5272. doi: 10.1016/j.cma.2005.01.007

Pagano NJ (1969) Exact solutions for composites laminates in cylindrical bending. J Compos Mater 3(3):398-411. doi: 10.1177/002199836900300304

Pagano NJ (1970) Exact solutions for rectangular bidirectional composites and sandwich plates. J Compos Mater 4:20-34.

Reddy JN (2002) Energy principles and variational methods in applied mechanics. New York: John Wiley and Sons.

Reinoso J, Blázquez A, Estefani A, París F, Cañas J, Arévalo E, Cruz F (2012) Experimental and three-dimensional global-local finite element analysis of a composite component including degradation process at the interfaces. Compos B Eng 43(4):1929-1942. doi: 10.1016/j.compositesb.2012.02.010

Salerno G (2009) Damage analysis of composite laminates subject to low velocity impacts (PhD thesis). Milan: Politecnico di Milano.

Schoeppner GA, Abrate S (2000) Delamination threshold loads for low velocity impact on composite laminates. Compos Appl Sci Manuf 31(9):903-915. doi: 10.1016/S1359-835X(00)00061-0

Sellitto A, Borrelli R, Caputo F, Riccio A, Scaramuzzino F (2011) Methodological approaches for kinematic coupling of non-matching finite element meshes. Procedia Engineering 10:421-426. doi: 10.1016/j.proeng.2011.04.071

Simo JC, Fox DD, Rifai MS (1989) On a stress resultant geometrically exact shell model. Part III: Computational aspects of the nonlinear theory. Comput Meth Appl Mech Eng 79(1):21-70. doi: 10.1016/0045-7825(90)90094-3

Simo JC, Rifai MS (1990) A class of mixed assumed strain methods and the method of incompatible modes. Int J Numer Meth Eng 29(8):1595-1638. doi: 10.1002/nme.1620290802

Stringfellow R, Paetsch C (2009) Modeling material failure during cab car end frame impact. Proceedings of the 2009 ASME Joint Rail Conference; Pueblo, USA.

Wang CM, Reddy JN, Lee KH (2000) Shear deformable beams and plates. New York: Elsevier Science.

13

PRO-ELICERE: A Hazard Analysis Automation Process Applied to Space Systems

Tharcius Augusto Pivetta[1], Glauco da Silva[1,2], Carlos Henrique Netto Lahoz[1,2,3], João Batista Camargo Júnior[4]

ABSTRACT: In the last decades, critical systems have increasingly been developed using computers and software even in space area, where the project approach is usually very conservative. In the projects of rockets, satellites and its facilities, like ground support systems, simulators, among other critical operations for the space mission, it must be applied a hazard analysis. The ELICERE process was created to perform a hazard analysis mainly over computer critical systems, in order to define or evaluate its safety and dependability requirements, strongly based on Hazards and Operability Study and Failure Mode and Effect Analysis techniques. It aims to improve the project design or understand the potential hazards of existing systems improving their functions related to functional or non-functional requirements. Then, the main goal of the ELICERE process is to ensure the safety and dependability goals of a space mission. The process, at the beginning, was created to operate manually in a gradual way. Nowadays, a software tool called PRO-ELICERE was developed, in such a way to facilitate the analysis process and store the results for reuse in another system analysis. To understand how ELICERE works and its tool, a small example of space study case was applied, based on a hypothetical rocket of the Cruzeiro do Sul family, developed by the Instituto de Aeronáutica e Espaço in Brazil.

KEYWORDS: ELICERE, Hazard analysis, Safety, Dependability, Quality attributes, Space systems.

INTRODUCTION

Critical systems or high-integrity systems are those in which a failure can lead to a severe consequence, such as economic, environmental or even human losses. In this context, aerospace systems can be highlighted, such as spacecraft, test facilities and ground equipment. One of the main activities of the safety engineering is performing hazard analysis, which aims to define potential hazards, consequent failures and defects into the system, identifying unplanned behaviours, problems related to exchanges information, wrong procedures execution, among others (Stark *et al.* 2004).

In 2009, a safety and dependability (S&D) analysis process called ELICERE was developed, whose intent was to improve the quality level of critical computer systems (Lahoz 2009). The "elicere" word is derived from infinitive of the Latin verb *"elicio"*, which means to elicit, to extract.

In general, the elicitation activity consists of the extraction and identification of the system and software requirements. Requirements can be classified into functional and non-functional. Basically, functional requirements describe the main features of the product under the user's perspective. Non-functional requirements describe various quality factors, or attributes, which affect the functional requirements, such as usability, dependability and safety. Dependability covers other safety-related features, as reliability, availability and maintainability and other factors related to the critical functioning of a product. These particular systems are known as safety critical or high-integrity systems.

1. Departamento de Ciência e Tecnologia Aeroespacial – Instituto Tecnológico de Aeronáutica – São José dos Campos/SP – Brazil. **2.** Departamento de Ciência e Tecnologia Aeroespacial – Instituto de Aeronáutica e Espaço – São José dos Campos/SP – Brazil. **3.** Massachusetts Institute of Technology – Department of Aeronautics and Astronautics – Cambridge/MA – USA. **4.** Universidade de São Paulo – Escola Politécnica – Departamento de Engenharia de Computação e Sistemas Digitais – São Paulo/SP – Brazil.

Author for correspondence: Tharcius Augusto Pivetta | Departamento de Ciência e Tecnologia Aeroespacial – Instituto Tecnológico de Aeronáutica | Praça Marechal Eduardo Gomes, 50 – Vila das Acácias | CEP: 12.228-900 – São José dos Campos/SP – Brazil | Email: tharcius@yahoo.com.br

ELICERE brings together goal-oriented requirements engineering technique — known as ISTAR (Yu 1995) — and features of safety engineering techniques such as Hazards and Operability (HAZOP) and Failure Mode and Effect Analysis (FMEA). The idea is to perform the hazard analysis over the system requirements model, in order to identify potential mitigation actions and improvements in the system. The process uses a questionnaire based on guidewords that are applied under the modelled system elements. The outcomes of the questionnaires are mitigation actions based on a set of quality attributes (factors) related with the technique to assure its integrity.

However, the ELICERE generates a large amount of information and requires a computational structure to deal with the relationship between the hazards and the quality attributes that could mitigate it and then to suggest prioritizations of hazard that should be treated. Besides, it is desirable that the results can be recorded as a database of knowledge in order to reuse it for future analysis in other projects, creating a statistical and historical database. To meet these needs, the PRO-ELICERE automated tool has been proposed to improve the process.

This paper introduces the main features of PRO-ELICERE, describing the architecture of the first prototype proposed, the automation of each step, how to run this tool and which results are expected, initially applied for a studied case related to space system.

In "The ELICERE Process" section, the process is described in summary, explaining how the ISTAR modelling language works, the approach for hazard analysis and, finally, the questionnaire submitted to the Analyst. "The PRO-ELICERE tool" section presents the tool that builds upon the ELICERE and shows the main features, mainly in terms of its questionnaire and how to present the mitigation options. "The Case Study Example" section is about a case study based on a hypothetical rocket called V-ALFA. The "Conclusions" section discusses the results and possible improvements for the next version of PRO-ELICERE.

THE ELICERE PROCESS

The ELICERE is an S&D process applied to critical computer systems and was created to support hazard analysis of space systems (Lahoz and Camargo Júnior 2011). ELICERE adopts the ISTAR framework (Yu 1995) for modelling the systems behaviour and the guidewords based on HAZOP and

FMEA to extract mitigations provisions and goals related to S&D. In general, this activity comprehends the establishment of the general business and technical goals, an outline description of the problem to be solved and the identification of the system constraints. ELICERE helps to define what the system cannot do, or what the system should do in order to minimize problems related to safety, security, reliability and so on, typically non-functional requirements. In addition, the process improves the product quality, mitigates problems such as ambiguity, risk behaviour, unclearness, besides omission of non-functional requirements. Figure 1 presents the two main activities of ELICERE.

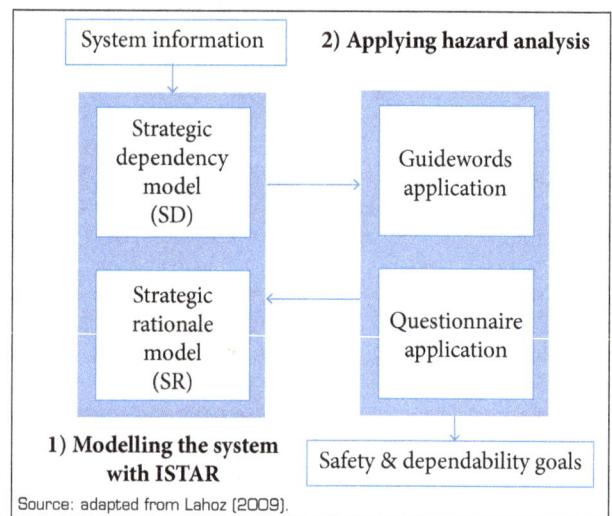

Figure 1. The ELICERE's two main activities.

ACTIVITY 1: MODELLING THE SYSTEM WITH ISTAR

The purpose of this activity is to create a system model, through the modelling language called ISTAR. The ISTAR (also called i* or i* framework) is an organizational requirements modelling technique suitable for use in early phase of system design in order to better understand the problem domain. This modelling language describes dependencies among actors through their four basic elements: goal, soft-goal, task and resource. Actors depend on each other for goals to be achieved, tasks to be performed and resources to be furnished. The ISTAR consists of two main modelling components: strategic dependency model (SD), which describes a network of dependency relationships among various actors in an organizational context, and the strategic rationale model (SR), which allows modelling the reasons associated with each actor and their dependencies and provides information about how actors achieve their goals and soft-goals. The ISTAR is used into ELICERE to represent a system in such a way that its hazards and vulnerabilities become evident, as well as the mission goals, the elements needed by the system to operate, and so on. The constraints should be identified and represented by the actors and its other four modelling elements. ELICERE apply the SD model to obtain a general overview of the relationship between actors, mainly concerned about goals and their soft-goals, tasks and resources. The SR model is used to represent a boundary of an actor, as well a view "in deep" about how each actor works to attend the goals and soft-goals. The main ISTAR features are represented in Table 1 (Yu and Grau 2006).

Figure 2 shows an example of ISTAR model as adapted to ELICERE. It shows that there is a relationship between two actors, where **GS** is the ground system and **OP_GS** is the operator of the ground system. This relationship shows, in a

Table 1. ISTAR features.

Symbol	Definition
Actor	Actor: active entity which leads to goals achievement, exercising their abilities
Goal	Goal: the affirmation or goal that the actor or system must meet or achieve
Soft-Goal	Soft-goal: the affirmation or goal related to non-functional requirements that an actor or system must meet or achieve
Task	Task: it concerns the activity that an actor or system must play in achieving a goal
Resource	Resource: system entity that provides some kind of information, product or service to the actor
→	Dependency link: representation of the dependency relationship ("Dependum") between two actors of the system: one is a "Depender" (consumer or depending actor on a dependency relationship) and the other is a "Dependee" (producer or the actor who is depended upon on a dependency relationship). In the graphical notation, the arrowhead points are presented from the "Depender" to the "Dependee"
—+—	Decomposition link: task element which is linked to its component nodes by decomposition links. A task can be decomposed into four types of elements: a subgoal, a subtask, a resource, and/or a soft-goal
⟶	Means-End link: ISTAR element that indicates a relationship between an end and a means for attaining it (the end). The "means" is expressed in the form of a task, since the notion of task embodies how to do something, while the "end" is expressed as a goal. In the graphical notation, the arrowhead points are presented from the means to the end.

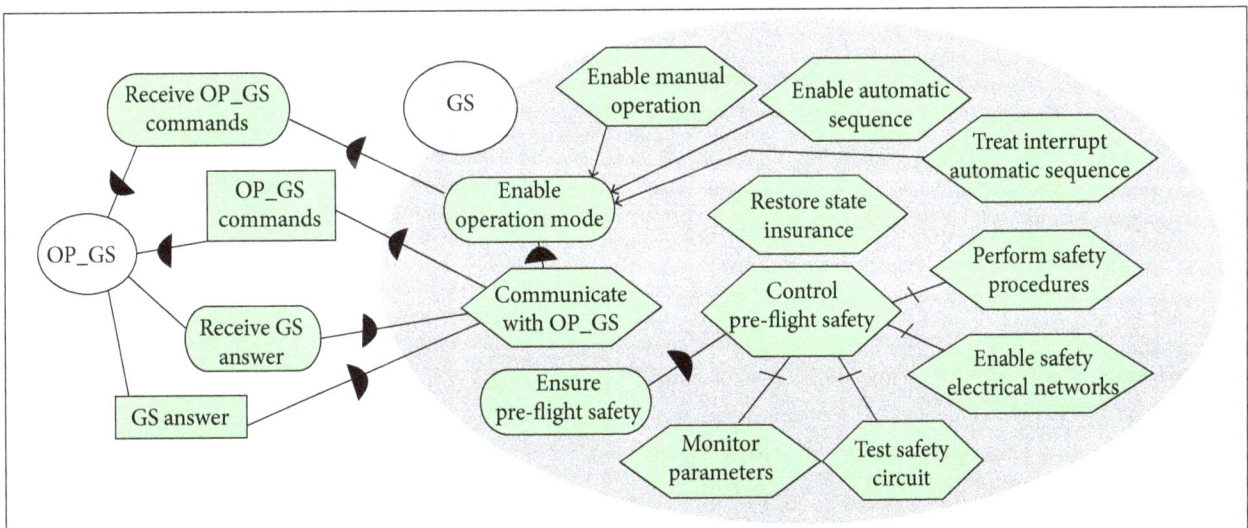

Figure 2. ISTAR SD and SR model example.

simply way, the message exchange between the operator and the system to enable the operational mode during the pre-launch, performing the tests of safety circuit and the change of the modes of operation, in a way to assure that the electrical network is ready to fly.

The relationship of the goals "Receive OP_GS commands" and "Receive GS answer" as well as the resources "OP_GS Commands" and "GS Answer" are part of the SD model, showing the external dependencies between two actors. On the other hand, all elements within the **GS** are part of the SR model and show the interoperability of this actor, *i.e.* how the elements relate internally.

ACTIVITY 2: APPLYING HAZARD ANALYSIS

The HAZOP study was used initially in the 1960s and, despite being based initially on the systematic examination of a chemical engineering plant, it is adopted for other areas and complex software systems (Crawley and Tyler 2015). It is based on guidewords to perform a qualitative analysis to each flow of a system, suggesting deviations operations, such as no, more, less or reverse (Souza 1995). These guidewords aim to identify deviations that may result in potential hazards to the system or function. The FMEA technique was developed in the 1940s as a US military security procedure to determine failures and effects on the system and in its equipment. Later, in the 1960s, the aerospace industry started to use the FMEA during the Apollo program. It aims to classify the flaws in relation to its impact on the mission success and staff safety. To do so, it individually investigates components or system functions, determining how and how often the components of a system can fail, and analyses the effects of this failures. After the analysis execution, it is made a verification of possible ways of reducing the probability of failures or effect analysed (Storey 1996).

The HAZOP and FMEA originated approaches such as Software Hazard Analysis and Resolution in Design (SHARD), Low-level Interaction Safety Analysis (LISA) (Pumfrey 1999) and Software FMEA (Lutz and Woodhouse 1996, 1997) that were used as reference for creating the ELICERE guidewords. While SHARD and LISA are more appropriated for hardware/software deviations, the SHARD technique examines the information flow deviations, initiating with the output system or its functions. LISA examines events of time deviations, such as interruptions, and physical resources used in the system operation. The Software FMEA approach of Lutz and Woodhouse (1997) is used to verify software requirements, specifically to analyse software requirements in space vehicles.

The ELICERE specific guidewords, strongly based on the HAZOP study nodes and FMEA, will be applied in the goal, resource, task and soft-goal of ISTAR components, observing their relationship dependency. The next step of the ELICERE hazard analysis is to apply guidewords over the components of the system modelled with ISTAR. The guidewords are used as a tool for the hazard analysis conduction, aiding the evaluation of the system components, anticipating their possible risks or failures. These guidewords represent the deviation of design intent, taking into consideration mainly the ISTAR components to be used in Programmable Electronic System (PES). Computer systems, communication systems, hardware devices (sensors and actuators), software or even human interface are some important actors considered during this step. This activity will allow obtaining a characterization of the hazard evidences that should be explored and prioritize the more critical components (resource, tasks, actors or even goals) that should be analysed. Then, it is necessary to fill each questionnaire chosen and the result in a set of soft-goals for the system.

Their settings are made for each type of ISTAR element and have a generic structure, which can be used to systems in a standard way, but can evolve and has new features for specific systems. Table 2 describes the guidewords used in the context of the ELICERE process.

Table 2. ELICERE generic guidewords.

Element	Guideword
Soft-goal	Soft-goal is not achieved
	Incorrect soft-goal
	Additional soft-goal
	Soft-goal out of time/order
Goal	Goal is not achieved
	Incorrect goal
	Additional goal
	Goal out of time/order
Task	Abnormal task termination
	Task omission
	Task is incorrect
	Task is out of time/order
Resource	Absent resource
	Incorrect resource
	Additional resource
	Resource out of time/order

Source: Lahoz (2009).

To accomplish this activity, it is necessary to apply a questionnaire for each guideword to conduct the hazard analysis that could result in a set of soft-goals for the system. This is an easy way to discover goals related to non-functional requirements (soft-goals) that allow mitigating the system hazards. More details about ELICERE guidewords and its questionnaire can be found in Lahoz (2009).

THE PRO-ELICERE TOOL

The PRO-ELICERE is a software tool that aims gathering the hazard analysis and its requirements for critical computer system from a system modelling perspective. Based on the ELICERE process, this new approach proposes an intelligent layer that allows performing the analysis, as possible, automatically. PRO-ELICERE is designed in several steps, as shown in Fig. 3.

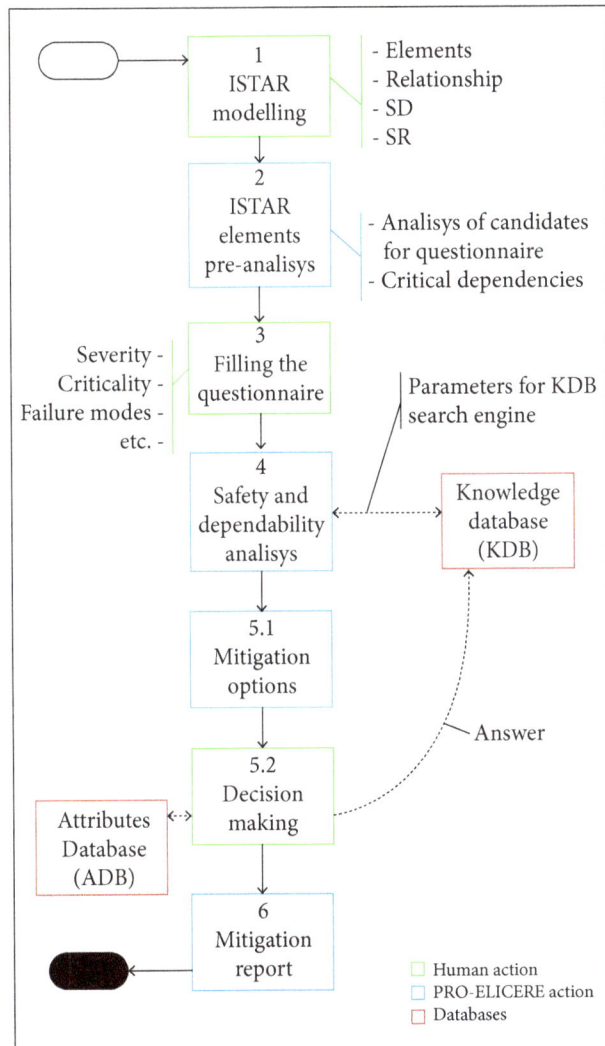

Figure 3. The PRO-ELICERE steps.

The PRO-ELICERE starts working with the ISTAR system models (inputs); after that, it creates the questionnaires based on guidewords related to ISTAR elements. Finally, it offers to the Analyst a menu of potential techniques and metrics (related to quality attributes) that could mitigate or even eliminate the hazards identified in the questionnaires. This set of techniques and metrics is an important contribution to the hazard analysis. Additionally, the knowledge database — capable of keeping the information of previous projects — become an important contribution to the Analyst's decision about which mitigation actions are more appropriated for the hazard.

Initially, it is made the system modelling (1), following the concepts of ISTAR. Thus, they shall be entered into the system (through manual entry or import from OpenOme, an open source tool) and separated by projects, identifying all the elements that make up the system and each of their relationships. After inserting the model in the system (2), a pre-analysis will be performed to identify candidates for the questionnaires. This filter is used to prioritize the elements, using criteria such as number of related elements or criticality in operation. After reviewing which questionnaires should be filled, they will be generated under the command of the Analyst.

With the questionnaires properly generated (3), the Analyst fills with the information that defines the task, resource, goal or soft-goal, to identify possible failure modes, severity, criticality, among other important factors. The automated questionnaire assists the analysis and helps to optimize the time.

The S&D analysis (4), an activity performed after the questionnaire, works with the data and recommendations (automatizing as much as possible) of the mitigation techniques to mitigate the hazards presented. It uses queries based on the attributes database (ADB) and the previous answered questionnaires. These answers are registered in the knowledge database (KDB) to find the best recommendations in futures hazard analysis queries. Originally, the ADB was created through a survey on the quality attributes of literature and techniques and methods to ensure these attributes. For another side, the KDB contains a repository of the mitigation suggestions coming from other analysis, of the same project or not.

At the end of the S&D analysis (5.1), the recommendations will be displayed and sorted by the degree of confidence or through the same parameters used in other projects. Despite the automation, the PRO-ELICERE does not choose the technique for hazard analysis. The Analyst (human inference) is still necessary (5.2), because he knows exactly the specificities and the scope of the problem and its variables; also he is free to consult others involved in the system. The PRO-ELICERE gives some possible answers to the Analyst determine the best way to mitigate the problem posed in the questionnaire. After that, the system will store that answer to the knowledge database that can be based on one PRO-ELICERE option chosen or an own free text of the Analyst. This information recorded in the knowledge database can be used in a future analysis of other systems. Finally, at the end of all questionnaires and answers (6), the PRO-ELICERE issues mitigation reports by several parameters, such as elements, models, guidewords, relationships or criticality, for example. This report can be used as a document to record the analysis and to suggest what actions can be taken to certain elements of the project.

Working with ISTAR Modelling

The PRO-ELICERE has two options to introduce the ISTAR models: manually, inserting each element of the system and their relationships, or importing the data from the OOD format (a XML format file), chosen from the OpenOme (Yu and Horkoff 2013).

The system model elements should include the Dependency Strengths, such as Open, Committed and Critical. This type of dependency strengths helps the PRO-ELICERE to prioritize the questionnaires to be generated.

When drawing or manually including the elements of the system, the Analyst should define the relationships between them, informing dependencies, task decomposition and other kinds of features available in the ISTAR modelling. The example in Fig. 4 shows a model definition used in PRO-ELICERE.

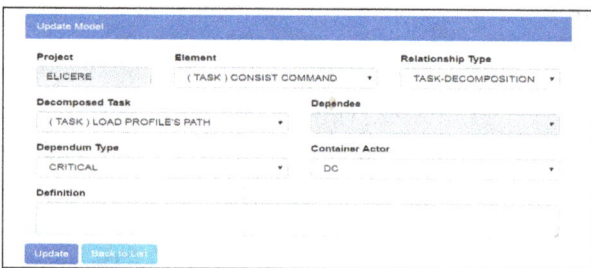

Figure 4. A PRO-ELICERE model definition.

In Fig. 4, **Relationship Type** is the field that describes the relationship between the element analysed and the system (another actor, resource, tasks etc.). This relationship could be **Task Decomposition**, **Means-End** and **Contribution Link**. Element is the element type and name analysed, such as task, resource, and goal. The **Decomposed Task** represents the task that will be fulfilled with the task analysed (in this case, the "Consist Command"). **Dependency**

type shows if the task is critical. The **Container Actor** is the actor who contains this task in the S&D model. The **Actor** is the field that contains the actor ("Dependee" or "Depender") in charge of the task or resource under analysis. Lastly, the **Definition** can be filled with additional information about the ISTAR model.

ISTAR Elements Pre-Analysis

The project elements description (and their relationships with other elements) is important to understand how they interact and to open the possibility of analysing the most critical items, using the dependency criteria, for example.

The PRO-ELICERE presents a set of guideword for each type of element helps to create the hazard questionnaires. With the models stored in the system, the Analyst will be able to identify which are the elements, and the related guidewords, that will have a critical importance. Figure 5 shows an example of the list of the potential questionnaires of an element.

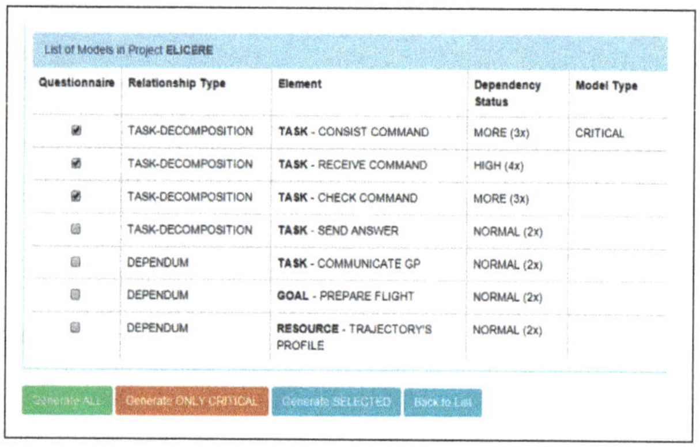

Figure 5. Example of the list of potential questionnaire generation.

One of the sorting options for choosing which questionnaires should be created is observing how many relationships (Dependency Status field) this element has in the whole system model.

For this, all design elements and relationships are seen, and the Analyst will check the items with more complexity, criticality or dependence. The PRO-ELICERE has the option to generate all questionnaires, to select only a few, or to choose just those suggested by the tool. For each element, the questionnaires will be generated according to 4 generic guidewords classified by ISTAR.

The Questionnaire Automation

The questionnaire proposed in ELICERE methodology had specific fields, but they were open; therefore, they did not have a fill pattern and would hardly be capable to be reused in other projects, like PRO-ELICERE aims. To avoid this, the PRO-ELICERE uses predetermined options from menus, aiming to enhance the Analyst's task, becoming the answer more direct and standardized. The example of the PRO-ELICERE questionnaire is presented in Fig. 6.

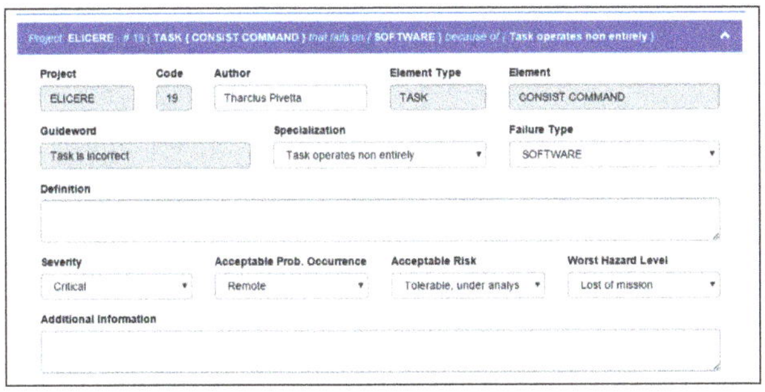

Figure 6. A PRO-ELICERE questionnaire example.

There are several fields to identify the hazard, but the most important are described as follows:

- Element Type: (1) resource, (2) task, (3) goal, and (4) soft-goal.
- Specialization: specific level of guideword to identify more objectively the deviation, such as "Saturated data", "Task occurs very early", "It was not provided the resource", "Sensor failure to send data", among others.
- Failure type: defines if the failure is (1) human, (2) software, (3) hardware or (4) environmental.
- Acceptable risk: defines the acceptable risk for failure. It is given by combining the Acceptable Probability of Occurrence and the Severity, being represented by the options: (1) intolerable, (2) undesirable, (3) tolerable under analysis, and (4) acceptable.

Safety and Dependability Analysis

To identify technical recommendations for the completed questionnaire, 2 possible approaches were applied: identify attributes and techniques directly from the ADB or through parameters from the questionnaire related to the KDB. Upon finished the questionnaire, the tool will suggest the recommendations, presented as a menu of options, from the ADB attributes. The option(s) will be chosen manually by the Analyst and will be recorded in the KDB. The PRO-ELICERE will create a database relationship table that matches some fields of the questionnaire (a hazard combination) such as "Element + Guideword + Specialization + Failure Type" related to the

mitigation techniques chosen by the Analyst. This relationship table will help the tool to present future recommendations, in case of some hazard combination occurs. It is considered that the hazard analysis is interactive (many interactions could happen) until finish the questionnaire with the final consideration about how to mitigate the hazard. The feedback from other projects about recommendations of the same kind of problem should be considered. These recommendations will be stored in a KDB, as well as the parameters filled of the hazard analysis questionnaire. With this, the tool can combine options based on knowledge, with techniques that can mitigate the hazard analysed, ensuring recommendations that are more reliable.

To illustrate how the questionnaire works, the PRO-ELICERE will analyse a failure related to the data transmission of an OP_GS and the GS (as illustrated in the model of Fig. 2). The resource "OP_GS Commands" and the task "Receive OP_GS commands" are related with 2 actors, OP_GS and GS, respectively. In the specific case of the **Element Type** "Resource", the questionnaire field **Guideword** is filled with "Resource Missing"; the **Specialization** of the hazard is "Loss or lack of message" for a **Failure Type** "Hardware". The tool will check in the KDB, in previous projects, if this similar scenario (communication problems) exists. Then, the tool could suggest a mitigation technique related to other questionnaires like "execute a Ping/Echo command", which recommends sending a standard signal to the sensor and waiting for the correct answer to check if the middle of transmission works well (Bass *et al.* 2003), based, for example, on the **Specialization** and **Element Type** fields. If the Analyst did not agree with this recommendation, the PRO-ELICERE can make other combinations to find other possible techniques disregarding parameters like **Specialization** (loss or lack of message). It can find 2 options, the first being already informed ("execute a Ping/Echo command") and a second, that is more generic, such as "create a passive redundancy", which is installing more than one sensor to read the same message and assure the availability of the data (Bass *et al.* 2003). As mentioned, the analysis is in charge of the final decision, although in many cases these options may be the best and the more reasonable solutions for the problem.

Mitigation Options versus Human Inference

The main goal of the PRO-ELICERE is to find mitigation recommendations in the databases and display the results to the Analyst. The Analyst will have the power to choose the options presented from the data from KDB. If the mitigation that fits the hazard analysis is not found, the Analyst can also manually choose among all techniques registered in the ABD or write his own mitigation action, creating a new input in the ABD and KDB.

Figure 7 shows a result that comes from the KDB for the questionnaire of the guideword "Task Incorrect".

Results

Combination	Technique	Method	
{Target Type: TASK } + { Specialization: Task operates non entirely } + { Failure Type: SOFTWARE }	(Testability) - Specialized Interfaces	Having specialized testing interfaces allows you to control or capture variable values for a component either through a test harness or through normal execution. Examples of specialized test routines include these: A set and get method for important variables, modes, or attributes, a report method that returns the full state of the object, a reset method to set the internal state to a specified internal state, a method to turn in verbose output, various levels of event logging, performance instrumentation, or resource monitoring.	Choose
{Target Type: TASK } + { Specialization: Task operates non entirely }	(Testability) - Record/Playback	The state that caused a fault is often difficult to re-create. Recording the state when it crosses na inteface allows that the state to be used to play the system back and to re-create the fault.	Choose

Figure 7. A PRO-ELICERE mitigation options example.

It is important to emphasize that the Analyst always takes the last decision about the best mitigation action. The PRO-ELICERE may suggest the recommendations through the criteria presented, but the final decision will always be of the Analyst, because he has the expertise about the project and the responsibility to determine the best solution to the problem presented.

Mitigation Report

Upon completion of the questionnaires, as well as the choice of recommendations for the presented analysis, the tool will allow the management of some reports, which can be used as formal documents of the hazard analysis. Among them, the "Questionnaires Report", which presents each performed analysis with details of hazards, criticality, failure mode and severity, as well as the recommendations of mitigation actions, may be cited. These reports will be described and exemplified in the next section.

THE CASE STUDY EXAMPLE

According to the Programa Nacional de Atividades Espaciais (PNAE, 2012) of the Agência Espacial Brasileira (AEB), a new launch vehicle program, called Cruzeiro do Sul, was established. The main goal is a continuation of the development carried out for the VLS-1 (Brazilian Satellite Launch Vehicle) — a medium-size solid propellant rocket motor, which is comprised of five new vehicles to be developed and qualified by the Instituto de Aeronáutica e Espaço (IAE). The first vehicle of the family is a 3-stage launch with an expected capability of transporting up to 400 kg payload into low inclination orbits of 400 km altitude (Moraes *et al.* 2006; Villas Boas 2006).

To understand how to perform the ELICERE hazard analysis and its PRO-ELICERE tool, some features of the VLS-1 were selected, related to the on-board to ground communication functionalities, potentially reused in the first prototype of the Cruzeiro do Sul family, denominated for this study V-ALFA.

THE V-ALFA MODEL EXAMPLE

For a better understanding of the model in ELICERE, it was created in the OpenOme tool a macro view of the goals models used by the actors/agents. The goal in ISTAR is an objective or function that the system should reach. To do so, it must perform tasks and use resources among actors and agents and even accomplish intermediate goals. These views are very important to be able to view which actors, goals, tasks and resources are critical for any project, noting their interdependencies. With the two ISTAR models created for the V-ALFA, it can be seen that one of the actors with the highest number of relationships is the Digital Controller (**DC**), so their tasks are critical for the V-ALFA meet the space mission. The role of the PRO-ELICERE is helping to identify soft-goals in order to assure this mission.

To illustrate the development analysis, a goal was chosen to show the model questionnaire, mitigation suggestion and finally the analysis report. Many goals can perform many tasks using many resources. It is advisable to check each goal separately, but after the analysis it must be validated the inter-relationship with the other goals and actors.

When the vehicle is in the launch pad, many activities are performed, such as testing the **DC** communication with the actuators and sensors, testing the pyrotechnic valves, checking of the destruction system, and loading the "profile of the trajectory", which contains the parameters relating to the V-ALFA flight profile. The file with the trajectory profile should be loaded correctly through the communication link between the Ground Support equipment (**GP**) and the on-board Digital Controller (**CD**), in the pre-flight phase. If this profile was loaded incorrectly, the goal "Prepare to Flight" will be not accomplished, and the mission probably will fail. This ISTAR model can be represented in Fig. 8.

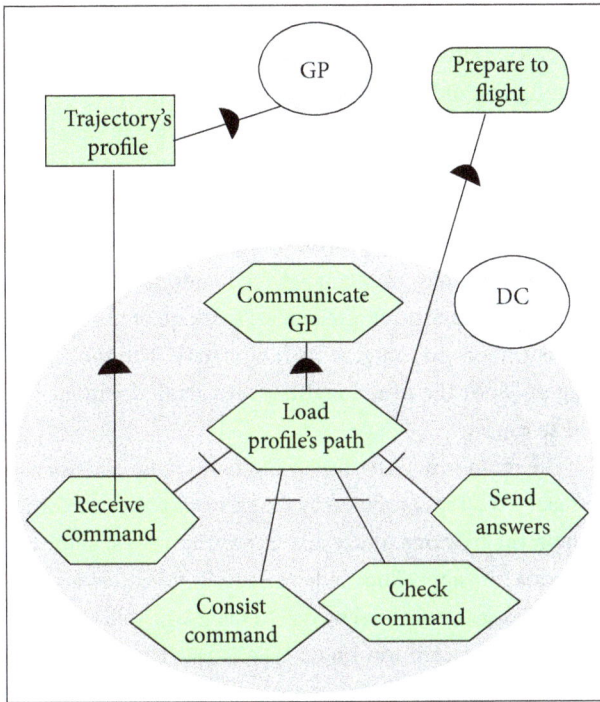

Figure 8. ISTAR model of the "Prepare to flight" goal.

The main elements of the specific "Prepare to flight" ISTAR model are:

- **GP** actor – ground support equipment: system responsible for the preparation to launch flight. It commands the tests and verifies the on-board networks, sending the trajectory profile with the parameters and inputs to the control algorithms.

- **DC** actor – digital controller: equipment responsible to communicate with the ground system. It performs all pre-flight tests and loads the trajectory profile to perform the control of the vehicle during the first stages of flight.

- **Trajectory's Profile** resource –contains the parameters for the V-ALFA flight profile. This resource must be loaded properly through the means of communication between the GP and DC, in the pre-flight phase. If any of the parameter is wrong, the goal will not be met, because the DC will not load the data with the specified values. There will be the need to carry this resource again with the corrected values.

- **Load Profile's Path** — this task is in charge of receiving data relating to the proposed flight trajectory. If the task sends wrong or absent values from the GP, the target will be missed and will create an incorrect table profile path leading

to loss of mission. This task is decomposed into 4 sub-tasks: "Communicate with GP", "Receive Command", "Consist Command", "Check Command" and "Send Answer".

THE V-ALFA QUESTIONNAIRE EXAMPLE

After the PRO-ELICERE generates all the possible questionnaires, the Analyst can select all or only some of them, following some criteria, such as the high criticality items, number of dependencies, level of risk etc. Figure 9 presents a questionnaire completed, analysing the "Receive Command" sub-task, decomposed from the "Load Profile's Path" task.

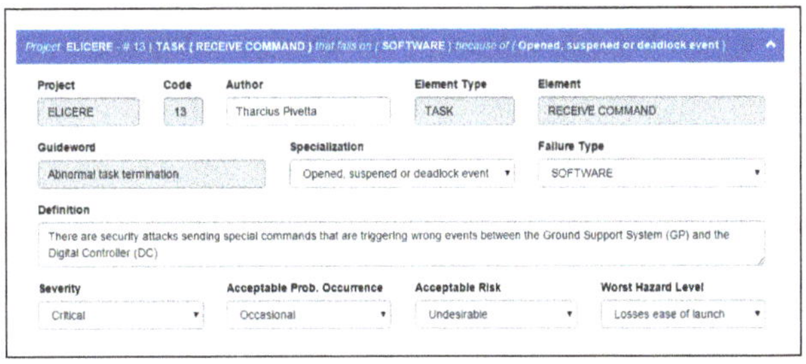

Figure 9. Example of questionnaire filled.

By performing all the hazard analysis, the system will allow the Analyst to extract certain system reports, such as the list of all filled questionnaires or the list of interaction (each other) of the elements in the system. Figure 10 presents an example of one report that informs the number of completed questionnaires, their risk analysis, and details about the mitigation action, such as the technique related, the quality attribute associated and the explanation about the method to achieve it.

In addition, it is shown a summary of the status of each project (in this context, V-ALFA), analysing quantitatively the questionnaires answered, such as:

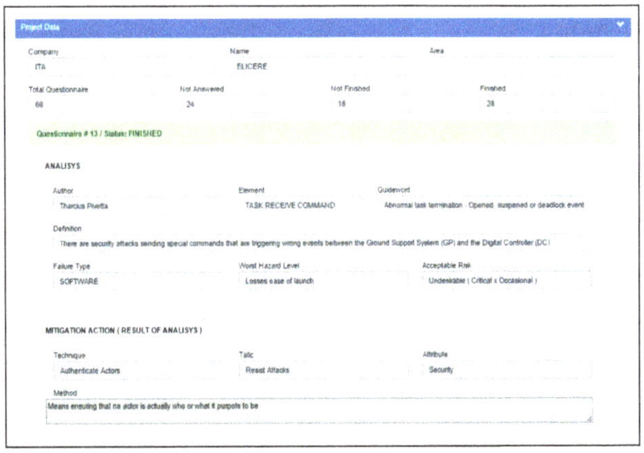

Figure 10. Questionnaire report example.

- **Total Questionnaire**: total of questionnaires generated by the elements in combination with de guidewords, selected in the step of Questionnaire Generation (Fig. 5).
- **Not Answered**: quantity of questionnaires created, but with information not filled like risk, probability occurrence, and other vital information for the understanding of the hazard.
- **Not Finished**: questionnaires fully answered, but without the mitigation technique chosen by the Analyst. It means the hazard needs one more step to finish.
- **Finished**: questionnaires fully answered and with proper mitigation technique chosen.

Figure 9 presents a report extracted from the Questionnaire #13 – "Receive command" task, which picked the mitigation action "Authenticate Actors". This technique suggests creating means to ensure that the actor access is authorized (Bass *et al.* 2003).

CONCLUSIONS

This paper described the main features of the automation of the ELICERE process, a methodology of hazard analysis through the system modelling, and its guidewords analysis. Due to the large amount of information to be manipulated and aiming to reuse the mitigation actions, the software tool called PRO-ELICERE was created. First, the system is described through the ISTAR modelling language. Then, the analysis of its elements is performed using guidewords, such as HAZOP technique and then a questionnaire is filled with the hazard information as proposed by the FMEA.

A database repository (ADB) has been created with a well-known set of quality attributes and related techniques identified in the literature, as well as another database (KDB) with the results of the previous hazard analysis and its mitigation actions. The first one was created using several literature references (Bass *et al.* 2003; Romani *et al.* 2010; Lahoz *et al.* 2012) and the other was created to record the results of the hazard analysis performed with the PRO-ELICERE. These quality attributes cover availability, interoperability, modifiability, performance, security, testability, usability, modularity, traceability, simplicity and robustness. Eighty-seven techniques are suggested to assure that elements under analysis meet this quality goal. For example, if the hazard analysis identified that an element

of the system needs to improve the Availability, the PRO-ELICERE could suggest applying techniques like "make reconfiguration", "include passive redundancy", "perform self-test" or "send Ping/echo" etc. If the element analysed needs to improve its Security, for instance, "detect Intrusion", "verify message integrity" and "check authorization login" are suggested. Another possible analysis performed by PRO-ELICERE is about problems related to development, when a guideword questionnaire detects that one of the system's element needs to improve its Modularity. The tool could suggest "Split the functionalities into small components" and so on.

The main function of the KDB is to create a history of mitigation actions proposed by the previous projects in order to help the new one under analysis with the best solution. The idea is to apply knowledge discovery techniques, using ontologies, generally presented in the language Web Ontology Language/Description Logic (OWL-DL; Horrocks *et al.* 2003) to infer knowledge about safety and dependability issues based the on-going PRO-ELICERE analysis, with the set of quality attributes and its techniques previously chosen by the last projects.

Finally, in the study case presented as a way to understand how PRO-ELICERE works, 83 ISTAR elements are created, such as 13 actors, 3 agents, 13 goals, 14 resources and 40 tasks. A total of 60 **Dependency** relationships (producer *versus* consumer), 27 **Decomposed-Task** and 4 **Means-End** relationships were produced, generating a total of 68 questionnaires. The next research step involves the extraction of more data from the V-ALFA, improving its ISTAR model, performing more hazard analysis and creating a proper ontology for the PRO-ELICERE.

The main benefits of the PRO-ELICERE are the organization in database of the huge amount of data obtained with the system models, with the guidewords questionnaires and with the mitigation options that come from ADB and KDB. Also, the PRO-ELICERE's database are capable of presenting, in terms of screen views and printable reports, all the information handled, such as the list of ISTAR actors related to their goals, tasks and resources necessary to perform a system goal, number of dependencies of an actor (extracted from SD or SR diagrams), and so on. The KDB is a strategic contribution for the hazard analysis due to its capacity of storing the previous hazards analysis, then presenting a potential solution for mitigation.

REFERENCES

Bass L, Clements P, Kazman R (2003) Software architecture in practice. Upper Saddle River: Addison-Wesley.

Crawley F, Tyler B (2015) HAZOP: guide to best practice. 3rd ed. Waltham: Elsevier. Chapter 1, Introduction; p. 1-3.

Horrocks I, Patel-Schneider PF, van Harmelen F (2003) From shiq and rdf to owl: the making of a Web Ontology Language. Web Semant Sci Serv Agents World Wide Web 1(1):7-26. doi: 10.1016/j.websem.2003.07.001

Lahoz CHN (2009) ELICERE — o processo de elicitação de metas de dependabilidade para sistemas computacionais críticos: estudo de caso aplicado à area espacial (PhD Thesis). São Paulo: Universidade de São Paulo. In portuguese.

Lahoz CHN, Camargo Júnior JB (2011) Introducing ELICERE guidewords for critical computer systems. Proceedings of the IEEE Fourth International Conference on Software Testing, Verification and Validation (ICST); Berlin, Germany.

Lahoz CHN, Romani MAS, Yano ET (2012) Dependability attributes for space computer systems: quality factors approach. Proceedings of the Space Operations Conference (SpaceOps); Stockholm, Sweden.

Lutz RR, Woodhouse RM (1996) Experience report: contributions of SFMEA to requirements analysis. Proceedings of the 2nd IEEE International Conference on Requirements Engineering (ICRE); Colorado Springs, USA.

Lutz RR, Woodhouse RM (1997) Requirements analysis using forward and backward search. Ann Software Eng 3:459-475.

Moraes Jr P, Carrijo DS, Garcia A, Costa LEL, Oliveira UC, Santana Jr A, Villas Boas DJF, Yamamoto MK (2006) An overview of the Brazilian launch vehicle program Cruzeiro do Sul. Proceedings of the 57th International Astronautical Congress, International Astronautical Congress (IAF); Valencia, Spain.

PNAE (2012). Programa Nacional de Atividades Espaciais: PNAE: 2012-2021 / Agência Espacial Brasileira. Brasília: Ministério da Ciência, Tecnologia e Inovação, Agência Espacial Brasileira. p. 36.

Pumfrey DJ (1999) The principled design of computer system safety analyses (PhD thesis). York: University of York.

Romani MAS, Lahoz CHN, Yano ET (2010) Identifying dependability requirements for space software systems. J Aerosp Technol Manag 2(3):287-300. doi: 10.5028/jatm.2010.02037810

Souza EA (1995) O treinamento industrial e a gerência de riscos – uma proposta de instrução programada (Master's thesis). Florianópolis: Universidade Federal de Santa Catarina. In portuguese.

Stark J, Swinerd G, Tatnall A (2004) Introduction. In: Fortescue P, Swinerd G, Stark J, editors. Spacecraft systems engineering. 3rd ed. Chichester: Wiley.

Storey N (1996) Safety-critical computer systems. Upper Saddle River: Addison-Wesley.

Villas Boas DJF (2006) O contexto histórico das atividades espaciais e a tecnologia dos foguetes. In: Ministério da Educação, Secretaria de Educação a Distância. Da Terra ao espaço: tecnologia e meio ambiente na sala de aula. Boletim 06/2006. p. 26-37; [accessed 2016 Jan 28]. http://cdnbi.tvescola.org.br/resources/VMSResources/contents/document/publicationsSeries/1426100949736.pdf

Yu ES (1995) Modelling strategic relationship for process reengineering (PhD thesis). Toronto: University of Toronto.

Yu ES, Grau G (2006) ISTAR quick guide; [accessed 2016 Jan 02]. http://istar.rwth-aachen.de

Yu ES, Horkoff J (2013) OpenOme beta; [accessed 2015 Dec 10]. http://sourceforge.net/projects/openome

14

A Comparative Study of Four Feedback Linearization Schemes for Motion Planning of Fixed-Wing Unmanned Aerial Vehicles

Hossein Bonyan Khamseh[1]

ABSTRACT: In this paper, different feedback linearization schemes are studied to address the motion planning problem of fixed-wing unmanned aerial vehicles. For a unmanned aerial vehicle model with second-order dynamics, several schemes are studied to make the vehicle (i) fly over and (ii) make a loitering around the objective position. For each scheme, comparisons are made to illustrate the advantages and disadvantages. Lyapunov stability analysis is used to prove the stability of the proposed schemes, and simulation results for some case studies are included to show their feasibility.

KEYWORDS: Feedback linearization, Unmanned aerial vehicle, Motion planning.

INTRODUCTION

In recent years, unmanned aerial vehicles (UAVs) have gained increasing attention for various missions such as remote sensing of agricultural products (Costa *et al.* 2012), forest fire monitoring (Casbeer *et al.* 2006), search and rescue (Almurib *et al.* 2011), transmission line inspection (Li *et al.* 2013) and border monitoring (Beard *et al.* 2006). To this date, various approaches have been employed to address the motion planning of UAVs to reach, fly over or loiter around an objective position. As an example, in Frew *et al.* (2008) and Lawrence *et al.* (2008), vector fields with a stable limit cycle centered on the target position were constructed. In the mentioned studies, the authors employed a Lyapunov vector field guidance (LVFG) law to bring the UAV to an observation "orbit" around the target. Also, in Gonçalves *et al.* (2011), a vector field approach was used to bring several non-holonomic UAVs to a static curve embedded in the 3-D space. In Gonçalves *et al.* (2010), vector fields were determined such that a robot converged to a time-varying curve in n-dimensions and circulated it. In Hsieh *et al.* (2008), decentralized controllers were proposed to bring a number of robotic agents to generate desired simple planar curves, while avoiding inter-agent collision. In Hsieh *et al.* (2007), the controllers were modified such that the robots converged to a star-shaped pattern and, once on the objective curve, circulated it. In Bonyan Khamseh *et al.* (2014), based on the concept of flight corridor, a decentralized coordination strategy was proposed to bring a team of fixed-wing UAVs to a circular orbit, while avoiding inter-UAV collision. In Hafez *et al.* (2013), model predictive control was used to create a

1.Universidade Federal de Minas Gerais – Escola de Engenharia Elétrica – Departamento de Engenharia Elétrica – Belo Horizonte/MG – Brazil.

Author for correspondence: Hossein Bonyan Khamseh | Universidade Federal de Minas Gerais - Escola de Engenharia Elétrica - Departamento de Engenharia Elétrica
Av. Pres. Antônio Carlos, 6627 – CEP: 31 270-901 - Pampulha - Belo Horizonte/MG – Brazil | Email: h.bonyan@gmail.com

dynamic circular formation around a given target. By means of simulations, it was shown that the system was stable, but formal stability analysis was not provided. In Marasco *et al.* (2012), the same approach was improved to address encirclement of multiple targets, without stability analysis.

It is also possible to employ feedback linearization to simplify the equations of motion in motion planning problems (Lawton *et al.* 2003; Fan and Zhiyong 2009; Kanchanavally *et al.* 2006). As an example, in Lawton *et al.* (2003), feedback linearization was employed to study the formation control of the end-effector position of a team of non-holonomic robots. Having obtained simpler double-integrator equations, control laws were designed and formation control was achieved. Stability of the system was proven by means of Lyapunov stability theory. In Fan and Zhiyong (2009), for a multi-agent system, the authors proposed a dynamic feedback linearization scheme to describe the equations of motion of each agent by third-order integrators. Then, a formation control law with inter-agent damping was developed, and asymptotic stability of the system was verified using Lyapunov stability analysis. In Kanchanavally *et al.* (2006), the specific problem of 3-D motion planning of UAVs via feedback linearization was studied. In that study, a non-holonomic UAV equipped with a fixed-angle camera was considered. The footprint of the camera was defined as the system output, and it was shown that it converged to an objective position. Similar to the previous papers (Lawton *et al.* 2003; Fan and Zhiyong 2009), the stability of the system was studied by means of Lyapunov stability theory. Yet, an important drawback of Kanchanavally *et al.* (2006) is that it did not include the constraints of minimum and maximum forward velocity of fixed-wing UAVs. Therefore, the UAVs came to rest, i.e. zero forward velocity, once the camera footprint converged to the target position.

In this paper, several feedback linearization schemes are studied to address the problem of motion planning of fixed-wing UAVs flying with constant forward velocity. The objective here is that the UAV (i) flies over or (ii) loiter around a static objective position, without coming to rest.

FEEDBACK LINEARIZATION SCHEMES

In the following subsections, a UAV with an on-board camera will be considered. For a fixed-angle forward-looking camera, the results were presented in Kanchanavally *et al.* (2006), where the UAV finally came to rest. Due to minimum

forward velocity constraint, that method is not applicable to fixed-wing UAVs. Here, we define several schemes such that the footprint of the on-board camera converges to an objective position while the UAV either flies over or loiters around the objective position with constant forward velocity.

UAV WITH FORWARD-LOOKING CAMERA, SCHEME #1

In the first scheme a fixed-wing UAV with variable-angle forward-looking camera is considered. In the control affine form, considering a simplified rigid-body model, the dynamic equations of motion of the system are given by:

$$
\begin{bmatrix} \dot{x}_1 \\ \dot{x}_2 \\ \dot{x}_3 \\ \dot{x}_4 \\ \dot{x}_5 \\ \dot{x}_6 \\ \dot{x}_7 \\ \dot{x}_8 \end{bmatrix} = \begin{bmatrix} \dot{r}_x \\ \dot{r}_y \\ \dot{\theta} \\ \dot{\omega} \\ \dot{r}_z \\ \dot{v}_z \\ \dot{\phi} \\ \dot{v}_\phi \end{bmatrix} = \begin{bmatrix} v \cos x_3 \\ v \sin x_3 \\ x_4 \\ 0 \\ x_6 \\ 0 \\ x_8 \\ 0 \end{bmatrix} + \begin{bmatrix} 0 & 0 & 0 \\ 0 & 0 & 0 \\ 0 & 0 & 0 \\ \frac{1}{I} & 0 & 0 \\ 0 & 0 & 0 \\ 0 & \frac{1}{m} & 0 \\ 0 & 0 & 0 \\ 0 & 0 & \frac{1}{J} \end{bmatrix} \begin{bmatrix} u_1 \\ u_2 \\ u_3 \end{bmatrix} = \boldsymbol{f} + g\boldsymbol{u} \quad (1)
$$

where: r_x, r_y and r_z represent x, y, and z positions of the UAV; v and v_z represent the constant velocity v in the $x - y$ horizontal plane and the velocity v_z in the z-direction; θ and ω represent the heading angle and angular velocity of the UAV; ϕ and v_ϕ represent the angle and angular speed of the camera, respectively. The UAV constants are given by m (UAV mass), I (UAV moment of inertia about z-axis) and J (camera moment of inertia about its rotation axis). The input vector is $\boldsymbol{u} = [u_1 \ u_2 \ u_3]^{\mathrm{T}}$.

In the model given by Eq. 1, it has been assumed that the forward velocity in the $x - y$ plane, i.e. v, is constant. With this simplification, if one initially chooses the forward velocity to satisfy $v_{min} \le v \le v_{max}$, one can conclude that the minimum and maximum forward velocity constraints will be automatically satisfied throughout the mission. Also, since the camera is not mounted with a fixed angle, one can come up with scenarios in which the UAV does not come to rest when the camera footprint converges to the objective position. As it can be seen from Eq. 1, the camera dynamics has been assumed to be second-order and completely decoupled from the UAV dynamics. For a forward-looking camera, one can consider the following output for the system (Kanchanavally *et al.* 2006):

$$\boldsymbol{\Gamma} = \begin{bmatrix} x_1 + L \cos x_3 \\ x_2 + L \sin x_3 \\ x_5 \end{bmatrix}, \tag{2}$$

where: $L = r_z \cot\phi$.

As schematically shown in Fig. 1, the first two elements of Γ represent x and y positions of the footprint of the forward-looking camera. Also, the third element represents the altitude of the UAV. An important advantage of this scheme, compared to Kanchanavally *et al.* (2006), is that ϕ is not constant here and is considered one of the system state variables.

Assuming relative degree of r_i for the i-th elements of Γ, after r_i times differentiation, one finds:

$$\begin{bmatrix} \Gamma_1^{(r_1)} \\ \Gamma_2^{(r_2)} \\ \Gamma_3^{(r_3)} \end{bmatrix} = \begin{bmatrix} L_f^{r_1}\Gamma_1 + \sum_{i=1}^{3} L_{g_i}(L_f^{r_1-1}\Gamma_1)u_i \\ L_f^{r_2}\Gamma_2 + \sum_{i=1}^{3} L_{g_i}(L_f^{r_2-1}\Gamma_2)u_i \\ L_f^{r_3}\Gamma_3 + \sum_{i=1}^{3} L_{g_i}(L_f^{r_3-1}\Gamma_3)u_i \end{bmatrix} = \boldsymbol{A} + \boldsymbol{B}\boldsymbol{u} \tag{3}$$

where: $L_f\Gamma_i$ is the Lie derivative, i.e. $L_f\Gamma_i = \partial\Gamma_i/\partial x$ and g_i is the i-th column of the matrix g. Also, $\boldsymbol{A} \in \boldsymbol{R}^{3\times1}$ and $\boldsymbol{B} \in \boldsymbol{R}^{3\times3}$, where the elements of $\boldsymbol{A}$ are given below. For the system given by Eq. 1 and with output given by Eq. 2, one can obtain the vector of relative degree as $[r_1\ r_2\ r_3] = [2\ 2\ 2]$.

$$A_1 = -vx_4 \sin x_3 - 2x_6 x_8 \frac{1}{\sin^2 x_7} \cos x_3 +$$
$$+ 2x_5(x_8)^2 \frac{\cos x_7}{\sin^3 x_7} \cos x_3 - 2x_6 \cot x_7\, x_4 \sin x_3 +$$
$$+ 2x_5 x_8 \frac{1}{\sin^2 x_7} x_4 \sin x_3 - L(x_4)^2 \cos x_3$$

$$A_2 = vx_4 \cos x_3 - 2x_6 x_8 \frac{1}{\sin^2 x_7} \sin x_3 +$$
$$+ 2x_5(x_8)^2 \frac{\cos x_7}{\sin^3 x_7} \sin x_3 + 2x_6 \cot x_7\, x_4 \cos x_3 -$$
$$- 2x_5 x_8 \frac{1}{\sin^2 x_7} x_4 \cos x_3 - L(x_4)^2 \sin x_3$$

and $A_3 = 0$.
Also:

$$B = \begin{bmatrix} \dfrac{-L}{I}\sin x_3 & \dfrac{1}{m}\cot x_7 \cos x_3 & -\dfrac{1}{J}\dfrac{x_5}{\sin^2 x_7}\cos x_3 \\[2mm] \dfrac{L}{I}\cos x_3 & \dfrac{1}{m}\cot x_7 \sin x_3 & -\dfrac{1}{J}\dfrac{x_5}{\sin^2 x_7}\sin x_3 \\[2mm] 0 & \dfrac{1}{m} & 0 \end{bmatrix}$$

One can verify that $\det(B) = -(x_5)^2 \cos x_7/m{\cdot}I{\cdot}J \sin^3 x_7$. This determinant can be zero if $x_5 = 0$ or $\cos x_7 = 0$. In order to show that $x_5 \neq 0$, we define the error as:

$$\boldsymbol{E} = (\boldsymbol{\Gamma} - \boldsymbol{R}) + (\dot{\boldsymbol{\Gamma}} - \dot{\boldsymbol{R}}) \tag{4}$$

where: $\boldsymbol{R}$ is the reference signal and, for a stationary reference signal, one will have $\dot{\boldsymbol{R}} = 0$. Assuming $E_3(0) > -R_3$, one can rewrite the third row of Eq. 4 as:

$$\dot{\Gamma}_3 = -\Gamma_3 + \underbrace{(E_3 + R_3)}_{u'} \tag{5}$$

Therefore:

$$\Gamma_3(t) = e^{-t}\Gamma_3(0) + \int_0^t e^{-(t-\tau)} u'(\tau) d\tau \tag{6}$$

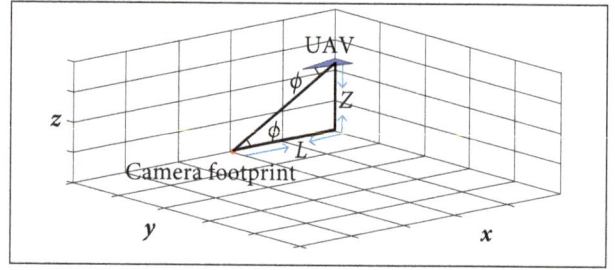

Figure 1. The UAV and its camera footprint in scheme #1.

In the following paragraphs it will be shown that the absolute value of E_3 monotonically decreases and converges to zero. Therefore, noting that $R_3 > 0$, it will be easy to see that $u'(t)$ is always positive. With $\Gamma_3(0) > 0$, from Eq. 6, one can conclude that $\Gamma_3(t) > 0$ and therefore $x_5 \neq 0$, i.e. the altitude cannot be zero. Also, for $\cos x_7 = 0$, one must have $x_7 = kn + \pi/2$. It means that the objective position is exactly underneath the UAV actual position. This is an important drawback which can lead to the failure of this scheme. Yet, due to discretization, control errors and other real-world phenomena, this is not a concern in practical situations. In our simulations, no problem was encountered due to this drawback. Also, $\det(B) \to \infty$ if $x_7 \to k\pi$. Yet, $x_7 \to k\pi$ means that $L \to \infty$. In practice, it is not a legitimate concern because $\Gamma_3(0) > 0$ and also the UAV cannot be infinitely far from its target ($x_7 \neq k\pi$). Therefore it is concluded that $\det(B) = -(x_5)^2 \cos x_7/m{\cdot}I{\cdot}J \sin^3 x_7 \neq 0$ for practical applications.

In order to study the error dynamics, we differentiate Eq. 4 to obtain:

$$\dot{E} = \dot{I} + \ddot{\Gamma} = \dot{I} + A + Bu \qquad (7)$$

Similar to Kanchanavally *et al.* (2006), if we define $u = B^{-1}(-A - \dot{I} + v)$ and , it is easy to see that:

$$\dot{E} = KE \qquad (8)$$

where: $K \in R^{3 \times 3}$ and eigenvalues of K have negative real parts.

In order to study the error dynamics given by Eq. 8, one may consider the following Lyapunov function:

$$V = \frac{1}{2} E^T E \qquad (9)$$

The time derivative of the above positive definite V is given by:

$$\dot{V} = \dot{E}^T E = E^T K^T E < 0 \qquad (10)$$

Therefore the error converges to zero. Regarding the internal dynamics with a vector of relative degree of [2 2 2], one needs to propose two more transformations to complete the diffeomorphism. The internal dynamics is given by Eq. 11:

$$\begin{cases} \eta_7 = x_3 \rightarrow \dot{\eta}_7 = x_4 \\ \eta_8 = x_4 \rightarrow \dot{\eta}_8 = \frac{1}{I} u_1 \end{cases} \qquad (11)$$

For a UAV with a forward-looking camera, the only possible motion where the error goes to zero is when the UAV flies on a straight line over the objective position. Therefore, as $t \to \infty$, $x_4 \to 0$. With $x_4 \to 0$, it is easy to conclude that x_3 will be bounded and therefore it is not going to cause undesirable effects. Simulation results verifying the feasibility of this approach will be given in "Simulations" section.

UAV WITH SIDE-LOOKING CAMERA, SCHEME #2

In this subsection, a UAV with variable-angle side-looking camera is considered. The equations of motion of this configuration are identical to those given in Eq. 1 and therefore are not repeated here. For a side-looking camera, the output is given by:

$$\Gamma = \begin{bmatrix} x_1 + L \sin x_3 \\ x_2 - L \cos x_3 \\ x_5 \end{bmatrix} \qquad (12)$$

where the first two elements of Γ represent x and y positions of the footprint of the camera (see Fig. 2).

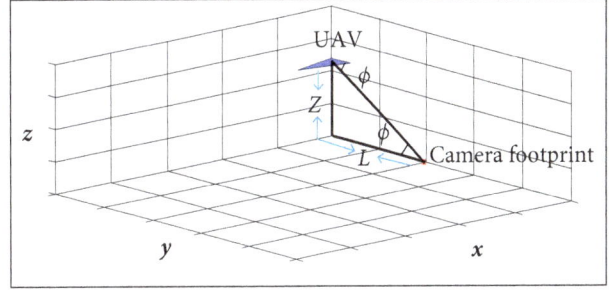

Figure 2. The UAV and its camera footprint in scheme #2.

For the given output, one can verify that $[r_1\ r_2\ r_3] = [2\ 2\ 2]$. Therefore, an equation identical to that given in Eq. 3 is obtained, in which:

$$A_1 = -vx_4 \sin x_3 - 2x_6 x_8 \frac{1}{\sin^2 x_7} \sin x_3 +$$
$$+ 2x_5 (x_8)^2 \frac{\cos x_7}{\sin^3 x_7} \sin x_3 + 2x_6 \cot x_7 x_4 \cos x_3 -$$
$$- 2x_5 x_8 \frac{1}{\sin^2 x_7} x_4 \cos x_3 - L(x_4)^2 \sin x_3$$

$$A_2 = vx_4 \cos x_3 + 2x_6 x_8 \frac{1}{\sin^2 x_7} \cos x_3 -$$
$$- 2x_5 (x_8)^2 \frac{\cos x_7}{\sin^3 x_7} \cos x_3 + 2x_6 \cot x_7 x_4 \sin x_3 -$$
$$- 2x_5 x_8 \frac{1}{\sin^2 x_7} x_4 \sin x_3 + L(x_4)^2 \cos x_3$$

$$A_3 = 0$$

$$B = \begin{bmatrix} \frac{L}{I} \cos x_3 & \frac{1}{m} \cot x_7 \sin x_3 & -\frac{1}{J} \frac{x_5}{\sin^2 x_7} \sin x_3 \\ \frac{L}{I} \sin x_3 & -\frac{1}{m} \cot x_7 \cos x_3 & \frac{1}{J} \frac{x_5}{\sin^2 x_7} \cos x_3 \\ 0 & \frac{1}{m} & 0 \end{bmatrix}$$

One can verify that $\det(B) = -(x_5)^2 \cos x_7 / m \cdot I \cdot J \sin^3 x_7$. This determinant can be zero if $x_5 = 0$ or $\cos x_7 = 0$. With the reasoning given in the previous subsection (see Eqs. 4 – 6), one can conclude that $x_5 \neq 0$, i.e. the altitude cannot be zero. For a UAV with constant forward velocity and a side-looking camera, the only possible motion where the error goes to zero is when the UAV loiters around the objective position, with a fixed loitering radius. Therefore, with a non-zero loitering radius, it is easy to conclude that $\cos x_7 \neq 0$. Also, $\det(B) \to \infty$ if $x_7 \to k\pi$. Yet, $x_7 \to k\pi$

means that $L \to \infty$, which is not common in practical scenarios. Therefore, it is concluded that $\det(B) = -(x_5)^2 \cos x_7 / m \cdot I \cdot J \sin^3 x_7 \neq 0$ for practical applications, and thus the system given by Eq. 1 with output given by Eq. 12 is input-output linearizable.

For this scheme, the Lyapunov stability analysis is identical to that given by Eqs. 7 – 10. Therefore, it can be concluded that the error dynamics asymptotically converges to zero. Regarding the internal dynamics with a vector of relative degree of [2 2 2], one needs to propose two more transformations to complete the diffeomorphism. The internal dynamics is given by and in Eq. 13:

$$\begin{cases} \eta_7 = x_3 \to \dot{\eta}_7 = x_4 \\ \eta_8 = x_4 \to \dot{\eta}_8 = \frac{1}{I} u_1 \end{cases} \quad (13)$$

For a UAV loitering around a given objective position with constant (finite) forward velocity, x_4 will be bounded and cannot go to infinity. Also, in a loitering motion, x_3, i.e. the heading angle, can be shown by $2k\pi + \theta'$ where θ' is a finite value and therefore the internal dynamics will not cause undesirable effects in our approach. Simulation results regarding this scheme will be given in "Simulations" section.

An important drawback of this method is that one cannot explicitly control the final loitering radius of the UAV. Therefore, in the next scheme, we try to explicitly define the loitering radius as one of the system outputs.

UAV WITH SIDE-LOOKING CAMERA, SCHEME #3

In this section, we modify the equations of motion given in Eq. 1 in a manner that a new useful scheme is obtained. In the control affine form, the new dynamic equations of motion are given by Eq. 14:

$$\begin{bmatrix} \dot{x}_1 \\ \dot{x}_2 \\ \dot{x}_3 \\ \dot{x}_4 \\ \dot{x}_5 \\ \dot{x}_6 \\ \dot{x}_7 \\ \dot{x}_8 \end{bmatrix} = \begin{bmatrix} \dot{r}_x \\ \dot{r}_y \\ \dot{\theta} \\ \omega \\ \dot{r}_z \\ \dot{v}_z \\ \dot{L} \\ \ddot{L} \end{bmatrix} = \begin{bmatrix} v \cos x_3 \\ v \sin x_3 \\ x_4 \\ 0 \\ x_6 \\ 0 \\ x_8 \\ 0 \end{bmatrix} + \begin{bmatrix} 0 & 0 & 0 \\ 0 & 0 & 0 \\ 0 & 0 & 0 \\ \frac{1}{I} & 0 & 0 \\ 0 & 0 & 0 \\ 0 & \frac{1}{m} & 0 \\ 0 & 0 & 0 \\ 0 & 0 & 1 \end{bmatrix} \begin{bmatrix} u_1 \\ u_2 \\ u_3 \end{bmatrix} = f + gu \quad (14)$$

Here, the main difference is that L and $\dot{L}$ are explicitly considered to be state variables. One may define the system output as:

$$\Gamma = \begin{bmatrix} x_1^2 + x_2^2 \\ x_5 \\ x_7 \end{bmatrix} \quad (15)$$

where the first element of Γ is the square of the distance of the UAV from the origin of the coordinate system, i.e. the stationary objective position.

It can be readily seen that $[r_1 \ r_2 \ r_3] = [3\ 2\ 2]$. Therefore, in a compact form, one can write:

$$\begin{bmatrix} \ddot{\Gamma}_1 \\ \ddot{\Gamma}_2 \\ \ddot{\Gamma}_3 \end{bmatrix} = A + Bu \quad (16)$$

where:

$$A = \begin{bmatrix} -2vx_4^2(x_1 \cos x_3 + x_2 \sin x_3) \\ 0 \\ 0 \end{bmatrix},$$

$$B = \begin{bmatrix} \frac{-2v}{I}(x_1 \sin x_3 - x_2 \cos x_3) & 0 & 0 \\ 0 & \frac{1}{m} & 0 \\ 0 & 0 & 1 \end{bmatrix}$$

One can verify that $\det(B) = -2v_c/m \cdot I\ (x_1 \sin x_3 - x_2 \cos x_3)$. An important disadvantage is that this determinant can be zero if $x_1 \sin x_3 - x_2 \cos x_3 = 0$, i.e. when the UAV is either flying radially inward or radially outward. Yet, if the heading of the UAV does not fall within this region, the UAV can converge to a loitering motion around the origin of the coordinate system. On the other hand, the advantage of this scheme is that, depending on the value of R_1, i.e. the first element of the reference signal, one can come up with scenarios in which the UAV converges to a loitering radius either smaller or greater than the initial one. Also, a second advantage is that one can explicitly control L, as R_3. Therefore, for $R_3 < 2\sqrt{R_1}$, $R_3 = 22\sqrt{R_1}$ and $R_3 > 2\sqrt{R_1}$, one can define scenarios in which the UAV loiters around the origin while the camera footprint sweeps a circle with the radius smaller than, equal to or greater than $\sqrt{R_1}$. This is schematically shown in Figs. 3a to 3c.

Regarding the stability of the system, if we define the error as:

$$\begin{bmatrix} E_1 \\ E_2 \\ E_3 \end{bmatrix} = \left(\begin{bmatrix} \Gamma_1 \\ \Gamma_2 \\ \Gamma_3 \end{bmatrix} - \begin{bmatrix} R_1 \\ R_2 \\ R_3 \end{bmatrix} \right) + \left(\begin{bmatrix} \dot{\Gamma}_1 \\ \dot{\Gamma}_2 \\ \dot{\Gamma}_3 \end{bmatrix} - \begin{bmatrix} \dot{R}_1 \\ \dot{R}_2 \\ \dot{R}_3 \end{bmatrix} \right) \quad (17)$$

for a stationary reference signal, one will have $\dot{R} = \ddot{R} = 0$. Differentiating Eq. 17, one has:

$$\begin{bmatrix} \dot{E}_1 \\ \dot{E}_2 \\ \dot{E}_3 \end{bmatrix} = \begin{bmatrix} \dot{\Gamma}_1 \\ \dot{\Gamma}_2 \\ \dot{\Gamma}_3 \end{bmatrix} + \begin{bmatrix} \ddot{\Gamma}_1 \\ \ddot{\Gamma}_2 \\ \ddot{\Gamma}_3 \end{bmatrix} \tag{18}$$

Now, if we define $u = B^{-1}(-A - \dot{\Gamma} + v)$ and $v = KE$, it is easy to see that:

$$\dot{E} = KE \tag{19}$$

where: K is a matrix with eigenvalues which have negative real parts.

In order to study the error dynamics given by Eq. 19, one may consider the following Lyapunov function:

$$V = \frac{1}{2}E^T E \tag{20}$$

The time derivative of the above positive definite V is given by:

$$\dot{V} = \dot{E}^T E = E^T K^T E < 0 \tag{21}$$

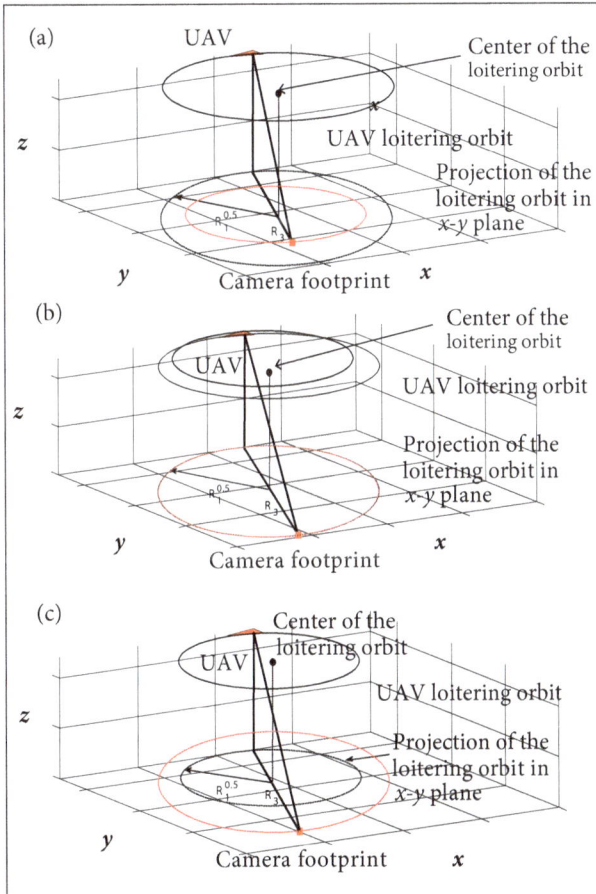

Figure 3. Three different scenarios in scheme #3.

Therefore the error dynamics is asymptotically stable. Regarding the internal dynamics with a vector of relative degree of [3 2 2], one needs to propose one more transformation to complete the diffeomorphism. The internal dynamics is given by $\dot{\eta}_8$ in Eq. 22:

$$\eta_8 = x_4 \rightarrow \dot{\eta}_8 = \frac{1}{l}u_1 \tag{22}$$

For a UAV loitering around a given objective position with constant (finite) forward velocity, x_4 will be bounded and cannot go to infinity. Therefore, the internal dynamics will not cause undesirable effects in our approach. Simulation results regarding this scheme will be given in "Simulations" section.

UAV WITH SIDE-LOOKING CAMERA AND ONE VIRTUAL FORWARD-LOOKING CAMERA, SCHEME #4

In this subsection, we modify the previous scheme in the sense that the UAV can loiter around the origin with a desirable radius while avoiding the singularity problem of scheme #3. Convergence to a loitering radius (i) smaller or (ii) greater than the initial radius is studied separately.

Convergence to a Loitering Radius Smaller than the Initial One

In this scenario, it is initially assumed that the UAV is equipped with a virtual forward-looking camera. From geometry, one can find two tangent lines (and their corresponding tangency points) between the initial position of the UAV and the circle with the reference radius. In the first phase, the UAV can choose one of the tangency points as its virtual objective position and fly over it, according to scheme #1, discussed earlier. Assuming that the UAV flies over the tangent line, its heading will be perpendicular to the radius of the objective circle as it reaches the virtual objective position. As the UAV reaches the tangent point, it switches to scheme #3. The advantage here is that, in the second phase, it is ensured that the heading of the UAV is far from inward-outward direction and therefore scheme #3 can bring the UAV to loiter around the objective position, with desirable radius. This is schematically shown in Fig. 4.

The details of feedback linearization, control laws and stability analysis of scheme #1 and scheme #3 were discussed in the previous subsections and are not repeated here. Simulation results of this scheme will be given in "Simulations" section.

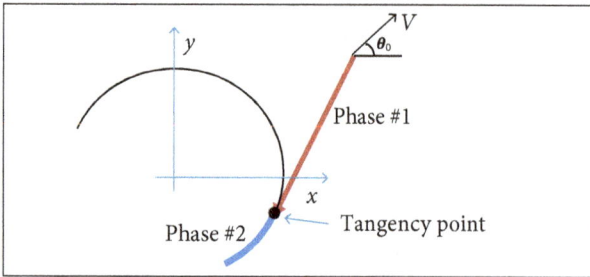

Figure 4. Scheme #3, convergence to a loitering radius smaller than the initial one.

Convergence to a Loitering Radius Greater than the Initial One

Similar to the previous subsection, we assume that the UAV is equipped with a virtual forward-looking camera and a side-looking camera. The scenario proposed here consists of three phases, as shown in Fig. 5.

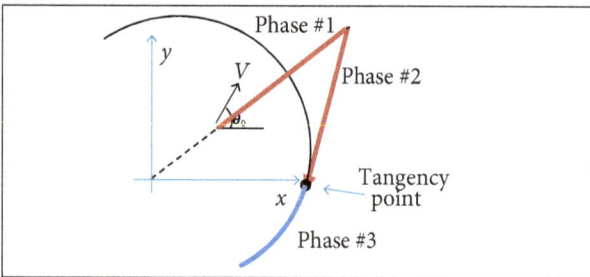

Figure 5. Scheme #3, convergence to a loitering radius greater than the initial one.

In the first phase, based on scheme #1, the UAV flies to a virtual objective position which is on the extension of the line connecting the origin to the initial position of the UAV. Let's denote the UAV distance from the origin by d and the desired loitering radius by R^*. At the end of the first phase, when d is relatively greater than R^*, the UAV finds the tangent lines from its current position to the circle with the radius R^*. With d relatively greater than R^*, one can assume that, in the second phase, based on scheme #1, the UAV flies on the tangent line to reach the tangency point (second virtual objective position). Once at this point, the UAV has reached the reference radius and switches to scheme #3. In the third phase, based on scheme #3, the UAV loiters around the origin with $R_1 = R^*$. The details of feedback linearization, control laws and stability analysis of scheme #1 and scheme #3 were discussed in the previous subsections and are not repeated here. Simulation results of this scheme will be given in "Simulations" section.

SIMULATIONS

In this section, some case studies are developed to verify the feasibility of the proposed schemes. A light fixed-wing UAV is considered, with its characteristics given in Table 1. In the simulations, where applicable, the initial condition of the UAV is assumed to be $[-1,800 \text{ m } 2,500 \text{ m } 240 \text{ deg } 1 \text{ deg/s } 300 \text{ m } 10 \text{ m/s } 10 \text{ deg } 1 \text{ deg/s}]^T$. For the first scheme, the reference signal is assumed to be $[100 \text{ m } -20 \text{ m } 500 \text{ m}]^T$. For the described case study, simulations were carried out and the results are shown in Fig. 6.

As it can be readily seen from Fig. 6, after the initial transition, the UAV has aligned its motion such that it flies almost over the objective position on a straight line. Once on this line, the objective position is monitored by merely controlling the angle of the camera (see Fig. 7).

As it was expected, in this scheme, the camera angle will approach zero as the UAV flies toward the objective position. As the UAV flies away from the objective position, the camera angle will approach π, as $t \to \infty$.

Table 1. Characteristics of the light fixed-wing UAV.

Mass (kg)	Moment of inertia — z-axis (kg·m²)	Moment of inertia of the camera (kg·m²)	Forward velocity (m/s)
1	0.01	0.001	10

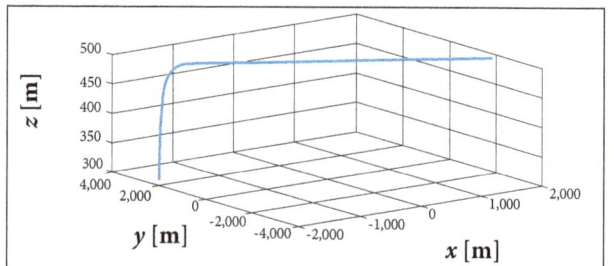

Figure 6. UAV trajectory obtained from scheme #1.

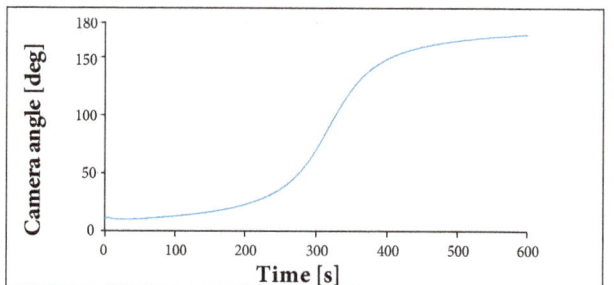

Figure 7. Forward-looking camera angle obtained from scheme #1.

For the second scheme, the reference signal is assumed to be $[-50 \text{ m} -50 \text{ m} \ 500 \text{ m}]^T$. For the described case study, simulations were carried out and the results are shown in Fig. 8.

As it can be seen from Fig. 8, the UAV has successfully converged to a loitering motion around the objective position. Also, as expected, the camera angle converges to a fixed value in the loitering motion (see Fig. 9).

The initial conditions of the third scheme are assumed to be identical to those of the first scheme. Here, the reference signal is assumed to be $[R_0 \ 500 \text{ m} \ 30 \text{ deg}]^T$, where R_0 is the square distance of the UAV from the origin, at the initial time. Similar to the previous scenarios, simulations were carried out and the results are shown in Fig. 10.

Also, in this scheme, it is possible for the UAV to converge to a loitering circle with radius smaller/greater than the initial one. For the loitering radius of 1,500 and 4,500 m, simulations were carried out and the results are shown in Figs. 11 and 12, respectively. It can be seen from Figs. 11 and 12 that the UAV has successfully converged to a loitering motion around the origin in both scenarios. Yet, it must be reminded that scheme #3 can fail if the UAV flies in the radial direction. Thus, it is recommended that one employs scheme #4 if a loitering motion is desirable.

It is important to note that scheme #4 includes scenarios where the loitering radius can be smaller/greater than the initial distance of the UAV from the origin.

To verify the feasibility of scheme #4, a case study is developed in which the UAV is at the same initial condition as before. In the first example, let's assume that the UAV is desired to loiter around the origin with a radius of 1,500 m, a value smaller than its initial distance to the origin. For this case study, simulation results are shown in Fig. 13.

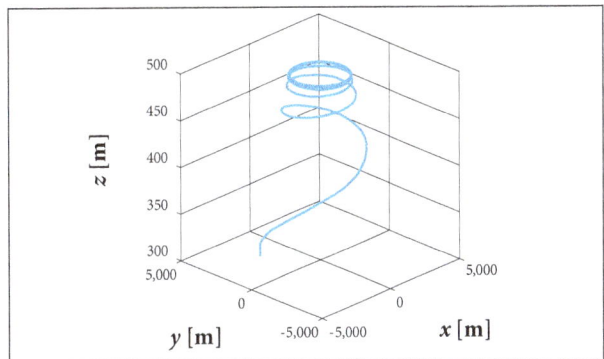

Figure 10. UAV trajectory obtained from scheme #3 — first example.

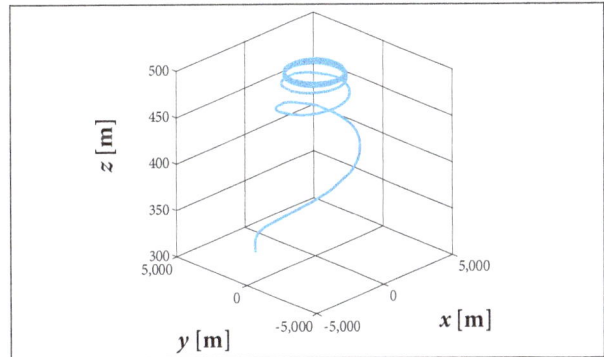

Figure 11. UAV trajectory obtained from scheme #3 — second example.

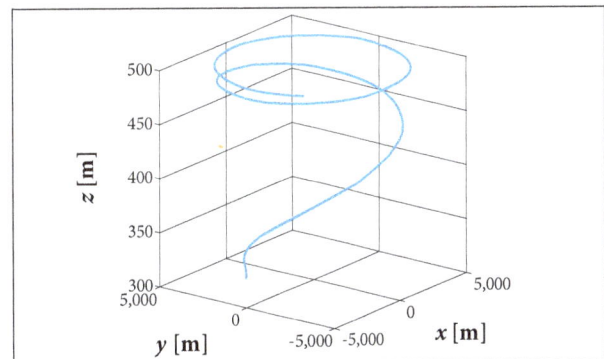

Figure 8. UAV trajectory obtained from scheme #2.

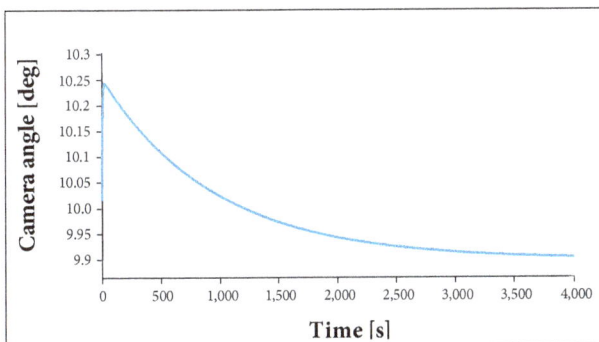

Figure 9. Side-looking camera angle obtained from scheme #2.

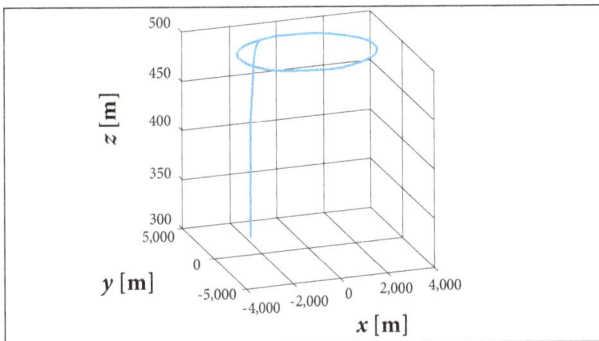

Figure 12. UAV trajectory obtained from scheme #3 — third example.

In Fig. 13, the first and the second phases of the path are shown in red and blue, respectively (see Fig. 4). As it can be seen from the figure, the UAV has successfully converged to the desired reference signal. In the second case study, assume that the UAV is desired to loiter around the origin with a radius of 4,000 m, a value greater than its initial distance to the origin. For this case study, simulation results are shown in Fig. 14, where the first and the second phases of the path are shown in red and the last phase is shown in blue (see Fig. 5). As it can be seen from Fig. 14, the UAV has successfully converged to a loitering motion around the origin with the desired loitering radius.

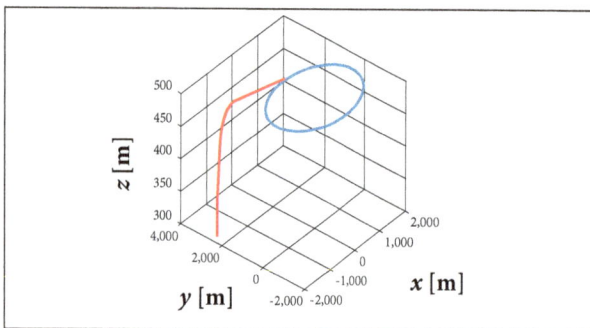

Figure 13. UAV trajectory obtained from scheme #4 — first case study.

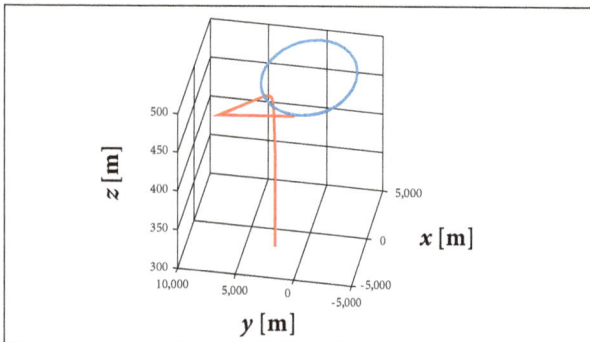

Figure 14. UAV trajectory obtained from scheme #4 — second case study.

CONCLUSION

In this paper, several feedback linearization schemes were studied to make a fixed-wing UAV, with constant forward velocity, (i) fly over or (ii) loiter around a stationary objective position. Throughout the paper, advantages and disadvantages of each scheme were discussed. A main drawback of the proposed schemes is that they do not take account of the maximum angular velocity constraint of fixed-wing UAVs. In scheme #1, the proposed method failed if the UAV was exactly above the objective position. Scheme #2 was disadvantageous in the sense that the loitering radius cannot be explicitly controlled. This was improved in scheme #3. Yet, the method in scheme #3 failed when the UAV had to fly radially inward or radially outward. However, if the heading of the UAV did not fall within this region, the UAV converged to a loitering motion around the origin. In scheme #4, the UAV had to be far enough from the objective circle. In this manner, one could assume that the UAV reaches the tangency point with its heading far from inward-outward direction, and therefore scheme #3 could bring the UAV to loiter around the objective position, with desirable radius. It is important to note that, in scheme #3 and scheme #4, the switching from a control strategy to another was given as an equality-type condition. Therefore, for real-world implementation, thresholds must be defined, and the equalities must be replaced by appropriate inequalities.

ACKNOWLEDGEMENTS

The authors gratefully acknowledge the financial support of Conselho Nacional de Desenvolvimento Científico e Tecnológico (CNPq), Financiadora de Estudos e Projetos (FINEP), Fundação de Amparo à Pesquisa do Estado de Minas Gerais (FAPEMIG) and Coordenação de Aperfeiçoamento de Pessoal de Nível Superior (CAPES), Brazil.

REFERENCES

Almurib HAF, Nathan PT, Kumar TN (2011) Control and path planning of quadrotor aerial vehicles for search and rescue. Proceedings of the IEEE SICE Annual Conference; Tokyo, Japan.

Beard RW, McLain TW, Nelson DB, Kingston D, Johanson D (2006) Decentralized cooperative aerial surveillance using fixed-wing miniature UAVs. Proc IEEE 94(7):1306-1324. doi: 10.1109/JPROC.2006.876930

Bonyan Khamseh H, Pimenta LCA, Tôrres LAB (2014) Decentralized coordination of constrained fixed-wing unmanned aerial vehicles: circular orbits. Proceedings of the IFAC World Congress; Cape Town, South Africa.

Casbeer DW, Kingston DB, Beard RW, McLain TW, Li SM, Mehra R (2006) Cooperative forest fire surveillance using a team of small unmanned air vehicles. Int J Syst Sci 37(6)351-360. doi: 10.1080/00207720500438480

Costa FG, Ueyama J, Braun T, Pessin G, Osório FS, Vargas PA (2012) The use of unmanned aerial vehicles and wireless sensor network in agricultural applications. Proceedings of the IEEE International Geoscience and Remote Sensing Symposium; Munich, Germany.

Fan W, Zhiyong G (2009) An approach to formation maneuvers of multiple nonholonomic agents using passivity techniques. Proceedings of the Chinese Control and Decision Conference; Guilin, China.

Frew E, Lawrence D, Morris S (2008) Coordinated standoff tracking of moving targets using Lyapunov guidance vector fields. J Guid Contr Dynam 31(2):290-306. doi: 10.2514/1.30507

Gonçalves MM, Pimenta LCA, Pereira GAS (2011) Coverage of curves in 3D with swarms of nonholonomic aerial robots. Proceedings of the IFAC World Congress; Milano, Italy.

Gonçalves VM, Pimenta LCA, Maia CA, Dutra BCO, Pereira GAS (2010) Vector fields for robot navigation along time-varying curves in n-dimensions. IEEE Trans Robot26(4):647-659. doi: 10.1109/TRO.2010.2053077

Hafez AT, Marasco AJ, Givigi AN, Beaulieu A, Rabbath CA (2013)

Encirclement of multiple targets using model predictive control. Proceedings of the American Control Conference; Washington, USA.

Hsieh MA, Kumar V, Chaimowicz L (2008) Decentralized controllers for shape generation with robotic systems. Robotica 26(5):691-701. doi: 10.1017/S0263574708004323

Hsieh MA, Loizou S, Kumar RV (2007) Stabilization of multiple robots on stable orbits via local sensing. Proceedings of the IEEE International Conference on Robotics and Automation; Rome, Italy.

Kanchanavally S, Ordonez R, Schumacher CJ (2006) Path planning in three dimensional environment using feedback linearization. Proceedings of the American Control Conference; Minneapolis, USA.

Li H, Wang B, Liu L, Tian G, Zheng T, Zhang J (2013) The design and application of SmartCopter: an unmanned helicopter based robot for transmission line inspection. Proceedings of the Chinese Automation Congress; Changsha, China.

Marasco AJ, Givigi SN, Rabbath CA (2012) Model predictive control for the dynamic encirclement of a target. Proceedings of the American Control Conference; Montreal, Canada.

15

Qualification of Magnesium/Teflon/Viton Pyrotechnic Composition Used in Rocket Motors Ignition System

Luciana de Barros[1,2], Afonso Paulo Monteiro Pinheiro[1], Josemar da Encarnação Câmara[1], Koshun Iha[2]

ABSTRACT: The application of fluoropolymers in high-energy-release pyrotechnic compositions is common in the space and defense areas. Pyrotechnic compositions of magnesium/Teflon/Viton are widely used in military flares and pyrogen igniters for igniting the solid propellant of a rocket motor. Pyrotechnic components are considered high-risk products as they may cause catastrophic accidents if initiated or ignited inadvertently. To reduce the hazards involved in the handling, storage and transportation of these devices, the magnesium/Teflon/Viton composition was subjected to various sensitivity tests, DSC and had its stability and compatibility tested with other materials. This composition obtained satisfactory results in all the tests, which qualifies it as safe for production, handling, use, storage and transportation.

KEYWORDS: MTV, Pyrotechnic, Rocket motor, Safety.

INTRODUCTION

Pyrotechnic compositions are mechanical mixtures of different components that combust to produce special effects such as heat, light, smoke or sound (Agrawal 2010). The compositions including magnesium are widely known by the pyrotechnics community for their efficiency and performance. Teflon (Fig. 1) is a polymer with highly-polarized flourine atoms (fluoropolymer) and, when mixed with magnesium, a highly energetic material is formed with application in flares and propulsion systems of rocket motors (Göçmez *et al.* 1999).

The application of the magnesium and Teflon compositions in propulsive systems is due to the quantity of heat produced by the oxidation of the magnesium by the gaseous fluorine released by the Teflon (Yong and Smit 1991). The combustion reaction for this composition can be simplified by the equation (Peretz 1982):

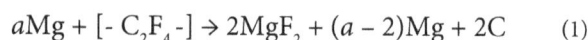

$$a\text{Mg} + [\text{- C}_2\text{F}_4\text{-}] \rightarrow 2\text{MgF}_2 + (a-2)\text{Mg} + 2\text{C} \quad (1)$$

To facilitate the processing, the magnesium and Teflon mixture is coated with another fluoropolymer, Viton (Fig. 2), which acts as a binder increasing the homogeneity of the mixture, the conception of the final product (pellets or granules), and protects the magnesium against oxidation from humidity during the storage period.

The magnesium/Teflon/Viton pyrotechnic composition (MTV) has good technical features for application in ignition systems because during combustion hot solid and liquid particles and condensable gaseous products are released, enhancing the energy transfer to the solid propellant surface (Göçmez *et al.* 1999).

1.Departamento de Ciência e Tecnologia Aeroespacial – Instituto de Aeronáutica e Espaço – Divisão de Propulsão Espacial – São José dos Campos/SP – Brazil.
2.Departamento de Ciência e Tecnologia Aeroespacial – Instituto Tecnológico de Aeronáutica – Departamento de Química – São José dos Campos/SP – Brazil.

Author for correspondence: Luciana de Barros | Departamento de Ciência e Tecnologia Aeroespacial – Instituto de Aeronáutica e Espaço – Divisão de Propulsão Espacial | Praça Marechal Eduardo Gomes, 50 – Vila das Acácias | CEP: 12.228-904 | São José dos Campos/SP – Brazil | Email: lucianabarros28@yahoo.com.br

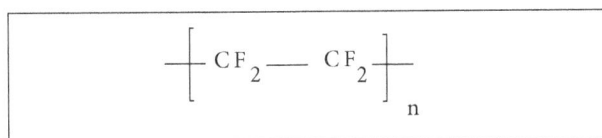

Figure 1. Teflon (polytetrafluoroethylene or PTFE).

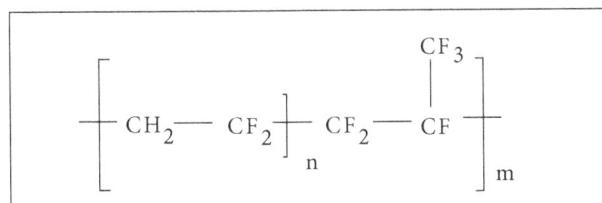

Figure 2. Viton (vinilydene fluoride/hexafluoropropylene copolymer).

However, pyrotechnic devices and rocket propellants have a potential for catastrophic accidents in case of an inadvertent initiation or ignition. In 1985, the US Army experienced an accident with the Pershing II missile, when a rocket motor made of Kevlar auto-ignited, killing three people. The auto-ignition of the motor was attributed to electrostatic discharge throughout the motor in cold and dry weather (Mellor and Boggs 1987).

Events like this made the global aerospace industry to evolve with respect to safety. In 2007, the *Agência Espacial Brasileira* (AEB) prepared the General Regulation of Space Safety, containing requirements to be applied in space activities that characterize Brazil as a launcher state to protect people, propriety and environment against potentially dangerous systems. The Regulation states that, to prevent potentially dangerous systems from activating unexpectedly by hardware failure or human error, barriers should be introduced. These barriers are referred as interception devices (AEB 2007).

The US military standard MIL-STD-1901A of 1992 states that only pyrotechnic compositions qualified by another standard, MIL-STD-1751A, now replaced by the North Atlantic Treaty Organization (NATO) standard AOP-7 (Edition 2) — Manual of Data Requirements and Tests for the Qualification of Explosive Materials for Military Use, may be used in a rocket motor ignition system without interruption (MIL-STD-1901A), that is, these compositions can be used after the interception device.

The AOP-7 standard provides mandatory tests that provide data to evaluate the performance and the safety of energetic materials, including pyrotechnic compositions. The mandatory tests to qualify these compositions are sensitivity to impact, to friction and to electrostatic discharge, stability, thermal analysis

and compatibility. The explosive hexogene (RDX) is used as a comparative for the sensitivity tests.

The MTV composition produced by the *Instituto de Aeronáutica e Espaço* (IAE) is used on the space vehicles developed by the Brazilian Space Program. The qualification of this pyrotechnic composition will increase personnel and patrimonial safety and reduce the costs of launch operations by providing physical-chemical data of this specific formulation.

Therefore, this study has the objective of certifying that the MTV pyrotechnic composition is safe for fabrication, handling, use, storage, and transportation, thereby meeting the safety requirements for launching space vehicles.

MATERIALS AND METHODS
MATERIALS

For the MTV pyrotechnic composition, Merch KGaA magnesium powder was used with purity higher than 97% and grain size less than 0.21 mm; Teflon 850A® (PTFE) of DuPont™ with grain size less than 0.42 mm; Viton B® (68% fluorine) of DuPont™ in pellets; and Acetone P.A. of Synth with minimum purity of 99.5%.

The RDX used was produced by SNPE Poudres et Explosifs, Groupe SNPE, France, class 5, with 97% of grain size less than 0.044 mm.

MTV COMPOSITION MANUFACTURE

The preparation of the MTV composition (58% magnesium and 38% Teflon) started with the complete dissolution of 4% Viton in acetone inside a proper recipient. A low-molecular-weight ketone swelled and vulcanized the Viton, creating a lacquer. The magnesium powder and the Teflon were added to the Viton lacquer and manually homogenized until the acetone completely evaporated. The mixture was passed through a sieve to classify the grain size less than 0.71 mm and was kiln-dried to remove traces of acetone.

IMPACT SENSITIVITY

The German Bundesanstalt fur Materialprufung (BAM) fall hammer is used to submit solid and liquid substances to an impact force that may cause detonation of the energetic material. The test set consists of two steel cylinders, one over another, and both cylinders are held together by a steel ring, as shown in Fig. 3 (AOP-7). A 20-mm^3 sample is placed between the two steel cylinders, and the test set is positioned at the base of the fall hammer. The weight is fixed at a chosen height and, guided by guide rails, the weight is released. The aim of the test is to estimate the energy that corresponds to a 50% probability of reaction. The reaction can be defined as flash, sparks, noise or explosion. The Bruceton analysis is used to evaluate the results in accordance with the French standard NF T70-500.

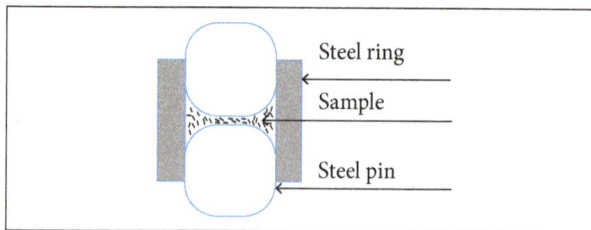

Figure 3. BAM impact test set.

FRICTION SENSITIVITY

The BAM friction tester is used to measure the response of energetic materials to a friction stimulus generated by two porous porcelain surfaces. A 10-mm^3 sample is placed in a porcelain plate attached to the sample table. The porcelain pin placed on the lever arm descends to come in contact with the sample, as shown in Fig. 4. Weights are set at different distances on the lever arm to adjust the force applied on the sample. An electric motor moves the table back and forth causing friction of the sample with the porcelain pin. The aim of the test is to estimate the energy in Newtons that corresponds to a 50% probability of reaction. A positive result is evidenced by gas liberation, sparks or noise. The Bruceton analysis is used to evaluate the results in accordance with the French standard NF T70-503.

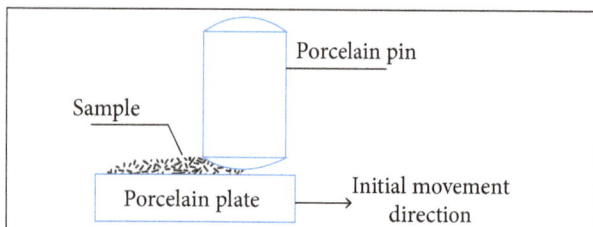

Figure 4. BAM friction test set.

ELECTROSTATIC DISCHARGE SENSITIVITY

This test determines the energy limit needed to ignite an explosive by electrostatic stimulus of various intensities. The test is run in an Energetic Materials Test Unit (Fig. 5), integrated to a Test Unit of 5 kV (Fig. 6) of the Electrostatic Discharge Sensitivity (ESD) Tester ESD-100 (Equatorial Sistemas). The initial energy is 0.25 J, charged by a capacitor of 0.02 μF connected to a discharge

circuit charged with 5 kV. The gap between the upper electrode (needle) and the metal sample holder is fixed at 0.018 mm. A sample of 30 mg is put inside a Teflon washer placed on the top of the grounded sample holder, and a mylar sheet (insulating material) is placed on the top of the washer to confine the powder. The needle is charged and manually moved down to the adjusted gap, piercing the mylar sheet and discharging into the material. A positive reaction is defined as a flash, spark, burn or noise. No reaction in 20 out of 20 trials is a pass (AOP-7).

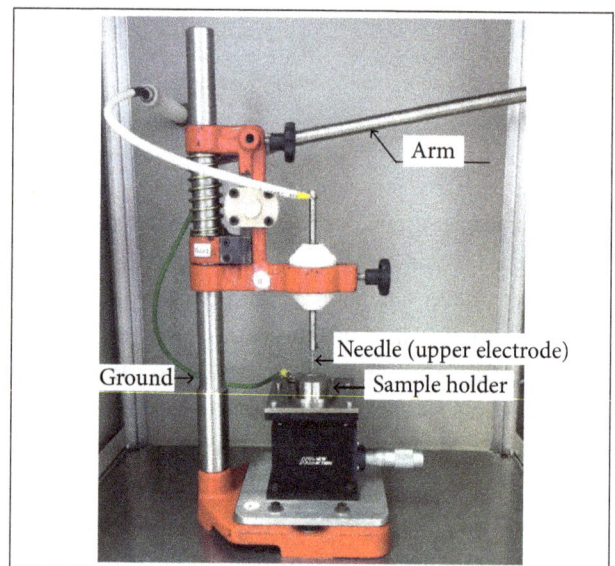

Figure 5. Energetic material test unit.

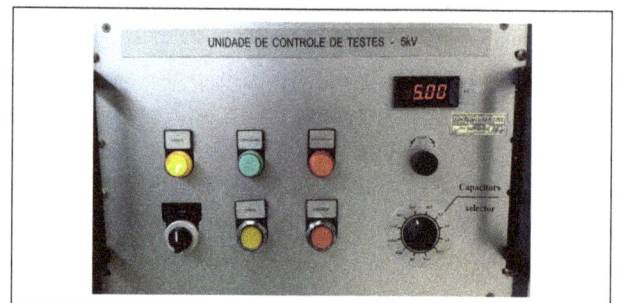

Figure 6. Test unit of 5 kV.

THERMAL ANALYSIS

The MTV composition was analyzed by the differential scanning calorimeter DSC-7 made by PerkinElmer, at a heating rate of 2 °C/min. Nitrogen gas was purged into the furnace at a rate of 50 mL/min. The temperature range was from 30 °C up to 500 °C, the maximum temperature allowed by this equipment. About 1 mg of sample was placed at a platinum crucible with perforated lid.

VACUUM STABILITY (CONSTANT TEMPERATURE)

The stability test was performed using STABIL Vacuum Stability Tester of OZM Research (Fig. 7), which consists in a heating block and glass tubes with temperature and pressure sensors. A 5-g mass was placed inside the glass tube and closed with a head that contains pressure and temperature transducers. A vacuum was created in the tube until the internal pressure reached values lower than 1 atm. The head is rotated to maintain the internal vacuum, and the tube is placed inside the heating block. The samples were heated at 100 °C for 40 h (AOP-7).

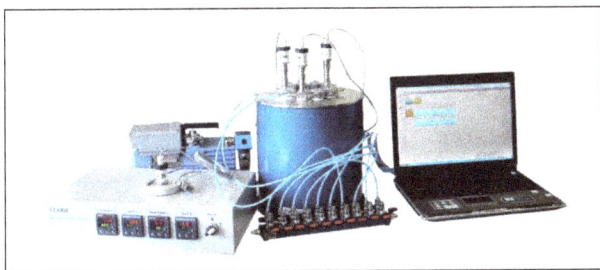

Figure 7. STABIL Vacuum Stability Tester.

COMPATIBILITY

The compatibility of the MTV composition was tested with the alkyl lacquer DOPE LP 453, a resin composed of nitrocellulose, aromatic solvents, alcohols, acetates and additives. This lacquer is used in the assembly of pyrotechnic devices that use the MTV composition as active material. During the assembly and sealing of the pyrotechnic device, the contact between the lacquer and the MTV composition could occur, so it was important to assess the compatibility between the two materials.

The test was conducted with a differential scanning calorimeter DSC-7 made by PerkinElmer, at a heating rate of 2 °C/min. Nitrogen gas was purged into the furnace at a rate of 50 mL/min (STANAG 4147). The temperature range was from 30 °C up to 500 °C, the maximum temperature allowed by this equipment. The samples were analyzed individually and in a mixture with a proportion of 1:1. About 2 mg was placed in a platinum crucible with a perforated lid.

RESULTS AND DISCUSSION

For the MTV pyrotechnic composition to be qualified and considered safe, all the tests must have satisfactory results. For sensitivity tests, the AOP-7 standard requires that RDX be tested contemporaneously with the composition, and the

MTV sensitivity levels should not be more sensitive than RDX, which is a military explosive with high stability and low levels of sensitivity.

IMPACT SENSITIVITY

The results are presented in Table 1. Using the "up and down" Bruceton method with a minimum of 30 tests, we can analyze the data statistically and find, with a 95% confidence level, the energy that causes a 50% probability of a positive result, that is, initiation or detonation, as well as the lower and upper limits. The RDX resulted in a range of initiation energy from 2.5 to 6.3 J with average of 5.0 J. Meyer *et al.* (2007) reported in their book "Explosives" an energy of 7.5 J for RDX, but they did not specify the grain size of the explosive, which has an influence on the sensitivity of the material. The results show that the MTV composition is less sensitive to impact than RDX, since the energy required to initiate the sample is 50 J. Koch (2012) states that large spherical grains, with 2 or 3 mm diameter, may not have a positive reaction at a 50-J energy. The grain size of the sample was less than 0.71 mm, which resulted in nine positive results in 30 trials, confirming the influence of the grain size on the sensitivity of the material.

In this test, the energetic material is subjected to a plastic work, which causes heating and initiation. The strain rate of the material in the edges is greater than in the center, causing an initiation reaction at the edges first (Mellor and Boggs 1987). This was seen when the MTV composition was tested. Positive reactions vary from small black dots at the edges to expulsion of the steel pin from the ring along with noise and sparks.

Table 1. Impact sensitivity results.

Product	Impact sensitivity results (J)
RDX	5.0 (2.5 to 6.3)
MTV	50

FRICTION SENSITIVITY

The results are presented in Table 2. Using once more the Bruceton method with 30 tests and analyzing the data statistically, the RDX tests resulted in a range of initiation energy from 105 to 183 N, with an average of 159 N. Meyer *et al.* (2007) reported an initiation energy level of 120 N but again they do not specify the grain size of the RDX. These results show that the MTV composition is less sensitive to friction than RDX, and the energy necessary to initiate the composition is greater than 353 N. Materials that contain PTFE are not susceptible to ignition by friction due to the low friction coefficient of the PTFE (m = 0.04). There is no reaction up to 360-N friction forces with the BAM apparatus (Koch 2012).

Friction is the hardest stimulus to eliminate, because it is present at handling, at manufacture and at the storage of the energetic material. Friction is found in sliding, rotation and scrape movements (Mellor and Boggs 1987), which often happens during the preparation of the composition and the assembly of pyrotechnic devices.

Table 2. Friction sensitivity results.

Product	Friction sensitivity results (N)
RDX	159 (105 to 183)
MTV	> 353

ESD SENSITIVITY

The results are presented in Table 3. As we can see in the tests results, the MTV composition is about ten times less sensitive to ESD than RDX. Low ESD sensitivity materials are in a range from 12.5 mJ to 12.5 J (TM 9-1300-214), therefore, both samples can be considered low ESD sensitive. The threshold energy is found with no positive results at 20 consecutive trials.

The Teflon on the composition justifies the low ESD sensitivity of the MTV composition. The Teflon has high dielectric strength, in other words, it is considered an insulating material. The hazard of initiation by ESD arises when the energetic material becomes charged to a potential where the breakdown of the material occurs, or when changes at the grounding allow the discharge of the existing charge. The discharge process reduces the material resistance, which increases the electric current passing through. This process can create an arc and can also establish discharge paths followed by the catastrophic initiation of the material (Mellor and Boggs 1987). Therefore, the use of insulating material in pyrotechnic compositions decreases the hazard of initiation due to ESD.

Table 3. ESD sensitivity results.

Product	ESD threshold energy [mJ]
RDX	22
MTV	250

THERMAL ANALYSIS

This analysis is conducted to characterize the thermal decomposition behavior of the MTV composition. This

characterization will be used afterwards to analyze the compatibility of the composition with other materials that it might come in contact with.

Figure 8 presents the DSC curve for the MTV composition. A small endothermic peak is identified at 340 °C and indicates the fusion point of the polymeric material in the mixture, because the Teflon fusion point is about 335 °C. The complete decomposition of the MTV composition at this heating rate occurs at temperatures above 500 °C, since there is no register of any exothermic event in the present curve.

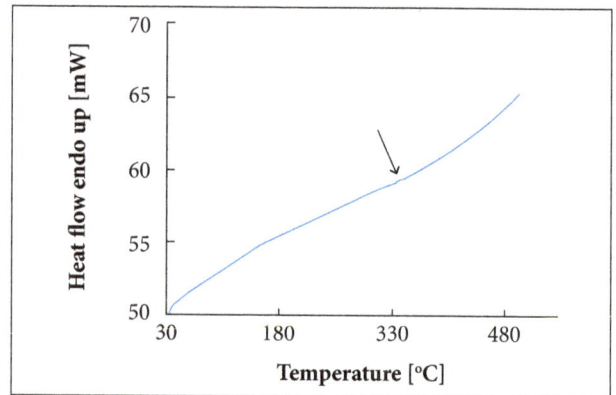

Figure 8. DSC curve of MTV composition — heating rate of 2 °C/min under nitrogen atmosphere from 30 to 500 °C.

VACUUM STABILITY (CONSTANT TEMPERATURE)

The gas volume released by the MTV composition after 40 h at 100 °C was of 0.61 mL/g. The AOP-7 standard requires a maximum gas release volume of 2 mL/g. The gas volume released in this test is a valid measure of the material stability (MIL-STD-1751A).

Usually, energetic materials remain in warehouses for a long period of time, and the environment conditions, such as temperature and humidity, depend exclusively on the storage location. The energetic material should maintain its performance, physical-chemical properties and safety during the entire storage time (Agrawal 2010).

The exposure of the MTV composition to the humidity transforms the magnesium in the composition into $Mg(OH)_2$ and MgO, as shown in Eqs. 2 and 3. These two products reduce the composition performance and increase the risk of accident due to the formation of gaseous hydrogen (Koc et al. 2009).

$$Mg(s) + 2H_2O \rightarrow Mg(OH)_2(s) + H_2(g) \qquad (2)$$

$$2Mg(s) + O_2(g) \rightarrow 2MgO(s) \qquad (3)$$

Besides helping in the conception of the final product, the Viton has the function of coating the magnesium grains, which protects the magnesium from oxidation by the humidity (Yong and Smit 1991).

COMPATIBILITY

When an energetic material has direct contact with a polymer or other type of material, the following conditions can occur: (a) one or more properties of the energetic material can be affected; (b) one or more properties of the contact material can be affected; and (c) none of the properties of the energetic material or the contact material can be affected.

This test provides evidence that certain materials can be used in a pyrotechnic device without any loss in safety or reliability (STANAG 4147). The alkyl lacquer DOPE was tested in a mixture with the MTV composition to verify if these two products are compatible, since they are used in the assembly of pyrotechnic devices.

Figure 9 presents the DSC curve for the MTV composition for the alkyl lacquer DOPE and a mixture of 1:1 proportion of the two materials. Table 4 presents the temperature peaks of the samples.

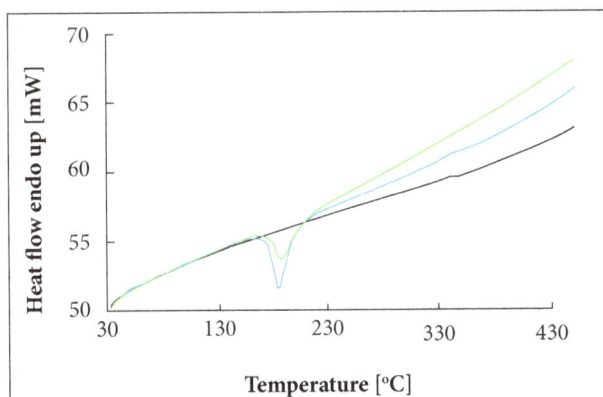

Figure 9. DSC curve of MTV, DOPE and MTV/DOPE 1:1 – heating rate of 2 °C/min at nitrogen atmosphere from 30 to 500 °C.

An exothermic peak with temperature of approximately 188 °C is identified in the DOPE curve, correspondent to the exothermic decomposition of the nitrocellulose contained in the lacquer (Andrade *et al.* 2007). It was verified before that the MTV composition has a small endothermic peak at 340 °C, relative to Teflon fusion point. Table 4 shows a small advance of 3 °C in the decomposition temperature of DOPE and the mixture MTV/DOPE. To evaluate the mixture behavior after

Table 4. Peak temperature of MTV, DOPE and MTV/DOPE.

Product	Peak temperature (°C)	Peak temperature (°C)
MTV	-	340.6 ± 0.3
DOPE	187.6 ± 0.7	-
MTV/DOPE	184.5 ± 0.4	340.0 ± 1.0

this exothermic peak, a run was performed until 240 °C, and the mixture was analyzed visually, which resulted in a small alteration of the polymeric material in the composition. It is assumed that the energy released by the nitrocellulose decomposition caused a thermal degradation of the polymeric material at the contact area, but it was not sufficient to initiate or ignite the MTV composition, because there are no exothermic reactions depicted in the curve before 500 °C. Therefore, at the analyzed temperature range by differential scanning calorimetry, we can say that the MTV composition and the DOPE alkyl lacquer are compatible (Pinheiro 2010).

CONCLUSION

The MTV pyrotechnic composition was subjected to a series of tests such as sensitivity to impact, friction and ESD, contemporaneously with RDX, and the results were compared. Stability, thermal analysis and compatibility were also performed, and the MTV composition obtained satisfactory results in all of them. The composition is less sensitive than RDX in sensitivity tests, it has good stability proprieties and is compatible with the DOPE lacquer used to assemble and seal pyrotechnic devices used in the ignition system of rocket motors.

Therefore, it is concluded that the MTV pyrotechnic composition developed by IAE is now qualified, being safe for fabrication, handling and application in the ignition systems of solid propellant rocket motors as well as storage and transportation.

ACKNOWLEDGMENT

I thank the Laboratory of Explosives in the IAE's Division of Defense Systems, especially the technician Nanci Miyeko Nakamura for performing the thermal analysis and the vaccum stability tests.

REFERENCES

Agência Espacial Brasileira (2007) Parte 1 – Regulamento técnico geral da segurança espacial. Brasília: Agência Espacial Brasileira.

Agrawal JP (2010) High energy materials: propellants, explosives and pyrotechnics. Weinheim: Wiley-VCH.

Andrade J, Iha K, Rocco JAFF, Suzuki N, Suárez-Iha MEV (2007) Determinação dos parâmetros cinéticos de decomposição térmica para propelentes BS e BD. Eclética Química 32(3):45-50. doi: 10.1590/S0100-46702007000300007

Göçmez A, Gürkan YA, Pekel F, Özkar S (1999) Development of MTV compositions as igniter for htpb/ap based composite propellants. Propellants Explos Pyrotech 24(2):65-69. doi: 10.1002/(SICI)1521-4087(199904)24:2<65::AID-PREP65>3.0.CO;2-8

Koc S, Erogul F, Tinaztepe HT (2009) Accelerated aging study for MTV ignition charges. AIAA-2009-5276. Proceedings of the 45th AIAA/ASME/SAE/ASEE Joint Propulsion Conference & Exhibit; Denver, USA.

Koch EC (2012) Metal-fluorocarbon based energetic materials. Weinheim: Wiley-VCH.

Mellor AM, Boggs TL (1987) Energetic materials hazard initiation: DoD assessment team final report. USA: U.S. Army Research Office.

Meyer R, Köhler J, Homburg A (2007) Explosives. 6th ed. Berlin: Wiley-VCH.

Peretz A (1982) Investigation of pyrotechnic MTV compositions for rocket motor igniters. Reston: American Institute of Aeronautics and Astronautics.

Pinheiro GFM (2010) Relatório interno de ensaio de compatibilidade de MTV com AERODOPE, SNOOP e LOCTITE 242. São José dos Campos: Instituto de Aeronáutica e Espaço.

Yong LV, Smit KJ (1991) A theoretical study of the combustion of magnesium/Teflon/Viton pyrotechnic compositions. Maribyrnong: Materials Research Laboratory.

16

Formation Flight Control of Multi-UAV System with Communication Constraints

Ruibin Xue[1], Gaohua Cai[2]

ABSTRACT: Three dimensional formation control problem of multi-UAV system with communication constraints of non-uniform time delays and jointly-connected topologies is investigated. No explicit leader exists in the formation team, and, therefore, a consensus-based distributed formation control protocol which requires only the local neighbor-to-neighbor information between the UAVs is proposed for the system. The stability analysis of the proposed formation control protocol is also performed. The research suggests that, when the time delay, communication topology, and control protocol satisfy the stability condition, the formation control protocol will guide the multi-UAV system to asymptotically converge to the desired velocity and shape the expected formation team, respectively. Numerical simulations verify the effectiveness of the formation control system.

KEYWORDS: Three dimensional formation control, Jointly-connected topologies, Multi-UAV system, Non-uniform time delays, Consensus protocol.

INTRODUCTION

Recently, with the development of computer control, sensors, communication network etc., many researches on the formation flight control have been performed. This is because various missions can be successfully completed by the formation flight, such as battlefield reconnaissance, multi-target attacking, environment monitoring and earthquake rescue and so on. Multi-UAV coordinated formation control has overwhelming superiority in high efficiency in performing tasks, low cost of fuel, strong robustness and more flexibility compared with single UAV (Ren and Beard 2008; Cao *et al.* 2012). Therefore, multi-UAV formation flight control has become a hot topic in UAV field.

In earlier years, typical approaches for formation control could be roughly categorized as leader-follower, behavioral, virtual leader/virtual structure. Most of the formation flight researches are performed based on the leader-follower approach, where some UAVs are designed as leaders while others are designed as followers (Ren 2007; Giulietti *et al.* 2000). In this approach, the leaders track the predefined trajectory, and the followers track the nearest leaders according to given schemes. It is easy to analyze and implement the leader-follower controller. However, the leader is a single point for the formation, and therefore this approach is not robust with respect to the leader failure.

In recent years, the problem of multi-UAV cooperative formation flight control based on consensus protocol has drawn substantial research effort from many studies (Kuriki and Namerikawa 2013; Menon 1989; Ren 2006; Seo *et al.* 2012). Ren (2007) extended a consensus protocol, which is introduced for systems modelled by second-order dynamics, to tackle multi-UAV formation control problems by appropriately choosing information states on which consensus is reached. Seo (2009)

1.Beijing Institute of Technology – School of Aerospace Engineering – Key Laboratory of Dynamics and Control of Flight Vehicle – Beijing – China. 2.Beijing Aerospace Automatic Control Institute – Beijing – China.

Author for correspondence: Ruibin Xue | Beijing Institute of Technology – School of Aerospace Engineering | Tiyu N Rd, Haidian | Beijing – China | Email: feirenlg@163.com

proposed a consensus-based formation flight control protocol and proved that the multi-UAV system can form and maintain a geometric formation flight with the network topology switching between a directed strongly-connected topology and a topology with a spanning tree. Dong *et al.* (2014) investigated the time-varying formation control problem by applying a consensus-based formation control protocol, and necessary and sufficient conditions are obtained for the stability of the system which contains a spanning tree in the fixed topology. Then a quadrotor formation platform was introduced to validate the theoretical results. However, most of the researches about consensus-based cooperative formation flight control are mainly focused on two systems: one is a fixed communication topology without time delays; the other is a switching communication topology without time delays as well. There are few results available to treat the formation control system with jointly-connected topologies and time delay. But, in reality, the time delay usually exists due to transmission rate and network congestion, and the communication topology of the multi-UAV system will be changed owing to communication jamming, complex terrain, limitation of communication distance etc. Therefore, it is of great significance in both theory and application to investigate cooperative formation flight control by considering time delay and changing topology.

The main contributions of the paper can be summarized as follows. First, to design a new formation flight control protocol considering two key-problems: one is the diverse and asymmetric time delays, and the other is the dynamically changing topologies. The topologies discussed here may not connect all the time but the union of the topologies is connected in each period of time. Second, the analysis of the complex topologies is turned to a simple research of connected component in each period of time according to the stability analysis, and a sufficient condition for the stability is obtained based on Lyapunov theory. The multi-UAV system can shape and maintain the expected formation with desired velocity, when it satisfies the sufficient condition.

MODEL OF THE MULTI-UAV SYSTEM

This paper considers a group system consisting of n autonomous UAVs, and the point-mass model is used to describe the motion of the UAV formation flying. The related variables are defined with respect to the inertial coordinate system and are shown in Fig. 1 (Wang and Xin 2012).

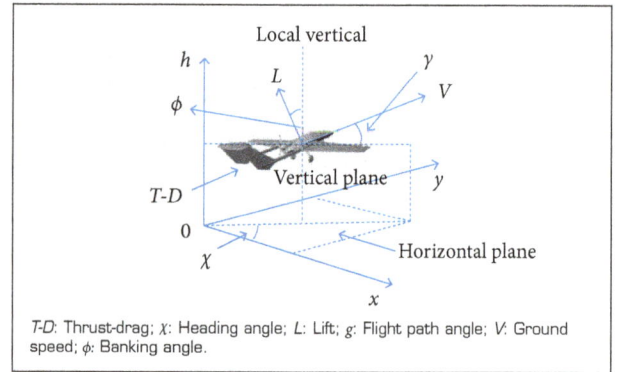

T-D: Thrust-drag; χ: Heading angle; *L*: Lift; *g*: Flight path angle; *V*: Ground speed; ϕ: Banking angle.

Figure 1. UAV model.

The model assumes that the aircraft thrust is directed along the velocity vector and that the aircraft always performs coordinated maneuvers. It is also assumed that the Earth is flat, and the fuel expenditure is negligible, *i.e.* the center of mass is time-invariant (Xu 2009). Under these assumptions, the motion equations of the ith UAV can be described as follows:

$$\begin{bmatrix} \dot{x}_i \\ \dot{y}_i \\ \dot{h}_i \\ \dot{v}_i \\ \dot{\gamma}_i \\ \dot{\chi}_i \end{bmatrix} = \begin{bmatrix} v_i \cos\gamma_i \cos\chi_i \\ v_i \cos\gamma_i \sin\chi_i \\ v_i \sin\gamma_i \\ (T_i - D_i)/m_i - g\sin\gamma_i \\ (g/v_i)(n_i \cos\phi_i - \cos\gamma_i) \\ L_i \sin\phi_i/(m_i v_i \cos\gamma_i) \end{bmatrix} \quad (1)$$

where: $i = 1, 2, …, n$ is the index of multiple UAVs under consideration. For UAV$_i$, x_i is the down-range; y_i is the cross range; h_i is the altitude; v_i is the ground speed; γ_i is the flight path angle; χ_i is the heading angle; T_i is the engine thrust; D_i is the drag; m_i is the mass; g is the acceleration due to gravity; ϕ_i is the banking angle; L_i is the vehicle lift.

The control variables in the UAVs are the g-load $n_i = L_i/gm_i$, controlled by the elevator, the banking angle ϕ_i, controlled by the combination of rudder and ailerons, and the engine thrust T_i, controlled by the throttle. Throughout the formation control process, the control variables will be constrained to remain within their respective limits.

Define $R^{m \times n}$ as a $m \times n$ real matrix set, $\xi_i = [x_i, y_i, h_i]^T \in R^3$, and $u_i = [u_{xi}, u_{yi}, u_{hi}]^T \in R^3$. Differentiating $\dot{v}_i, \dot{\gamma}_i, \dot{h}_i$ with respect to time twice and substituting x_i, y_i, χ_i, one has the transformed dynamic models of the ith UAV as follows:

$$\ddot{\xi}_i = u_i \quad (2)$$

where: ξ_i is the position of UAV$_i$; u_i is a new control variable, and the relationship between ui and the actual control variable U_i is given by the expressions (Xu 2009):

$$\phi_i = \tan^{-1}\left[\frac{u_{yi}\cos\chi_i - u_{xi}\sin\chi_i}{\cos\gamma_i(u_{zi}+g) - \sin\gamma_i(u_{xi}\cos\chi_i + u_{yi}\sin\chi_i)}\right] \quad (3)$$

$$n_i = \frac{\cos\gamma_i(u_{zi}+g) - \sin\gamma_i(u_{xi}\cos\chi_i + u_{yi}\sin\chi_i)}{g\cos\phi_i} \quad (4)$$

$$T_i = \left[\sin\gamma_i(u_{zi}+g) + \cos\gamma_i(u_{xi}\cos\chi_i + u_{yi}\sin\chi_i)\right]m_i + D_i \quad (5)$$

FORMATION CONTROL PROTOCOL DESIGN OF THE MULTI-UAV SYSTEM

The multi-UAV system and its behavior are described in graph theory. It is supposed that the multi-UAV system under consideration consists of n UAVs and $G(\Gamma, E, A)$ is an undirected graph of the multi-UAV system, where $\Gamma = \{s_1, s_2, ..., s_n\}$ is the set of nodes, $\ell = (1, 2, 3, ..., n)$ is the set of the number of nodes, and $E = \{(s_i, s_j) \in \Gamma \times \Gamma, i \neq j\}$ is the set of edges. At each time, each UAV updates its current state based upon the information received from its neighbors. Undirected graphs are used to model communication topologies. Each UAV is regarded as a node. Each edge (s_i, s_j) or (s_j, s_i) corresponds to an available information link between UAV$_i$ and UAV$_j$. A communication topology is formed when the UAVs begin to communicate to each other at any time. In reality, the communication topology usually switches due to link failure brought by communication blocking, external disturbance, hardware failure etc. To describe the variable topologies, a piecewise constant switching function $\sigma(t): [0, \infty \to p = \{1, 2, ..., N\}$($\sigma$ in short) is defined, where N denotes the total number of all possible communication undirected graphs. The communication graph at time t is denoted by G_σ and the corresponding Laplacian, by L_σ. This paper investigates the design of the control protocol of the multi-UAV system under jointly-connected communication graph.

The state-space form of the dynamics of the ith UAV is obtained from Eq. 2, as follows:

$$\begin{cases} \dot{\xi}_i(t) = \zeta_i(t) \\ \dot{\zeta}_i(t) = u_i(t) \end{cases} \quad (6)$$

where: $\xi_i(t) \in R^3$ is the position state; $\zeta_i(t) \in R^3$ is the velocity state; $u_i(t) \in R^3$ is the control input.

We say that the control protocol $u_i(t)$ solves the formation control problem if the states of UAVs satisfy $\lim_{t\to+\infty}[\xi_i(t) - \xi_j(t)] = r_{ij}$ and $\lim_{t\to+\infty}\zeta_i(t) = \zeta_i(t) = \zeta^*$ ($r_{ij} = -r_{ji}$ is the expect distance between UAV$_i$ and UAV$_j$ in formation and $\zeta^* \in R^3$ is the expect velocity), i.e. the multi-UAV system can shape and maintain an expected formation with a desired velocity under the control protocol $u_i(t)$.

In this paper, a formation flight control protocol for the multi-UAV system is designed, and the two key-problems of non-uniform time delays and jointly-connected topologies are considered. To solve this problem, a linear control protocol for the ith UAV is firstly presented, as follows:

$$u_i(t) = \sum_{v_j \in N_i(t)} a_{ij}(t)\left\{k_1[\xi_j(t-\tau_{ij}(t)) - \xi_i(t-\tau_{ii}(t)) - r_{ji}] + \right.$$
$$\left. + \frac{2}{k_2}[\zeta_j(t-\tau_{ij}(t)) - \zeta_i(t-\tau_{ii}(t))]\right\} + \zeta^* - k_3(\zeta_i(t) - \zeta^*) \quad (7)$$

where: $a_{ij}(t)$ is the adjacency weight of the communication graph G_σ; $N_i(t)$ is the neighbor set of the ith UAV; $k_1 > 0, k_2 > 0$, and $k_3 = k_1 k_2$; $\tau_{ii}(t)$ is the time-varying self-delay of the ith UAV that may be caused by measurement or computation, and $\tau_{ij}(t)$ is the time-varying delay for the ith UAV to get the state information of the jth UAV.

Here, it is not required that $\tau_{ij}(t) = \tau_{ji}(t)$. It is supposed that there are altogether M different time delays, denoted by $\tau_m(t) \in \{\tau_{ii}(t), \tau_{ij}(t), i, j, \in \ell\}$, $m = 1, 2, ..., M$, satisfying the following assumptions 1 and 2.

Assumption 1: the time-varying delays $\tau_m(t)$, $m = 1, 2, ...,$ M (τ_m in short), satisfy $0 \leq \tau_m(t) \leq h_m$ and $\dot{\tau}_m(t) \leq d_m < 1$ for specified constants $h_m > 0$ and $d_m > 0$.

A model transformation is made to analyze the close-loop control performance of the multi-UAV system. Therefore, the concept of formation center is introduced, which is a formation centroid of the multi-UAV system. A formation of "regular pentagon" is considered as an example for convenient and easy understanding of the formation problem, as shown in Fig. 2, where O is the origin of Cartesian coordinates, O_C is the formation center, $\xi_i(t)$ and $\xi_j(t)$ are positions of UAV$_{i,j}$ in plane coordinate system, respectively, and $\xi_0(t)$ is the formation center. The distance between UAV$_{i,j}$ and the formation center are r_i and r_j, respectively.

Consequently, the control protocol (Eq. 7) can be transformed into:

$$u_i(t) = \dot{\zeta}^* + \sum_{v_j \in N_i(t)} a_{ij}(t)\left\{k_1[(\xi_j(t-\tau_{ij}(t)) - r_j) - (\xi_i(t-\tau_{ii}(t)) - \right.$$
$$\left. - r_i)] + \frac{2}{k_2}[\zeta_j(t-\tau_{ij}(t)) - \zeta_i(t-\tau_{ii}(t))]\right\} - k_3(\zeta_i(t) - \zeta^*) \quad (8)$$

where: $r_{ji} = r_j - r_i$.

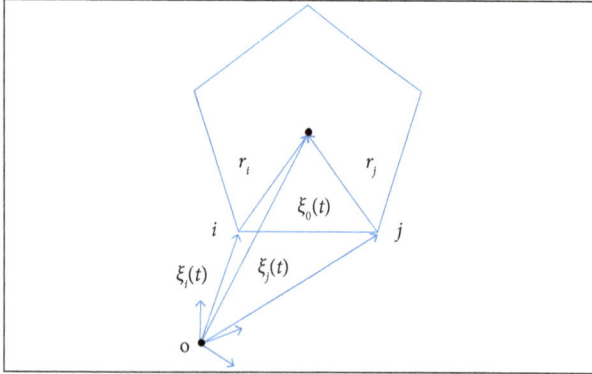

Figure 2. Graph of "regular pentagon" formation structure.

According to the position and velocity of the expected formation of the multi-UAV system, $\bar{\xi}_i(t) = \xi_i(t)\xi_0(t) - r_i$ and $\bar{\zeta}_i(t) = \zeta_i(t)\zeta^*$ are denoted, then control protocol (Eq. 8) can be transformed into:

$$u_i(t) = \sum_{v_j \in N_i(t)} a_{ij}(t)\left\{k_1[\bar{\xi}_j(t - \tau_{ij}(t)) - \bar{\xi}_i(t - \tau_{ii}(t))] + \right.$$
$$\left. + \frac{2}{k_2}[\bar{\zeta}_j(t - \tau_{ij}(t)) - \bar{\zeta}_i(t - \tau_{ii}(t))]\right\} + \dot{\zeta}^* - k_3\bar{\zeta}_i(t) \quad (9)$$

It is denoted:

$$\hat{\zeta}_i(t) = 2\bar{\zeta}_i(t)/k_1k_2 + \bar{\xi}_i(t); \quad \varepsilon(t) = [\bar{\xi}_1(t), \hat{\zeta}_1(t), \dots, \bar{\xi}_n(t), \hat{\zeta}_n(t)]^T$$

$$B = \begin{bmatrix} -k_3/2 & k_3/2 \\ k_3/2 & -k_3/2 \end{bmatrix}; \quad Q = \begin{bmatrix} 0 & 0 \\ 0 & 2/k_2 \end{bmatrix} \quad (10)$$

Under the protocol (Eq. 9), the closed-loop dynamics of the multi-UAV system is:

$$\dot{\varepsilon}(t) = (I_n \otimes B)\varepsilon(t) - \sum_{m=1}^{M}(L_{\sigma m} \otimes Q)\varepsilon(t - \tau_m) \quad (11)$$

where: I_n is the n-dimensional unit matrix; $\otimes$ denotes the Kronecker product; $L_{sm} \in R^{n \times n}$; $L_{\sigma m} \otimes Q$ is the coefficient matrix of the variable $\varepsilon(t - t_m)$ for $m = 1, 2, \dots, M$. It is clear that $L_\sigma = \sum_{m=1}^{M} L_{\sigma m}$ and $L_\sigma^T = L_\sigma$.

Evidently, if $\lim_{t \Rightarrow +\infty} \varepsilon(t) = 0$, then $\lim_{t \Rightarrow +\infty} \bar{\xi}_i(t) = 0$ and $\lim_{t \Rightarrow +\infty} \bar{\zeta}_i(t) = 0$, i.e. $\lim_{t \Rightarrow +\infty} \xi_i(t) - \xi_i(t) = r_{ji}$ and $\lim_{t \Rightarrow +\infty} \zeta_i(t) = \zeta^*$, that is, the multi-UAV system can shape and maintain the expected formation with a desired velocity under the formation control protocol. In the following, we prove that the multi-UAV system can realize $\lim_{t \Rightarrow +\infty} \varepsilon(t) = 0$ under the protocol (Eq. 7).

STABILITY ANALYSIS OF FORMATION FLIGHT CLOSE-LOOP CONTROL SYSTEM

Definition of switching topology and related lemmas

Some preliminary definitions and results need to be presented before the stability analysis. The concept of switching topology is introduced first. It is considered an infinite sequence of non-empty, bounded, and contiguous time intervals $[t_k, t_{k+1})$, $k = 0, 1, \dots$, with $t_0 = 0$ and $t_{k+1} - t_k \leq T_1 (k \geq 0)$ for some constant $T_1 > 0$. It is supposed that, in each interval $[t_k, t_{k+1})$, there is a sequence of non-overlapping subintervals

$$[t_{k_0}, t_{k_1}), \dots, [t_{k_b}, t_{k_{b+1}}), \dots, [t_{k_{m_k-1}}, t_{k_{m_k}})$$
$$t_k = t_{k_0}, t_{k+1} = t_{k_{m_k}} \quad (12)$$

satisfying $t_{kb+1} - t_{kb} \geq T_2, 0 \leq b \leq m_k$ for some integer $m_k \geq 0$ and a given constant $T_2 > 0$ such that the communication topology G_σ switches at t_{kb} and it does not change during each subinterval $[t_{kb}, t_{kb+1})$.

Assumption 2: the collection of graphs in each interval $[t_k, t_k+1)$ is jointly-connected.

With the switching topologies defined above, it is supposed that the time-invariant communication graph G_σ in the subinterval $[t_{kb}, t_{kb+1})$ has $d_\sigma (d_\sigma \geq 1)$ connected components with the corresponding sets of nodes denoted by $\psi_{kj}^1, \psi_{kj}^2, \dots, \psi_{kj}^{d\sigma}; f_\sigma^i$ denotes the number of nodes in ψ_{kj}. Then there exists a permutation matrix $P_\sigma \in R^{n \times n}$ such that $P_\sigma^T L_\sigma P_\sigma = \text{diag}\{L_{\sigma m}^1, L_{\sigma m}^2, \dots, L_{\sigma m}^{d\sigma}\}$,

$$P_\sigma^T L_{\sigma m} P_\sigma = diag\left\{L_{\sigma m}^1, L_{\sigma m}^2, \dots, L_{\sigma m}^{d\sigma}\right\} \quad (13)$$

and

$$\varepsilon^T(t)(P_\sigma \otimes I_2) = [\varepsilon_\sigma^{1\,T}, \varepsilon_\sigma^{2\,T}, \dots, \varepsilon_\sigma^{d_\sigma\,T}]$$

where each block matrix $L_\sigma^i \in R^{f_\sigma^i \times f_\sigma^i}$ is the Laplacian of the corresponding connected component, $L_{\sigma m}^i \in R^{f_\sigma^i \times f_\sigma^i}$ and $L_\sigma^i = \sum_{m=1}^{m} L_{\sigma m}^i$. Then, in each subinterval $[t_{kb}, t_{kb+1})$, the system (Eq.11) can be decomposed into the following d_σ subsystems:

$$\dot{\varepsilon}_\sigma^i(t) = (I_{f_\sigma^i} \otimes B)\varepsilon_\sigma^i(t) - \sum_{m=1}^{M}(L_{\sigma m}^i \otimes Q)\varepsilon_\sigma^i(t - \tau_m)$$
$$i = 1, 2, \dots, d_\sigma \quad (15)$$

where: $\varepsilon_\sigma^i(t) = [\varepsilon_{\sigma 1}(t), \dots, \varepsilon_{\sigma 2f_\sigma}(t)] \in R^{2f_\sigma}$.

Lemma 1 (Lin and Jia 2010): consider the matrix $C_n = nI_n - \mathbf{1}\mathbf{1}^T$ ($\mathbf{1}$ represents $[1, 1, \dots, 1]^T$ with compatible dimensions), then there exists an orthogonal matrix $U_n \in R^{n \times n}$ such that $U_n^T D U_n = diag\{nI_{n-1}, 0\}$ and the last column of U_n is $1\sqrt{n}$. Given a matrix $D \in R^{n \times n}$ such that $1^T D = 0$ and $D1 = 0$, then $U_n^T D U_n = diag\{\bar{U}^T D \bar{U}_n, 0\}$, where $\bar{U}_n$ denotes the first $n-1$ columns of U_n.

Lemma 2 (Lin and Jia 2011): for any real differentiable vector function $x(t) \in R^n$, any differentiable scalar function $\tau(t) \in [0, h]$, and any constant matrix $0 < H = H^T \in R^{n \times n}$, the following inequality can be obtained:

$$\frac{1}{h}[x(t)-x(t-\tau(t))]^T H[x(t)-x(t-\tau(t))] \le \int_{t-\tau(t)}^t \dot{x}^T(s)H\dot{x}(s)ds, \ t \ge 0$$

where $h > 0$ is a specified scalar value.

Sufficient conditions
for the multi-UAV close-loop
control system

Theorem 1: Cconsider a multi-UAV system with non-uniform time delays and switching topologies, for each subinterval $[t_{kb}, t_{kb+1})$, if there is a common constant $\gamma > 0$ and $F_\sigma^i \in R^{f_\sigma^i \times f_\sigma^i}$, $i = 1, 2, ..., d_\sigma$ such that

$$F_\sigma^{i\,T} \Xi_\sigma^i F_\sigma^i < 0 \tag{16}$$

then $\lim\limits_{t\to+\infty} \xi_j(t) - \xi_i(t) = r_{ji}$ and $\lim\limits_{t\to+\infty} \zeta_i(t) = \zeta^*$ that is, the multi-UAV system can finally shape an expected formation with the desired velocity

$F_\sigma^i = diag\{U_{2f_\sigma^i}, I_{2Mf_\sigma^i}\}$ and $U_{2f_\sigma^i}$ is defined as in Lemma 1, where

$$\Xi_\sigma^i = \begin{bmatrix} \Xi_{11} & \Xi_{12} \\ \Xi_{12}^T & \Xi_{22} \end{bmatrix},$$

$$\Xi_{11} = 2\gamma(I_{f_\sigma^i} \otimes B) + \sum\nolimits_{m=1}^M h_m (I_{f_\sigma^i} \otimes B)^T (I_{f_\sigma^i} \otimes B) -$$

$$- \sum\nolimits_{m=1}^M \frac{1-d_m}{h_m} I_{2f_\sigma^i}$$

$$\Xi_{12} = [-\gamma(L_{\sigma 1}^i \otimes Q) + \frac{1-d_1}{h_1} I_{2f_\sigma^i} - \sum\nolimits_{m=1}^M h_m (I_{f_\sigma^i} \otimes B)^T$$

$$(L_{\sigma 1}^i \otimes Q), ..., -\gamma(L_{\sigma M}^i \otimes Q) + \frac{1-d_M}{h_M} I_{2f_\sigma^i} -$$

$$- \sum\nolimits_{m=1}^M h_m (I_{f_\sigma^i} \otimes B)^T (L_{\sigma M}^i \otimes Q)]$$

$$\Xi_{22} = [-diag\left\{\frac{1-d_1}{h_1} I_{2f_\sigma^i}, \frac{1-d_2}{h_2} I_{2f_\sigma^i}, ..., \frac{1-d_M}{h_M} I_{2f_\sigma^i}\right\} +$$

$$+ \sum\nolimits_{m=1}^M h_m [(L_{\sigma 1}^i \otimes Q), (L_{\sigma 2}^i \otimes Q), ... (L_{\sigma M}^i \otimes Q)]^T$$

$$[(L_{\sigma 1}^i \otimes Q), (L_{\sigma 2}^i \otimes Q), ... (L_{\sigma M}^i \otimes Q)]$$

Theorem 1 is proven in the following.

Proof: Define a Lyapunov-Krasovskii function for the system (Eq. 11) as follows:

$$V(t) = \gamma \varepsilon^T(t)\varepsilon(t) + \sum\nolimits_{m=1}^M \int_{-\tau_m}^0 \int_{t+a}^t \dot{\varepsilon}^T(s)\dot{\varepsilon}(s)dsda \ , \ \gamma > \tag{17}$$

It is easy to see that $V(t)$ is a positive definite decrescent function. Calculating $\dot{V}(t)$, it can be obtained:

$$\dot{V}(t) = 2\gamma\varepsilon^T(t)[(I_{f_\sigma^i} \otimes B)\varepsilon(t)] - 2\gamma\varepsilon^T(t)\sum\nolimits_{m=1}^M [(L_{\sigma m} \otimes Q)\varepsilon(t-\tau_m)] +$$

$$+ \sum\nolimits_{m=1}^M \tau_m \dot{\varepsilon}^T(t)\dot{\varepsilon}(t) - \sum\nolimits_{m=1}^M (1-\dot{\tau}_m)\int_{t-\tau_m}^t \dot{\varepsilon}^T(s)\dot{\varepsilon}(s)ds$$

Moreover, from (Eq. 14) and Assumption 1, $\dot{V}(t)$ can be rewritten as:

$$\dot{V}(t) \le \sum\nolimits_{i=1}^{d_\sigma} \left\{2\gamma\varepsilon_\sigma^{i\,T}(t)[(I_{f_\sigma^i} \otimes B)\varepsilon_\sigma^i(t) - 2\gamma\varepsilon_\sigma^{i\,T}(t)\sum\nolimits_{m=1}^M [(L_{\sigma m}^i \otimes Q)\right.$$

$$\varepsilon_\sigma^i(t-\tau_m)] + \sum\nolimits_{m=1}^M h_m \dot{\varepsilon}_\sigma^{i\,T}(t)\dot{\varepsilon}_\sigma^i(t) - \sum\nolimits_{m=1}^M (1-d_m)$$

$$\int_{t-\tau_m}^t \dot{\varepsilon}^T(s)\dot{\varepsilon}(s)ds$$

Applying Lemma 2, it can be obtained:

$$\dot{V}(t) \le \sum\nolimits_{i=1}^{d_\sigma} \left\{2\gamma\varepsilon_\sigma^{i\,T}(t)[(I_{f_\sigma^i} \otimes B)\varepsilon_\sigma^i(t) - 2\gamma\varepsilon_\sigma^{i\,T}(t)\sum\nolimits_{m=1}^M [(L_{\sigma m}^i \otimes Q)\varepsilon_\sigma^i(t-\tau_m)] + \right.$$

$$+ \sum\nolimits_{m=1}^M h_m \dot{\varepsilon}_\sigma^{i\,T}(t)\dot{\varepsilon}_\sigma^i(t) - \sum\nolimits_{m=1}^M \frac{1-d_m}{h_m}[\varepsilon_\sigma^{i\,T}(t)\varepsilon_\sigma^i(t) - \varepsilon_\sigma^{i\,T}(t)\varepsilon_\sigma^i(t-\tau_m) -$$

$$- \varepsilon_\sigma^{i\,T}(t-\tau_m)\varepsilon_\sigma^i(t) + \varepsilon_\sigma^{i\,T}(t-\tau_m)\varepsilon_\sigma^{i\,T}(t-\tau_m)]\right\} = \sum\nolimits_{i=1}^{d_\sigma} \delta_i^T \Xi_\sigma^i \delta_i$$

where: $\delta = [\varepsilon_\sigma^{iT}(t), \varepsilon_{\sigma 1}^{i\,T}(t-\tau_1), \varepsilon_{\sigma 2}^{i\,T}(t-\tau_2), ..., \varepsilon_{\sigma M}^{i\,T}(t-\tau_M)]$

Considering $\eta = [\varepsilon_\sigma^{iT}(t) - h1, \varepsilon_{\sigma 1}^{iT}(t), \varepsilon_{\sigma 2}^{iT}(t), ..., \varepsilon_{\sigma M}^i(t)]$, where $h > 0$ is a constant, it is obvious that $\Xi_\sigma^i(\delta_i - \eta) = 0$. Therefore:

$$\delta_i^T \Xi_\sigma^i \delta_i = \eta^T \Xi_\sigma^i \eta \le \lambda_{\Xi_\sigma^i} \|\eta\|^2 \le \lambda_{\Xi_\sigma^i}\left[\left\|\varepsilon_\sigma^i(t) - h1\right\|^2 + \sum\nolimits_{m=1}^M \sum\nolimits_{K=1}^{f_\sigma^i} (\varepsilon_{\sigma mk})^2(t)\right]$$

where: $\lambda_{\Xi_\sigma^i} < 0$ denotes the largest non-zero eigenvalue of Ξ_σ^i. Therefore:

$$\dot{V}(t) \le \lambda_{\Xi_\sigma^i} \sum\nolimits_{i=1}^{d_\sigma}\left[\left\|\varepsilon_\sigma^i(t) - h1\right\|^2 + \sum\nolimits_{m=1}^M \sum\nolimits_{K=1}^{f_\sigma^i} (\varepsilon_{\sigma mk}^i)^2(t)\right] \le 0 \tag{18}$$

From the analysis above, system (Eq. 11) is stable (Gu *et al.* 2003), *i.e.* $\lim_{t \to +\infty} V(t) = 0$, thus $\lim_{t \to +\infty} \varepsilon(t) = 0$; consequently, $\lim_{t \to +\infty} \xi_j(t) - \xi_i(t) = r_{ji}$ and $\lim_{t \to +\infty} \zeta_i(t) - \zeta$, that is, the multi-UAV system can shape and maintain the expected formation with an desired velocity under the formation control protocol (Eq. 7).

MULTI-UAV CONTROL SYSTEM SIMULATION

Numerical simulations will be given to verify the designed control protocol and illustrate the theoretical results obtained in the previous section. In this paper, the drag in the UAV model (Eq. 1) is calculated by (Xu 2009):

$$D_i = 0.5\rho(v_i - v_{w_i})^2 S_i C_{D0} + 2kk_n n_i^2 W_i^2 / \left[\rho(v_i - v_{w_i})^2 S_i\right] \quad (19)$$

where: the wing area $S_i = 37.16$ m²; the zero lift drag coefficient $C_{D0} = 0.02$; the load factor effectiveness $k_n = 1$; the induced drag coefficient $k = 0.1$; the gravitational coefficient $g = 9.81$ kg/m²; the atmospheric density $r = 1.2207$ kg/m³; the weight of the UAV $W_i = m_i g = 14{,}515$ N. The gust model is $v_{wi} = v_{wi, n} + v_{wi, t}$ and varies according to the altitude h. In the simulated gust, the normal wind shear $v_{wi, n} = 0.215U\log10(h_i)$, where $U = 22.7$ m/s is the mean wind speed at an altitude of 5,000 m. The turbulence part of the wind gust $v_{wi, t}$ has a Gaussian distribution with a zero mean and a standard derivation of $0.09\,U$.

The six UAVs system will complete the task of formation climbing, level flight, and gliding. The communication topology graph of the UAVs and the expected formation structure are shown in Figs. 3 and 4, respectively.

The communication topology in Fig. 3 switches every 0.1 s in the sequence of $(G_I, G_{II}, G_{III}, G_I)$. All graphs in this figure are not connected, and the weight of each edge is 1.0, but the

union of the graphs is jointly-connected. It is supposed that there are altogether three different time delays, denoted by $\tau_1(t)$, $\tau_2(t)$, and $\tau_3(t)$: $\tau_{ii}(t) = \tau_{ij}(t) = \tau_1(t)$ for any $i \neq j$; $\tau_{12}(t) = \tau_{23}(t) = \tau_{34}(t) = \tau_{45}(t) = \tau_{56}(t) = \tau_{61}(t) = \tau_2(t)$; and $\tau_{21}(t) = \tau_{32}(t) = \tau_{43}(t) = \tau_{54}(t) = \tau_{65}(t) = \tau_{16}(t) = \tau_3(t)$. The time delays satisfy $0 \leq \tau_1(t) \leq 0.01$, $0 \leq \tau_2(t) \leq 0.02$, $0 \leq \tau_3(t) \leq 0.03$ and $\dot{\tau}_1(t), \dot{\tau}_2(t), \dot{\tau}_3(t) \leq 0.3$.

It is supposed that all initial conditions of position, velocity, and flight path angle are randomly set. The desired $v_1 = (50 + 10\sin(0.08t))$ *m/s* and $\chi = 45°$. It is solved that (Eq. 16) is feasible for $k_1 = 0.6$, $k_2 = 1.1$, $k_3 = 0.66$. The trajectories of position, velocity, flight path angle, heading angle, and the formed formation are shown in Figs. 5 to 11.

It is clear that the multi-UAV system can complete the maneuver formation flight task with the expected velocity and heading angle as well as maintain the desired formation during the flight.

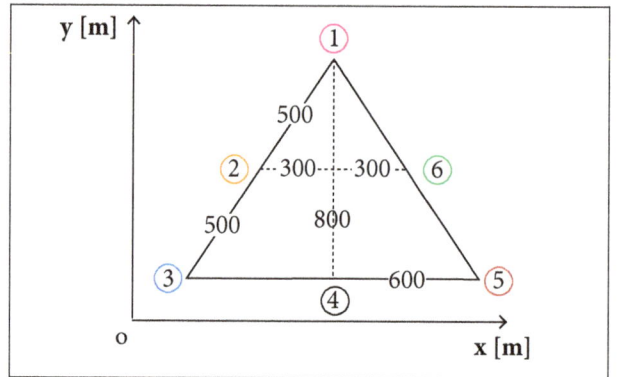

Figure 4. Expected "triangle" formation diagram.

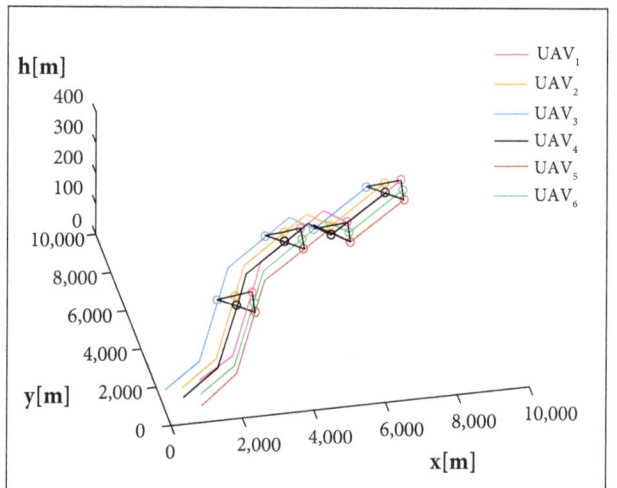

Figure 5. 3-D trajectories of UAVs' formation flying.

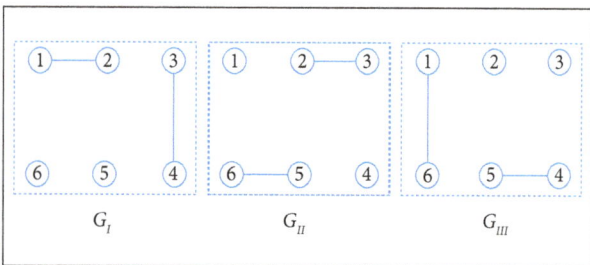

Figure 3. Communication topology of UAVs.

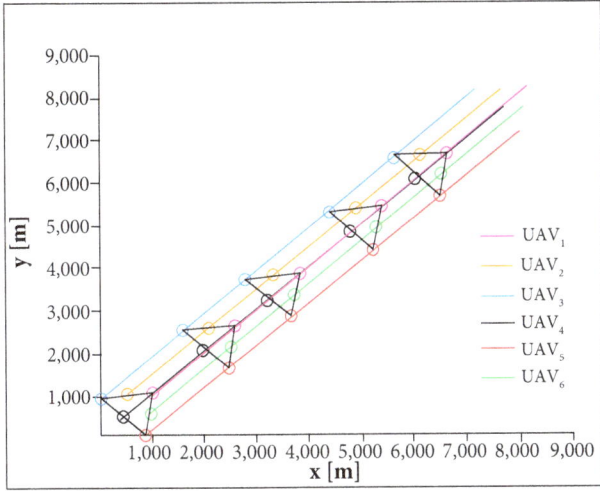

Figure 6. Top view of UAVs' formation flying.

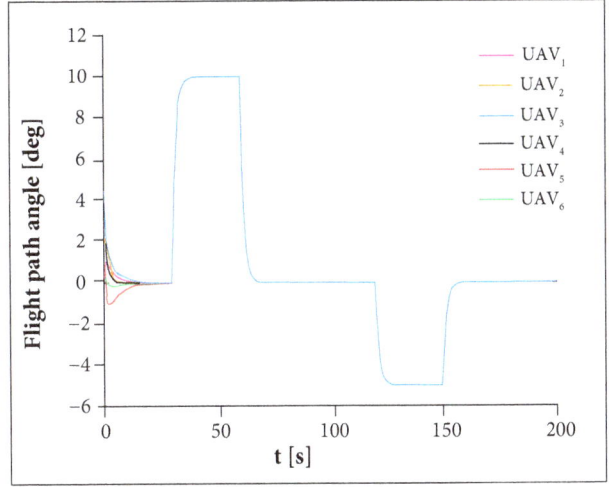

Figure 11. Time histories of the flight path angle.

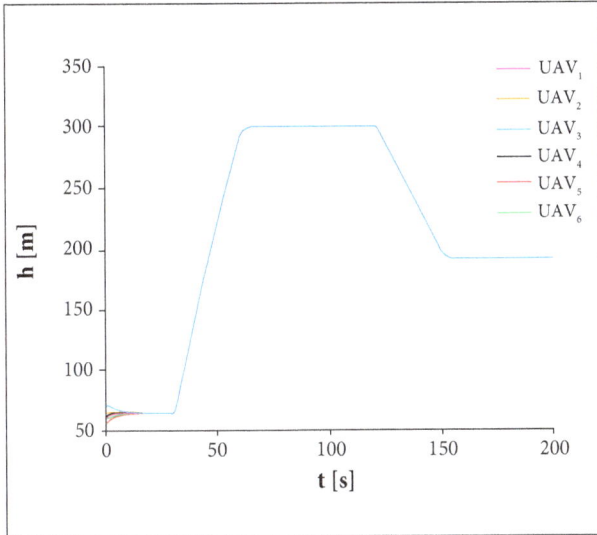

Figure 7. Time histories of the height.

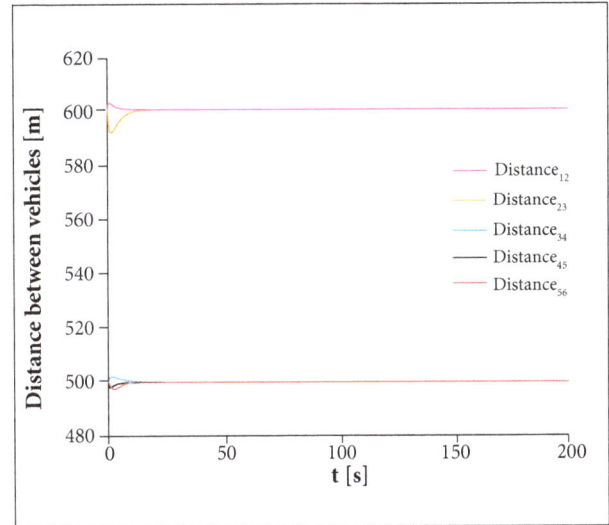

Figure 9. Time histories of the distance between the UAVs.

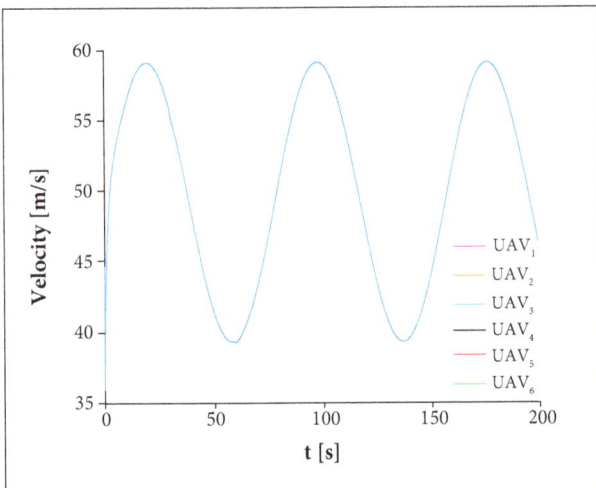

Figure 8. Time histories of the velocity.

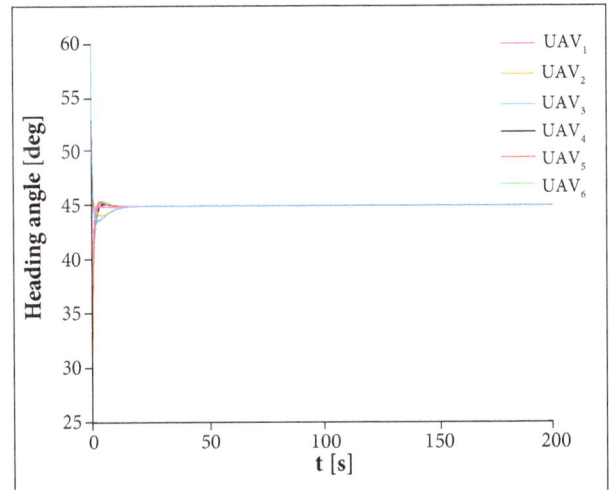

Figure 10. Time histories of the heading angle.

CONCLUSION

Three dimensional formation flight control problems are investigated, considering the constraints of jointly-connected topologies and non-uniform time delays, where each UAV has a self-delay, and all delays are independent of each other. A consensus-based formation control protocol is designed, and the stability problem of the multi-UAV formation control system is turned into the problem that looks for a feasible solution by solving the linear matrix inequality. In reality, it is only necessary to study the connected components with different topology structures, making it possible to simplify the analysis of the whole topology structures. Numerical examples are included to illustrate the obtained results in addition. If the communication topology is jointly-connected and the non-uniform time delays satisfy the designing requirements, then the multi-UAV system can shape the desired formation and also maintain the expected velocity, heading angle, and expected flight path angle.

The problems of collision avoidance constraint and the size of the UAVs are not considered here. These challenging and meaningful problems will be presented in future studies.

REFERENCES

Cao Y, Yu W, Ren W, Chen G (2012) An overview of recent progress in the study of distributed multi-agent coordination. IEEE Trans Ind Inf 9(1):427-438. doi: 10.1109/TII.2012.2219061

Dong X, Yu B, Shi Z (2014) Time-varying formation control for unmanned aerial vehicles: theories and applications. IEEE Trans Control Syst Technol 23(1): 340-348. doi: 10.1109/TCST.2014.2314460

Giulietti F, Pollini L, Innocenti M (2000) Autonomous formation flight. IEEE Control Syst 20(6): 34-44. doi: 10.1109/37.887447

Gu K, Kharitonov VL, Chen J (2003) Stability of time-delay systems. Boston: Birkhäuser.

Kuriki Y, Namerikawa T (2013) Consensus-based cooperative control for geometric configuration of UAVs flying in formation. Proceedings of the SICE Annual Conference; Nagoya, Japan.

Lin P, Jia Y (2010) Consensus of a class of second-order multi-agent systems with time-delay and jointly-connected topologies. IEEE Trans Autom Control 55(3):778-785. doi: 10.1109/TAC.2010.2040500

Lin P, Jia Y (2011) Multi-agent consensus with diverse time-delays and jointly-connected topologies. Automatica 47(4):848-856. doi: 10.1016/ j.automatica.2011.01.053

Menon PKA (1989) Short-range nonlinear feedback strategies for aircraft pursuit-evasion. J Guid Contr Dynam 12(1): 27-32. doi: 10.2514/3.20364

Ren W (2006) Consensus-based formation control strategies for multi-vehicle systems. Proceedings of the American Control Conference; Minnesota, USA.

Ren W (2007) Consensus strategies for cooperative control of vehicle formations. IET Control Theory Appl 1(2):505-512. doi: 10.1049/iet-cta:20050401

Ren W, Beard RW (2008) Distributed consensus in multi-vehicle cooperative control; London: Springer.

Seo J (2009) Controller design for UAV formation flight using consensus-based decentralized approach. Proceedings of the AIAA Aerospace Conference; Seattle, USA.

Seo J, Kim Y, Kim S, Tsourdos A (2012) Consensus-based reconfigurable controller design for unmanned aerial vehicle formation flight. Proc IME G J Aero Eng 226(7):817-829. doi: 10.1177/0954410011415157

Wang J, Xin M (2012) Integrated optimal formation control of multiple Unmanned Aerial Vehicles. Proceedings of the AIAA Guidance, Navigation, and Control Conference; Minnesota, USA.

Xu Y (2009) Nonlinear robust stochastic control for Unmanned Aerial Vehicles. J Guid Contr Dynam 32(4): 1308 - 1319. doi: 10.2514/1.40753

17

Comparative Assessment of Stabilised Polybutadiene Binder under Accelerated Ageing

Luiz Felipe Cannaval Sbegue[1], Luciene Dias Villar[2]

ABSTRACT: Polybutadiene elastomers are versatile materials, being employed at several applications from rocket propellant binder to adhesives and sealants. The elastomers derived from hydroxyl-terminated polybutadiene are usually stabilised with antioxidants to prevent degradation. In this study, a comparative assessment among 2,2'-methylene-bis (4-methyl-6-*tert*-butylphenol) (AO2246), 2,6-di-*tert*-butyl-4-methylphenol (BHT), *p*-phenylenediamine (*p*PDA), and triphenylphosphine (TPP) regarding stabilisation of hydroxyl-terminated polybutadiene binder under accelerated ageing (six months at 65 °C) was carried out. Evaluation of antioxidants effectiveness was examined through Oxidation Induction Time, sol/gel extraction, swelling and mechanical testing, dynamic mechanical analysis, and mass variation measurement. AO2246 yielded the best performance, meanwhile BHT was poorly protective. TPP acted as prooxidant, causing a severe degradation of the binder, and *p*PDA was not manageable to be assessed due to the lower curing degree of the resulted polyurethane.

KEYWORDS: Ageing, Antioxidant, HTPB propellants, Stabilisers.

INTRODUCTION

Despite the ongoing trend into development of energetic polymeric materials, hydroxyl-terminated polybutadiene (HTPB) elastomers are still widely used as polymeric binders for solid rocket propulsion due to the excellent mechanical properties they provide. Moreover, polyurethanes based on HTPB are commonly used as coatings, adhesives and sealants (Gupta and Adhikari 2003), which does make studies about stabilisation of HTPB elastomers of broad interest. In fact, HTPB sensitivity to oxidation through the olefinic double bonds is well known (Coquillat *et al.* 2007a, 2007b, 2007c), thus rendering a need for using stabilisers, namely antioxidants (AO), to trap the radical species and to avoid an undesired crosslinking (Désilets and Côté 2000).

In general, AO are classified as primary or secondary, depending upon their mechanisms of action. Primary AO act as H-atoms donators to free-radicals, meanwhile secondary AO act as hydroperoxide decomposers, both acting as to prevent propagation of chain reaction (Vulic *et al.* 2002). The use of primary and secondary AO in conjunction may create a synergy between these compounds (Vulic *et al.* 2002). The sterically hindered phenols and secondary aromatic amines are the most representative classes of primary AO. Among the secondary AO, phosphine and phosphonite compounds are regarded as the most effective.

Since n-phenyl-2-naphthylamine (PBNA) has been banned for its use in propellant formulation, due to carcinogenic effects (Chhabra *et al.* 1993), the development

1.Universidade de São Paulo – Escola de Engenharia de Lorena – Departamento de Engenharia de Materiais – Lorena/SP – Brazil. **2.**Departamento de Ciência e Tecnologia Aeroespacial – Instituto de Aeronáutica e Espaço – Divisão de Química – São José dos Campos/SP – Brazil.

Author for correspondence: Luciene Dias Villar | Departamento de Ciência e Tecnologia Aeroespacial – Instituto de Aeronáutica e Espaço – Divisão de Química | Praça Marechal Eduardo Gomes, 50 – Vila das Acácias | CEP: 12.228-904 – São José dos Campos/SP – Brazil | Email: lucieneldv@iae.cta.br

of solid propellant technology has extensively relied on 2,2'-methylene-bis(4-methyl-6-*tert*-butylphenol), usually named as AO2246 for HTPB stabilisation (Coquillat *et al.* 2008). Despite of that, some studies (Celina *et al.* 2006; Désilets and Côté 2000) have pointed through a partly attachment of AO2246 to the polymeric HTPB network through the reaction with isocyanate. The practical consequence of this finding is the lower availability of this antioxidant to act as scavengers of radical species. Désilets and Côté (2000) also found that this attachment is dependent upon the [NCO]/[OH] ratio, being quite pronounced at ratios higher than the stoichiometric one.

Since AO are used with the ultimate target to enhance long-term shelf life, the aim of this research was to conduct an assessment of the effectiveness of AO2246 to stabilise HTPB binder under accelerated ageing (six months at 65 °C) compared to other AO usually employed to stabilise polybutadiene, namely 2,6-di-*tert*-butyl-4-methylphenol (BHT), *p*-phenylenediamine (*p*PDA), and triphenylphosphine (TPP). Even thought it was not the main concern of this study, important findings were raised about the relationship between the chemical structure and the properties of unaged HTPB elastomers formulated with the AO investigated.

MATERIALS AND METHODS
MATERIALS

Uncrosslinked HTPB prepolymer, M_n 2,900 g/mol, of predominant trans 1,4 structure was supplied by Petroflex (Brazil) containing 1% w/w BHT. AO were used as received. BHT, *p*PDA, and TPP were supplied by Sigma-Aldrich, meanwhile AO2246 was supplied by Cyanox. Chemical name, structure and classification according to antioxidant mechanism of action are presented in Table 1. The preparation of HTPB elastomers was accomplished by reaction with diisocyanate of isophorone (IPDI, Veba Chemie) in the presence of ferric acetylacetonate ($Fe(acac)_3$, Merck), both of them used as supplied.

OIT MEASUREMENT IN HTPB PREPOLYMER

The assessment of antioxidant activity in the improvement of HTPB prepolymer oxidative resistance was carried out by measurement of Oxidation Induction Time (OIT). Samples of HTPB prepolymer added with 1% w/w AO were analysed. A control sample, without addition of AO, was also provided. OIT was measured in TA5008 (TA Instruments), DSC-Q1000 module. The samples were heated at 20 °C/min under nitrogen atmosphere up to 170 °C and kept in isothermal temperature for 5 min. Chamber atmosphere was changed afterwards to

Table 1. Chemical name, structure and classification of investigated antioxidants.

Acronym	Chemical name	Structural formulae	Classification
AO2246	2,2'-methylene-bis(4-methyl-6-*tert*-butyl-phenol)		Primary
BHT	2,6-di(*tert*-butyl)hydroxytoluene		Primary
*p*PDA	*p*-phenylenediamine		Primary
TPP	triphenylphosphine		Secondary

synthetic air. The time required to reveal the first oxidation peak after the change of nitrogen to synthetic air is assumed to be OIT at a constant temperature (ASTM D3895-14).

PREPARATION OF HTPB BINDER

Polyurethane binders were prepared from polymerization of HTPB with IPDI, using Fe(acac)$_3$ (0.012% w/w) as catalyst. Stoichiometric [NCO]/[OH] ratio was used. Formulations were prepared containing 1% w/w AO. A control formulation (CF), without addition of AO, was also prepared. Sheets of 2 mm thickness were cast and cured for seven days at 50 °C.

THERMAL AGEING ASSAY

Thermal ageing of the HTPB binders were carried out in type IIB forced ventilation ovens (ASTM E145-11) under atmospheric pressure and relative humidity (RH) less than 20%. The formulations were aged in separate ovens, in order to avoid cross-contamination by AO volatilisation. Ageing conditions were set up by following ASTM D3045-10 recommendations. The ageing assay was carried out at 65 °C having withdrawn after zero, one, two, four, and six months of storage.

The evaluation of antioxidant performance during thermal ageing was achieved by submitting withdrawn samples to sol/gel extraction, swelling and mechanical testing, dynamical mechanical analysis (DMA), as well as measurement of mass changes. Strands of HTPB binders were hanging inside the oven and used for DMA (Fig. 1a), sol/gel extraction, swelling and mechanical testing (Fig. 1b). Specimens for hardness and mass variation measurements were stored at the bottom of the oven (Figs. 1c and 1d, respectively).

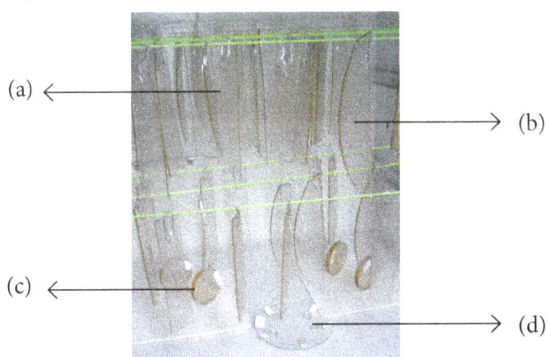

Figure 1. Storage of HTPB binders inside type IIB forced ventilation oven during accelerated ageing. (a) Strand for DMA; (b) Strand for sol/gel extraction, swelling and mechanical testing; (c) Specimen for hardness measurement; (d) Traveller for mass measurement.

SOL/GEL EXTRACTION

Solvent extraction was performed by using continuous Soxhlet extraction with chloroform (120 mL) for 24 h to determine sol/gel fractions. Samples of approximately 0.5 g of grounded HTPB elastomer were analysed in duplicate. Soluble fraction was submitted to evaporation in a rotary evaporator. Sol and gel fractions were dried at 60 °C until constant mass.

SWELLING TESTING

Swelling testing was performed by using HTPB binder discs (10 mm diameter, 2 mm thickness, 0.2 ± 0.02 g). Dried sample disks were placed, in triplicates, into chloroform at room temperature. Replicates were placed in separated vials, under occasionally agitation, until equilibrium was reached (approximately, seven days). The swollen sample disks were weighted ($m_{1,2}$) and placed to dry until constant mass (m_2). Molecular mass between chemical crosslinks (M_c) and crosslink density (ρ_e) was determined by Eqs. 1 and 2, respectively, following Flory-Rehner's equations (Flory and Rehner 1943).

$$M_c = -\frac{\rho_2 V_1 \left(v_2^{1/3} - v_2/2\right)}{\ln(1-v_2) + v_2 + \chi \cdot v_2^2} \tag{1}$$

$$\rho_e = \frac{\rho_2}{M_c} \tag{2}$$

where: ρ_2 is the density of the gel; V_1 is the molar volume of the solvent; v_2 is the volume fraction of the gel; χ is the Flory-Huggins interaction parameter between solvent and polymer.

In this study, ρ_2 was 0.84 g/mL, V_1 was 81.21 mL/mol, v_2 was calculated for each HTPB elastomer formulation, as defined in Eq. 3, and χ was 0.24 (Eroğlu 1998) for HTPB-chloroform pair at 45 °C. Hence, some loss of accuracy is expected due to the fact that, in the present study, swelling tests were performed at room temperature.

$$v_2 = \frac{1}{Q} = \left[1 + \frac{\rho_2}{\rho_1} \times \left(\frac{m_{1,2} - m_2}{m_2}\right)\right]^{-1} \tag{3}$$

where: Q is the swelling ratio by volume; ρ_1 is the density of the solvent at 25 °C (1.47 g/mL for chloroform).

MECHANICAL TESTING

Dumbbell-shaped specimens (ASTM D412-13 model C) were cut from the HTPB binder 2 mm-strands and assayed for uniaxial tensile tests, according to ASTM D412-13, in a Zwick 1474 testing machine at 23 ± 2 °C, and a cross-head speed of 500 mm/min. Actual strain was measured with an optical extensometer. The initial modulus of elasticity (Young modulus) was defined as the secant modulus between 3 and 10% of elongation. Reported results are mean values of ten specimens. Hardness was measured in four specimens with 10 mm thickness (Fig. 1c) by using Zwick Shore A tester according to ASTM D2240-10.

DMA ANALYSIS

DMA measurements were carried out in flexion mode with single cantilever clamps by using a DMA-Q800 analyzer (TA Instruments). Specimens (35 × 11 × 2 mm) were cooled at −130 °C, with the aid of liquid nitrogen, and heated up to +80 °C at 3 °C/min. Tests were conducted at a frequency of 1 Hz and oscillation amplitude of 30 μm. The samples were analysed in duplicate, and the results were expressed as mean values.

MASS VARIATION

Discs of HTPB binders, with the same dimensions as the ones used for swelling test, were employed as travellers for gravimetrically measurement of mass change during thermal ageing (Fig. 1d). Measurements were made on a regular basis (once a week, usually), and, for each formulation, five travellers were used. Results are expressed as mean values.

RESULTS AND DISCUSSION
OIT OF HTPB PREPOLYMER

Thermal analysis to determine OIT is usually employed to verify the readiness of materials to degrade through oxidation. In this study, antioxidant performance was firstly assessed by measuring OIT in HTPB prepolymer samples added with 1% w/w AO. OIT values are presented in Table 2, which includes a control sample without the addition of any further antioxidant.

Due to stabilisation of HTPB prepolymer from its manufacturer by using 1% w/w BHT, samples were expected to have some residual content of this antioxidant, including the control sample. However, this would not interfere in the assessment of antioxidant activity, since the analysis was made on a comparison basis.

OIT results showed greater stabilisation activity for AO2246 (Table 2), with a time span of 96 min to appear the first oxidation peak. This performance was followed by pPDA and TPP, with 35.6

and 16.76 min, respectively. BHT showed a lower performance, with only 3.6 min, just a little higher than the control sample (2.79 min). The delay observed in the control sample to initiate oxidation has demonstrated that BHT residual content in HTPB prepolymer was significant, thus additional BHT resulted in only a slight improvement of stabilisation around 30%.

From the OIT results, the most promising decreasing order of antioxidant performance to prevent oxidation of HTPB was AO2246 >> pPDA > TPP > BHT. Since TPP is a secondary antioxidant, it is likely that some synergetic effect has occurred with the residual BHT present in HTPB prepolymer. In spite of this, the primary antioxidants AO2246 and pPDA presented higher OIT values.

Table 2. OIT values for HTPB prepolymer in the presence of the investigated antioxidants.

Sample	OIT (min)
HTPB (control)	2.79 ± 0.08
HTPB + AO2246	96 ± 2
HTPB + BHT	3.6 ± 0.6
HTPB + pPDA	35.6 ± 0.6
HTPB + TPP	16.76 ± 0.08

UNAGED HTPB BINDER

HTPB binder containing pPDA resulted in a very soft material, unable to be properly extracted from the mould. Similar result was obtained by increasing the [NCO]/[OH] ratio from 1.0 to 1.25 in order to take into account the reaction between pPDA and IPDI. The formation of short-chain polyurethanes from pPDA/IPDI reaction is prone to be the most likely reason for the softening behaviour observed. In fact, it is well known that the reaction between isocyanates and primary aryl amines presents comparable relative rates to primary hydroxyl groups (Szycher 1999). Since the elastomer obtained was not able to be assayed by the tests employed in this study, pPDA was excluded from the further characterization of unaged HTPB binders.

Tables 3 and 4 summarise the results obtained for sol/gel extractions, swelling testing, mechanical testing and DMA for unaged HTPB elastomers. Mean values were evaluated by one-way Analysis of Variance (ANOVA). When appropriate, Tukey's test was applied for comparison between the means. In Tables 3 and 4, identical letters placed after mean values indicate no significant difference at 95% confidence level.

Table 3. Sol/gel fractions and swelling testing parameters for unaged HTPB binders.

AO	Soxhlet extraction			Swelling testing	
	Sol fraction [%]	Gel fraction [%]	Sol-Gel Recovery [%]	M_c [g/mol]	ρ_e [mol/mL × 10^{-4}]
CF	8.5 ± 0.2	83.6 ± 0.4	92.1	4,500a ± 50	1.86a ± 0.02
AO2246	9.5 ± 0.7	83.3 ± 0.7	92.8	6,200c ± 300	1.37c ± 0.07
BHT	8.0 ± 0.6	84.7 ± 0.3	92.7	5,370b ± 50	1.56b ± 0.01
TPP	9.2 ± 0.2	82.9 ± 0.2	92.1	5,680d ± 20	1.48d ± 0.01

Table 4. Mechanical property at rupture and dynamic mechanical property for unaged HTPB binders.

AO	Mechanical property at rupture			Dynamic mechanical property	
	σ_r (MPa)	ε_r [%]	E (MPa)	E' at −125 °C (MPa)	T_g (max E") [°C]
CF	1.37a ± 0.06	280ac ± 20	1.80a ± 0.03	3,450 ± 150	−74.2 ± 0.3
AO2246	1.2b ± 0.1	290a ± 40	1.56c ± 0.05	3,420 ± 30	−73.7 ± 0.1
BHT	1.14b ± 0.08	200b ± 30	1.78ab ± 0.02	3,430 ± 90	−73.7 ± 0.1
TPP	1.19b ± 0.06	230bc ± 30	1.73b ± 0.02	3,540 ± 30	−73.2 ± 0.4

σ_r: Strength at rupture; ε_r: Elongation at rupture; E: Young modulus; E': Storage modulus; T_g: Glass transition temperature; $E"$: Loss modulus.

Sol/gel analysis presented a recovery around 90% (Table 3), which can be considered a quite satisfactory recovery for a gravimetric analysis. From ANOVA, sol fraction resulted not significantly different for all the formulations studied at 95% confidence level (Table 3). Similar behaviour was observed for gel fraction. Sol/gel extraction showed that the chemical structure of unaged HTPB elastomers was dominated by gel (approximately 80%).

On the other hand, swelling tests have pointed through a great difference in crosslink architecture of the gel fraction for HTPB elastomers prepared with different AO. The results showed a steady increase in crosslink density for the elastomers in the following sequence: AO2246 < TPP < BHT < CF (Table 3). As a consequence, decrease in molecular mass between chemical crosslinks (M_c) followed the opposite sequence. In general, when formulating a HTPB binder, there is little concern about the possible effects that AO may have over the gel structure which is going to be formed. However, these findings have shown that, although the sol/gel fractions of these formulations can be not significantly different, their chemical structure may have some expressive variations, which may interfere in the mechanical behaviour of the binders.

Actually, in agreement with the swelling results, HTPB binder containing AO2246 presented the lowest stiffness among the formulations evaluated, as shown by the values obtained for Young modulus and elongation at rupture (Table 4). As indicated by the Tukey's test, the other formulations presented some similarities in mechanical results, depending on the property

considered. From DMA results, values of storage modulus (E') at −125 °C and glass transition temperature (T_g) obtained from $E"$ (loss modulus) peak have presented no significant difference at 95% confidence level for all the elastomeric binders studied. Thus, it seems that the gel structure of stabilised HTPB is more sensitive to the loads applied during unaxial stress-strain testing than to the loads under DMA testing.

AGED HTPB BINDER

Samples of HTPB binders withdrawn during accelerated ageing were submitted to Soxhlet extraction and swelling testing with the aim of evaluating the possible changes on their structure. For conciseness reasons, only sol fraction (Fig. 2) and crosslink density (Fig. 3) are presented, since gel fraction and molecular mass between chemical crosslinks are data complementary to the ones provided herein.

Sol fraction presented a continuous decrease during thermal ageing for all the formulations studied (Fig. 2). In addition, formulations containing AO2246 and TPP have presented the highest (6.0%) and the lowest (4.7%) ultimate sol content, respectively.

Crosslink density may be defined as the number of chain segments between the covalent or ionic bonds present in the polymeric chains. Higher crosslink density means higher interlacing, thus resulting in lower mobility. As unaged material, CF has presented the highest crosslink density, meanwhile AO2246 has presented the lowest value (Table 3). After ageing, AO2246 has not presented significant change in the crosslink density (Fig. 3). BHT and CF have, otherwise, shown exponential increase, after a lag phase of approximately two months. On the other hand, TPP followed an asymptotically increase from the early beginning of the thermal ageing. After six months, samples containing BHT, TPP or without AO (CF) became very brittle, and, hence, it was not possible to continue the swelling testing due to the loss of accuracy.

Sol/gel analysis and swelling testing have indicated that the most important changes in the chemical structure of polybutadiene binders have occurred in gel fraction, due to the major increase in crosslink density of HTPB chains. Changes in sol fraction also indicate that some loosely chains, such as low molecular mass polyurethane and unreacted HTPB prepolymer, may have been linked to gel fraction, thus decreasing the sol content. The most remarkable changes have occurred for binders containing TPP, followed by CF and BHT. Being a secondary antioxidant, TPP was expected to behave synergistically with

residual BHT from HTPB prepolymer, which, nevertheless, was not observed, despite the predictions obtained from OIT results (Table 2). As a matter of fact, Pauquet *et al.* (1993) have pointed out limitations of OIT predictions for secondary AO, when results are extrapolated to long-term stabilisation.

Ageing of the formulation containing AO2246 has shown little effect on the mechanical properties (Figs. 4 to 6), with a smooth tendency of elongation decrease and modulus increase. On the other hand, the other formulations were not able to be evaluated during the entire ageing assay due to the high level of stiffness reached by the samples. TPP presented the worst scenario, being not able to be mechanically tested after the first month of accelerated ageing. BHT and CF were assayed up to the second month, which corresponds to the lag phase observed before exponential increase in crosslink density (Fig. 3).

Following mechanical characterisation of the aged binders, hardness measurements indicated an asymptotically increase

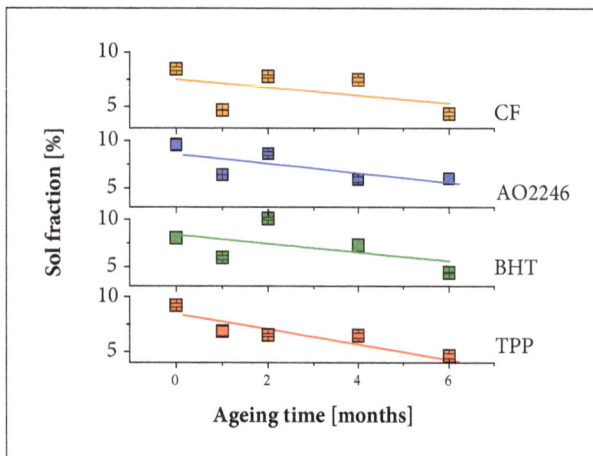

Figure 2. Sol fraction of HTPB binder under accelerated ageing.

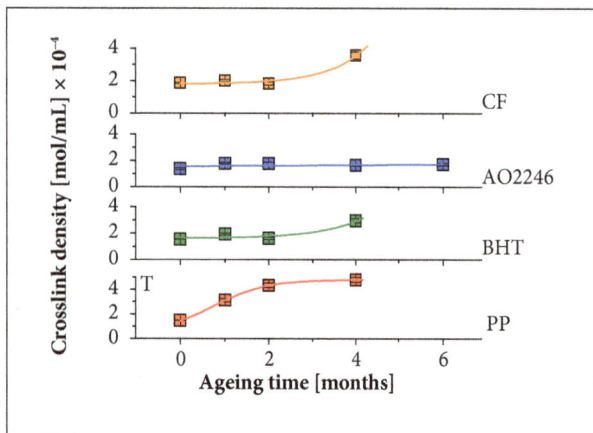

Figure 3. Crosslink density of HTPB binder under accelerated ageing.

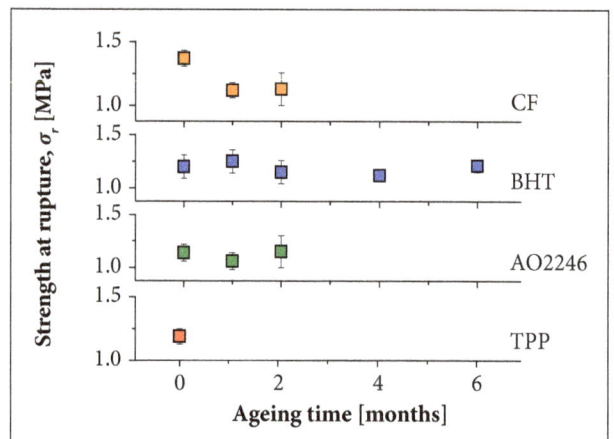

Figure 4. Strength at rupture of HTPB binder under accelerated ageing.

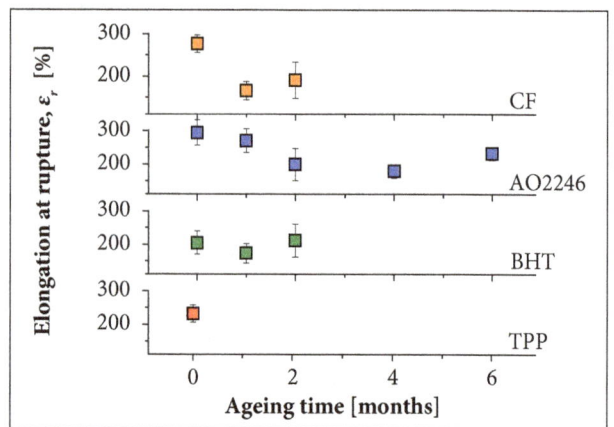

Figure 5. Elongation at rupture of HTPB binder under accelerated ageing.

for TPP from the early beginning of the ageing assay (Fig. 7). A tendency of hardness increase was very low pronounced for BHT and control, being not significant for AO2246 (Fig. 7).

Aged samples withdrawn after six months were also submitted to DMA. However, only AO2246 was able to be assayed, due to the high level of degradation observed for other samples, as reported for mechanical testing (Figs. 3 to 6). Storage modulus (E') of AO2246 resulted in 3,430 ± 30 MPa, meanwhile T_g resulted in −72.6 ± 0.1 °C. By applying mean comparison, only T_g showed to be significantly different from the initial value (Table 4) at 95% confidence level. The change to a slightly higher T_g indicates that some degree of crosslink did occur in the presence of AO2246, although it cannot be realized by the measurement of crosslink density, as shown in Fig. 3.

Measurement of net gain or loss of mass by using traveller specimens resulted in the behaviour described in Fig. 8, where m_0 corresponds to initial traveller mass and m, to the mass at ageing time t. At the beginning of the ageing assay, all the

formulations showed some loss of mass, which can be attributed to volatilisation of AO, and it was more pronounced for the formulation containing BHT, due to lower molar mass. Gain of mass due to the oxidation of the HTPB backbone was readily noticed for TPP formulation, followed by control and BHT samples. In the presence of AO2246, there was no expressive gain of mass throughout the assay.

With no shadow of doubt, the results from mechanical testing, DMA and mass variation have pointed to a negligible effect of ageing on AO2246 formulation, and, in opposite, a very significant effect on TPP. Although some loss of mass at the beginning of ageing assay indicates that unattached AO2246 was lost by volatilisation, the remained content of this antioxidant was very effective in protect HTPB binder against oxidation.

On the other hand, although TPP was reported earlier (Hinney and Murphy 1989) to efficiently avoid HTPB prepolymer oxidation, thus preventing its increase in functionality, this antioxidant was not suited for HTPB binder protection. In fact, TPP has acted as a prooxidant, promoting a fast loss of mechanical properties and an expressive gain of mass by oxygen uptake through oxidation.

Finally, the results from BHT formulation suggested that this antioxidant may be used for stabilisation of HTPB binder. Although, if used previously for stabilisation of BHT prepolymer, the additional protective effect may be negligible. In fact, loss of BHT through volatilisation was more expressive than the observed for CF (Fig. 8). In addition, it was expected that BHT would have a reduced performance compared to AO2246, due to the presence of only one active hydroxyl group and to the lower molar mass, thus being more susceptible to volatilisation.

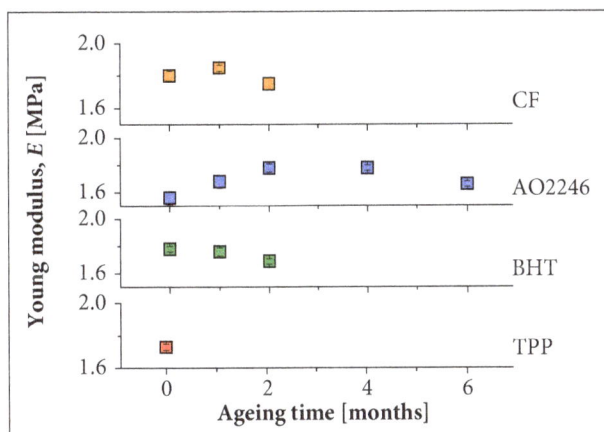
Figure 6. Young modulus of HTPB binder under accelerated ageing.

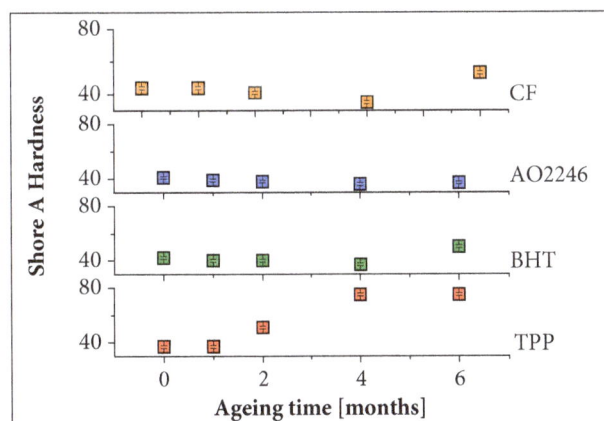
Figure 7. Hardness of HTPB binder under accelerated ageing.

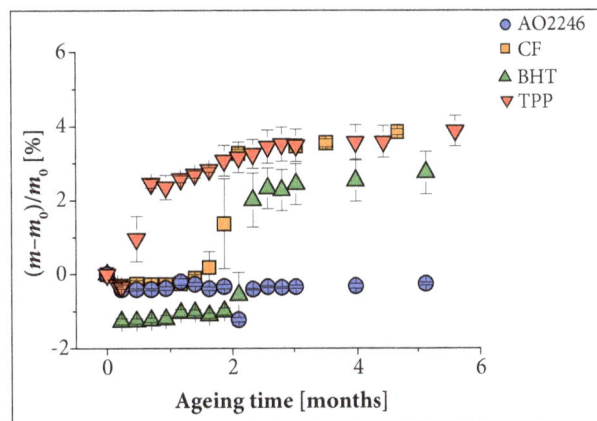
Figure 8. Mass variation of HTPB binder under accelerated ageing.

CONCLUSION

The main objective of this investigation was to compare the effectiveness of AO2246, a phenolic hindered antioxidant, with those of BHT, *p*PDA, and TPP in the stabilisation of HTPB binders for solid rocket propellants. In order to estimate the readiness of the investigated AO to prevent HTPB oxidation, the measurement of OIT was carried out, and the results revealed as the most promising decreasing order of antioxidant performance to be AO2246 >> *p*PDA > TPP > BHT. Otherwise, a more complete evaluation has confirmed OIT results only partially. Hence, in the presence of AO2246, HTPB binder presented the lower crosslink density, sol fraction decrease, mechanical and thermo-mechanical (DMA) changes and mass (loss/gain) variations. *p*PDA has to be withdrawn from this study, since its reaction with isocyanate (IPDI) resulted in a very soft material, unable to be assayed. Finally, the relative order between TPP and BHT was proved to be the opposite envisaged by OIT results. In fact, BHT

showed to be an alternative antioxidant for HTPB stabilisation, meanwhile TPP has actually acted as a prooxidant additive.

The findings described herein raised an important point about using analysis of one component, in this case, HTPB prepolymer, to predict the ageing behaviour of a formulation containing this component, as in the HTPB binder. Being a more complex system, AO proved to have interactions with the HTPB backbone and with the curing agent, thus resulting in a different comparative performance of stabilisation. In addition, these interactions may, eventually, interfere with mechanical properties of the resulted HTPB binders.

ACKNOWLEDGEMENTS

The author Luiz Felipe C. Sbegue acknowledges Conselho Nacional de Desenvolvimento Científico e Tecnológico (CNPq - PIBIC) for his scholarship (Grant 126365/2010-5).

REFERENCES

Celina M, Elliot JMS, Winters ST, Assink RA, Minier LM (2006) Correlation of antioxidant depletion and mechanical performance during thermal degradation of an HTPB elastomer. Polym Degrad and Stabil 91(8):1870–1879. doi: 10.1016/j.polymdegradstab.2005.11.006

Chhabra JS, Athar J, Agrawal JP, Singh H (1993) Comparative study of various antioxidants for HTPB prepolymer. Plast Rub Compos Process Appl 20:305-310.

Coquillat M, Verdu J, Colin X, Audouin L, Celina M (2008) A kinetic evaluation of the thermal oxidation of a phenol stabilised polybutadiene. Polym Degrad Stabil 93 (9):1689–1694. doi: 10.1016/j.polymdegradstab.2008.05.031

Coquillat M, Verdu J, Colin X, Audouin L, Nevière R (2007a) Thermal oxidation of polybutadiene. Part 1: effect of temperature, oxygen pressure and sample thickness on the thermal oxidation of hydroxyl-terminated polybutadiene. Polym Degrad Stabil 92(7):1326–1333. doi: 10.1016/j.polymdegradstab.2007.03.020

Coquillat M, Verdu J, Colin X, Audouin L, Nevière R (2007b) Thermal oxidation of polybutadiene. Part 2: mechanistic and kinetic schemes for additive-free non crosslinked polybutadiene. Polym Degrad Stabil 92(7):1334-1342. doi: 10.1016/j.polymdegradstab.2007.03.019

Coquillat M, Verdu J, Colin X, Audouin L, Nevière R (2007c) Thermal oxidation of polybutadiene. Part 3: molar mass changes of additive-free non crosslinked polybutadiene. Polym Degrad Stabil 92(7):1343–1349. doi: 10.1016/j.polymdegradstab.2007.03.018

Désilets S, Côté S (2000) Chemical bond between stabilizers and HTPB binders in propellants. Propell Explos Pyrotech 25(4):186–190. doi: 10.1002/1521-4087(200009)25:4<186::AID-PREP186>3.0.CO;2-B

Eroğlu MS (1998) Characterization of the network structure of hydroxyl terminated poly(butadiene) elastomers prepared by different reactive systems. J Appl Polym Sci 70(6):1129–1135. doi: 10.1002/(SICI)1097-4628(19981107)70:6<1129::AID-APP9>3.0.CO;2-Q

Flory PJ, Rehner J (1943) Statistical mechanics of swelling of crosslinked polymer networks. Chem Phys 11:521-526.

Gupta T, Adhikari B (2003) Thermal degradation and stability of HTPB-based polyurethane and polyurethaneureas. Thermochim Acta 402(1-2):169–181. doi: 10.1016/S0040-6031(02)00571-3

Hinney HR, Murphy JR (1989) Suppression of reactivity changes in PolyBD R-45M resin. Paper presented at ICT 1989. Proceedings of the 20th International Annual Conference of Fraunhofer ICT; Karlsruhe, Germany.

Pauquet JR, Todesco RV, Drake WO (1993) Limitations and applications of oxidative induction time (OIT) to quality control of polyolefins. Paper presented at IWCS 1993. Proceedings of the International Wire and Cable Symposium; Saint Louis, USA.

Szycher M (1999) Szycher's handbook of polyurethanes. Boca Raton: CRC Press.

Vulic I, Vitarelli G, Zenner JM (2002) Structure-property relationships: phenolic antioxidants with high efficiency and low colour contribution. Polym Degrad Stabil 78(1):27–34. doi: 10.1016/S0141-3910(02)00115-5

18

Multidisciplinary Design Optimization of UAV Under Uncertainty

Majid Hosseini[1], Mehran Nosratollahi[1], Hossein Sadati[1]

ABSTRACT: Uncertainty-based multidisciplinary design optimization considers probabilistic variables and parameters and provides an approach to account for sources of uncertainty in design optimization. The aim of this study was to apply a decoupling uncertainty-based multidisciplinary design optimization method without any dependence on probability mathematics. Existing approaches of uncertainty-based multidisciplinary design optimization are based on probability mathematics (transformation to standard space), calculating an approximation of the constraint functions in standard space and finding the most probable point, which is the best possible one. The current approach used in this paper was inspired on interval modeling, so it is good when there is insufficient data to develop a good estimate of the probability density function shape or parameters. This approach has been implemented for an existing Unmanned Aerial Vehicle (UAV, Global Hawk) designed for purposes of comparison and validation. The advantages of the provided approach are independence of probability mathematics, appropriate when there is insufficient data to approximate the uncertainties variables, appropriate speed to calculate the best reliable response, and proper success rate in the presence of uncertainties.

KEYWORDS: Uncertainty-based multidisciplinary design optimization, MDO, Systemic design, Unmanned Aerial Vehicles design.

INTRODUCTION

Uncertainty existing at the early phases of the design process influences the system reliability. It is important to manage error early in the design process to decrease the redesign likelihood. Designing complicated and large systems such as aerospace vehicles requires appropriate compromise for gaining balance between multiple coupled targets. The targets include high performance and low costs. The sooner these compromises are understood in design process, the more technology, programming, and cost-related risks can be minimized. There are complicated relationships existing between assignment requirements, constraints, design sub-systems and contradicted targets, which could be coordinated using a suitable strategy of optimization. Multidisciplinary design optimization (MDO) or coordination between multidisciplinary analyses makes understandable more effective solutions during design and optimization of complicated systems. This allows system engineers to look for a vast scope of compromise in a systematic and thoughtful way and to consider more structures in conceptual design phase and before concentrating on final design.

Preliminary application of optimization in aerospace industries is accompanied by optimization of sub-systems or components such as aerodynamic shape, orbital path, as well as optimization of sub-systems altogether. Anyway, an optimized systemic compound will not be always created through optimization of sub-systems. In aerospace engineering, MDO was first applied in design of airplanes and to date the method has been used in academic papers and manufactured airplanes (Geethaikrishnan 2003).

In Olds (1993), González *et al.* (2005), Morris and Kroo (1990), Raymer (2002), Çavuş (2009), Neufeld (2010), Hendrich

1.Malek Ashtar University of Technology – Faculty of Aerospace – Department of Aerospace – Tehran/Tehran – Iran.

Author for correspondence: Majid Hosseini | Malek Ashtar University of Technology – Faculty of Aerospace – Department of Aerospace | Babaii Highway | Zip code: 15875-1774 – Tehran/Tehran – Iran | Email: Mjdhosseini@mut.ac.ir

(2011), Giunta (1997), Buonanno (2005), Iqbal (2009), Mattos and Secco (2013), Eisler (2003), Rowell *et al.* (1999), Goraj (2005), Lee *et al.* (2007), Tianyuan and Xiongqing (2009), Jaeger *et al.* (2013), Ahn *et al.* (2002), Perez *et al.* (2004), Sóbester and Keane (2006), Lee *et al.* (2009), Choi *et al.* (2010), Zill *et al.* (2011), and Tekinlap and Cavus (2012), the airplane conceptual designing is done by taking advantage of MDO. In these references, all fighter, passenger, and conventional planes, as well as Unmanned Aerial Vehicles (blended wing body and conventional) are considered. In a systemic attitude towards previous studies, the applied design algorithms are: MDOs in different frameworks, such as All at Once (AAO), Multiple Discipline Feasible (MDF), Collaborative (CO), and Bi-level Integrated System Synthesis (BLISS), which are made compatible to different optimization methods like evolutionary algorithms and Steepest Descent. Mostly, design criteria are minimum-cost, minimum-weight, and minimum-drag under constraints of scenario performance and functional capability. New researches started to study and develop novel design methods, applying optimized and efficient frameworks in various fields. Aerospace science is included in a way that the application of MDO methods with various single- or multi-level frameworks in aerospace vehicles — such as airplane, launch vehicles, and satellites — is seen referring to reliable papers, being considered as a current theme up to date. Aircraft design under uncertainty has been the subject of some recent studies too. In Ahn and Kwon (2006), it was introduced a BLISS based on Reliability-Based Design Optimization (RBDO) framework to design a simplified supersonic transport problem (Ahn and Kwon 2006; Sobieszczanski-Sobieski *et al.* 2000). The study assumed normal distributions with coefficients of variation equal to 0.3 (the ratio of the mean to standard deviation) on each of the 10 design variables considered such as wing area, span, and others describing aircraft geometry. In Smith and Mahadevan (2003), it was solved a spacecraft conceptual optimization problem using RBDO to consider uncertain design variables, reflecting the possibility of minor design changes later in the design process. Probabilistic error terms were added to the responses of the aerodynamics and structural analysis output with assumed values of 10%. The optimization problems were solved with several MDO architectures and First-Order Reliability Method (FORM) based reliability analysis methods. The aforementioned studies consider uncertainties in the design variables or parameters such as atmospheric conditions or material properties.

This study aimed to introduce the decoupling of Uncertainty-based Multidisciplinary Design Optimization (UMDO) method, applying it to design UAVs as a case study. The following section introduces UMDO methods. Then, it is developed UAV decoupling UMDO algorithm, based on MDO method in single-level frameworks (MDF), using genetic algorithm optimization and sequential quadratic programming (SQP). In the section "Implementation of Uncertainty-Based Multidisciplinary Design Optimization Methods and Their Comparison", the redesign of Global Hawk UAV was made using prepared algorithm, and the comparison was carried out between MDO and UMDO while validating the results.

UMDO METHODS

MDO is a branch of engineering science that applies optimization methods for solving design problems with multiple contexts and themes with coupled parameters. The method is called multidisciplinary optimization or Multidisciplinary System Design Optimization (MSDO). The method allows designers to consider related themes simultaneously (Olds 1993). Any deviation in the designer's assumptions (*i.e.*, the material strength or the manufacturing precision of a structural member) or approximate analysis methods may result in the failure of the optimized design because the results take place in feasibility bound.

Any design variable, parameter, or any output from analysis codes in a given optimization problem can be considered as uncertain quantities provided that the uncertainty can be mathematically represented. Uncertainty at the early phases of the development exists due to the limited knowledge concerning the system characteristics and due to the low-fidelity simulations and analyses performed. UMDO methods are recent, still under development, and partially applied in conceptual phases. Many methods currently exist for quantifying the behavior of aleatory and epistemic uncertainty. The methods most often applied in design optimization are interval analysis, fuzzy numbers, and probability theory. The choice of the method is driven by the quantity of information available to the designer about the source of uncertainty. In general, sources of uncertainty in which there is insufficient data to accurately estimate a probability density function (PDF), interval analysis or fuzzy numbers are preferred (Hu and Qiu 2010; Schueller and Jensen 2008; Hajela 2002; Vittal and Hajela 2003). UMDO process relies on 2 steps (Fig. 1): uncertain system modeling and UMDO procedure.

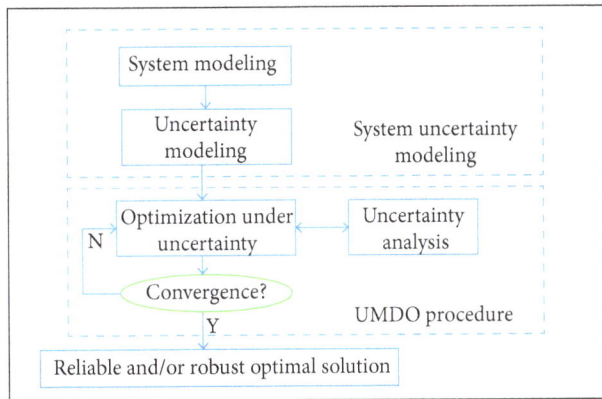

Figure 1. UMDO process (Yao *et al*. 2011).

The first phase of the process consists on desired system and mathematical uncertainty modeling; the second phase includes optimization in the presence of uncertainty, resistance, and uncertainty analysis, as well as the reliability of the answer. UMDO should match the mathematical uncertainty modeling (probability theory, possibility, etc.) and type of design algorithm (AAO, MDF, CO, etc.). In an optimum problem designing, the failure of the plan will be determined by the problem constraints. If these constraints are identified by uncertainty variable, the output of the constraints will have some uncertainty. There are 2 ways to resolve the issue through reliability strategies: nested and decoupling approaches.

In the first method (nested approach), reliability analysis is carried out through optimization cycles for feasibility in the presence of uncertainty. For this purpose, at each iteration of optimizer, reliability analysis is carried out through uncertainty modeling in the parameters in mind. One of the theories available can be used for uncertainty modeling (probability theory, possibility, etc.). Depending on the use of existing theories for uncertainty modeling, there are 2 cases (Rao and Cao 2002):

- RBDO: in this method, the probability theory is used for uncertainty modeling. This technique allows the propagation of uncertainty (given probability density function of variables) in the design process to determine their influence on the final answer (Rao 1992). For uncertainty analysis in RBDO strategy, there are various methods such as first- and second-order analysis (first-order reliability method — FORM and second-order reliability method — SORM). In this method, all design variables are mapped to normal distribution space, then the minimum

distance between the design point and feasibility boundary is calculated. The point in the border of feasibility nearest to the design point is known as MPP. The distance between the design point and MPP is called reliability (β) of the desirable value which is determined by the designer and must have a minimum value. Several methods are used to calculate the MPP and β such as performance measure approach (PMA) and reliability index approach (RIA). These methods are based on FORM but differ in methods of finding MPP (Yu *et al*. 1997; Tu *et al*. 1999).

- Possibility-Based Design Optimization (PBDO): when there is not enough information about the uncertainty of design variables, PBDO is appropriate. In this method, the interval or fuzzy theory is used for uncertainty modeling, and interval or FORM-based optimization methods are applied. In general, the results of this method compared to RBDO are the more stable (Du *et al*. 2006).

Finally, to check the reliability of the obtained optimal solution, uncertainty analysis can be used.

In the second method (decoupling approach), optimization cycles and reliability analysis are separated. For this purpose, 4 steps are carried out:

1. Optimization without uncertainty.
2. Uncertainty analysis on the found answer from step 1 by applying the uncertainty on the desired parameters of the issue.
3. Convergence: if the uncertainties violate the possibility of the answer, a reliable optimal solution is obtained.
4. Lack of convergence: if the uncertainties violate the possibility of the answer, a shifting vector of the answers must be found for feasibility. Then step 3 is carried out again.

The important thing in this method is finding the shifting vector of answers in a way that the least number of repetitions are needed. One of the ways to find shifting vector is using PMA method to find MPP for each constraint. In this case, the difference between optimal solution without uncertainty and the answer located at the MPP is considered as a shifting vector, and this cycle continues to achieve reliable optimal solution (Du and Chen 2004). In finding shifting vector, the designer's experience can be very effective.

UAV DECOUPLING UMDO ALGORITHM

Before presenting the design algorithm, it is necessary that various parts of the UAV design be identified, modeled, and transformed into a software code. Other subjects involved in design are: aerodynamic, structure, propulsion, and path simulation.

INTEGRATED DESIGN ALGORITHM WITH UMDO METHOD

UAV design includes, respectively, 11, 15, and 11 common, uncommon, and coupled parameters (a total of 37) in addition to 33 constraints. The above design algorithm, mission definition, and flight scenario exist within which the following parameters are determined: payload geometrical specification and mass; cruise phase speed, altitude, and range; loiter phase speed, altitude, and duration; in addition to stall speed. Optimization methods applied in system level is a combination of genetic algorithm and SQP, taking the number of design parameters and related constraints into consideration. Combining these 2 methods makes the optimized point resulted from genetic algorithm using SQP method more accurate.

Figure 2 presents UAV's multidisciplinary design optimization algorithm in MDF structure, in which an optimizer is located at system level within which UAV design parameters (a total of 37) are achieved in a way that, observing the problem's constraints (a total of 33), optimization criterion (overall mass of the UAV) becomes minimum.

In order to expedite design algorithm and maintain a better convergence, constraints are prioritized, and a certain value is added to the criterion function against any of these priorities not being observed. This way, optimization algorithm will satisfy constraints while minimizing criterion function. Constraints are prioritized according to their importance: 1) geometrical constraints; 2) constraints related to minimum required thrust; 3) constraints related to minimum require lift; 4) static stability constraints; 5) load coefficient constraints; 6) propulsion mass constraints; and 7) constraints related to performing an mission.

In this framework, disciplines would be feasible through multidisciplinary analyses and internal cycles between coupled sub-systems. In other words, internal cycles between coupled sub-systems are continued till all disciplines are feasible in each optimizer evaluation.

The uncertainty of some of these parameters, such as the aerodynamic ones, according to the theory used, is available in the references; however, for the rest of the parameters, there is no authoritative reference. Thus, according to the authors' experience, sources of uncertainty and their values (3σ) are considered in this work as follows:

- Environmental uncertainties include gravity (5% of nominal value) and density (5% of nominal value).
- Uncertainties of the appliance model include mass (5% of nominal value), fuel consumption (5% of nominal value), aerodynamic coefficients (10% of nominal value [5, 17]), and thrust (5% of nominal value).

To provide design algorithm in the presence of uncertainties, it is necessary to identify the sources of the uncertainty. According to the given description, all design methods are based on reliability on

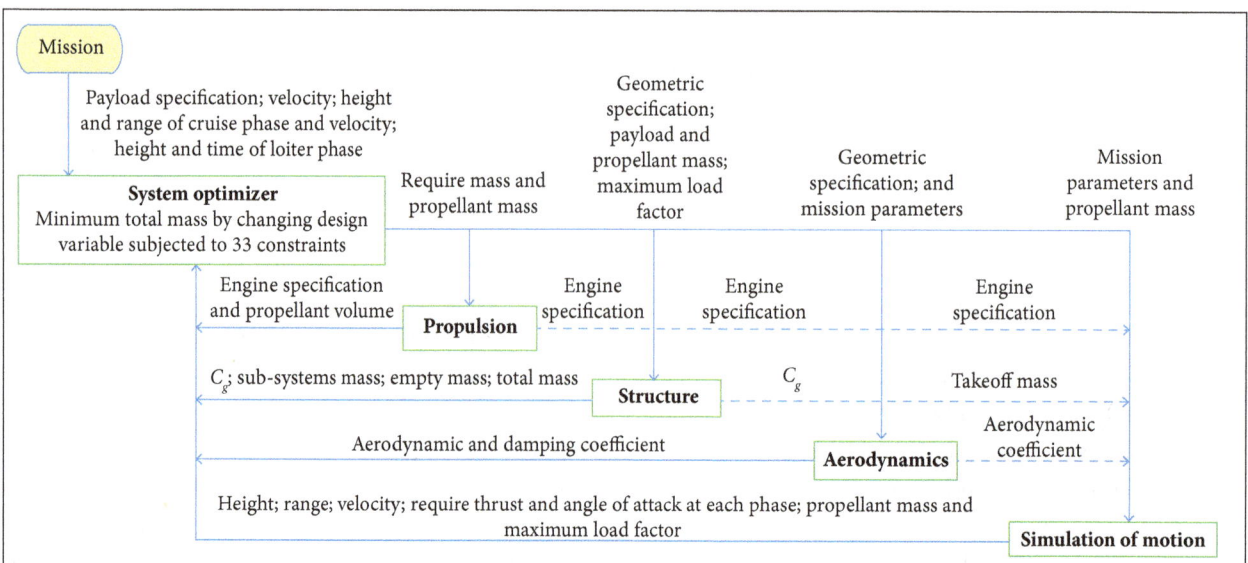

Figure 2. Multidisciplinary design optimization of UAV.

the basis of the mathematics of uncertainty (mapping to a normal distribution space), calculating the border of feasibility and the nearest point from it with a point (MPP). Since the constraints defined in the current problem of UAV are in a way that there is no clear feasibility border, using such methods is very difficult. Here it is presented the method for UAV multidisciplinary optimal design in the presence of uncertainties, which is based on the evolutionary algorithm. The advantage of this method compared to introduction ones is its independence from mathematics of uncertainty and no need for calculating the feasibility border. To perform the design in the presence of uncertainties, the following steps should be taken separately:

1. Optimization without uncertainty (Table 1-3).
2. Uncertainty modeling by the found answer from step 1 as follows: the worst case of uncertainty in a definite range for any constraint is calculated by an optimization algorithm as below. The parameters of optimization algorithm are uncertainty values, and optimization criteria maximize the impossibility of any constraint (for example, minimizing the duration of the flight or an increase in instability). The optimization algorithm in this method, like the original optimization algorithm, is a combination of genetic algorithm and SQP. Therefore, its output will be the set of the worst uncertainties for each constraint):
   Maximize constraint(i), (i = No. of constraints)
   By changing (7 parameters): $U_T, U_{SFC}, U_{We}, U_{CD}, U_{CL}, U_g, U_p$
3. Convergence: uncertainty analysis by applying the set of the worst uncertainties on the parameters of the problem. If the uncertainties do not violate the possibility of the answer, a reliable optimal solution is obtained.

Table 1. Optimization without uncertainty (step 1). Fitness function.

Fitness function	Function name
Minimize takeoff weight in order to below mission parameters	
V_{Cruise}	Cruise velocity
H_{Cruise}	Cruise height
R	Range
V_{Loiter}	Loiter velocity
H_{Loiter}	Loiter height
E	Endurance
Payload specifications	
V_S	Stall speed
R_N	Nose radius

Calculate & check constraint(i), (i = No. of constraints) in order to mission parameters ($V_{Cruise}, H_{Cruise}, R, V_{Loiter}, H_{Loiter}, E$, Payload Specifications, V_S, R_N) & Uncertainty value ($U_i = [U_T, U_{SFC}, U_{We}, U_{CD}, U_{CL}, U_g, U_p]$) for i = 1, 2, ..., n_c & Optimum design variable: $X_W, C_{rW}, C_{tW}, b_W, R_B, L_B, \Lambda_W, C_{rH}, C_{tH}, b_H, \Lambda_H, C_{rV}, C_{tV}, b_V, \Lambda_V, i_P, Z_W, i_W, i_H, L_N, L_A, \Theta_W, \Gamma_W, \Theta_H, \Gamma_H, Z_{CG}, W_F, T, n$

Table 2. Optimization without uncertainty (step 1). Design variable.

Design variable	Variable name
C_{rW}	Wing root chord
C_{tW}	Wing tip chord
b_W	Wing span
R_B	Body radius
L_B	Body length
Λ_W	Wing sweep
C_{rH}	Horizontal tail root chord
C_{tH}	Horizontal tail tip chord
b_H	Horizontal tail span
Λ_H	Horizontal tail sweep
C_{rV}	Vertical tail root chord
C_{tV}	Vertical tail tip chord
b_V	Vertical tail span
Λ_V	Vertical tail sweep
i_P	Propulsion system incidence
Z_W	Wing vertical position
i_W	Wing incidence
i_H	Horizontal tail incidence
L_N	Nose body length
L_A	Aft body length
Θ_W	Wing twist
Γ_W	Wing dihedral
Θ_H	Horizontal tail twist
Γ_H	Horizontal tail dihedral
Z_{CG}	Vertical position of gravity center
W_F	Fuel weight
T	Thrust
N	Load factor

4. Lack of convergence: if the uncertainties violate the possibility of the answer, a shifting vector of the answers, appropriate for the violated constraint, must be found for feasibility — for example, if the constraints related to flight time requirements is not met. The difference between the required fuel and the available one will be added to the latter (required fuel is the output of movement simulation), then step 3 is carried out again.

After designing in the presence of uncertainty, its analysis using the Monte Carlo method is performed to check the final answer. The UAV multidisciplinary optimal design algorithm in the presence of uncertainties is presented in Fig. 3.

UAV design algorithm in the presence of uncertainties was transformed into a software code; through it, the results of re-designing UAV — Global Hawk (RQ-4B) — in the presence of the uncertainties are presented. The results of UAV design have been compared without the presence of uncertainties.

Table 3. Optimization without uncertainty (step 1). Constraints.

Constraints
$V_{PL} + V_F < V_W + V_b$
$b_W > 2R_B$
$b_H > 2R_B$
$b_V > 2R_B$
$C_{tW} < C_{rW}$
$C_{tH} < C_{rH}$
$C_{tV} < C_{rV}$
$Z_W < R_B$
$X_W + C_{rW} < 0.8L_B$
$L_N < 0.5L_B$
$L_A < 0.5L_B$
$C_{m_\alpha} < 0$ without fuel and external payload
$C_{n_\beta} > 0$ without fuel and external payload
$C_{m_\alpha} < 0$ without fuel and with external payload
$C_{n_\beta} > 0$ without fuel and with external payload
$C_{m_\alpha} < 0$ with fuel and without external payload
$C_{n_\beta} > 0$ with fuel and without external payload
$C_{m_\alpha} < 0$ with fuel and external payload
$C_{n_\beta} > 0$ with fuel and external payload
$W_{F\,Opt} > W_{F\,Required}$
$n_{Opt} > n_{Calculated}$
Lift $> W_{TO}$ in stall speed
Lift $> W_{TO}$ in cruise speed
Lift $> W_{TO}$ in loiter speed
$T > T_{Calculated}$ in stall speed
$T > T_{Calculated}$ in cruise speed
$T > T_{Calculated}$ in loiter speed
$V_{Cruise} = V_{Cruise}$ calculated in simulation
$H_{Cruise} = H_{Cruise}$ calculated in simulation
$R = R_{Calculated}$ in Simulation
$V_{Loiter} = V_{Loiter}$ calculated in simulation
$H_{Loiter} = H_{Loiter}$ calculated in simulation
$E = E_{Calculated}$ in simulation

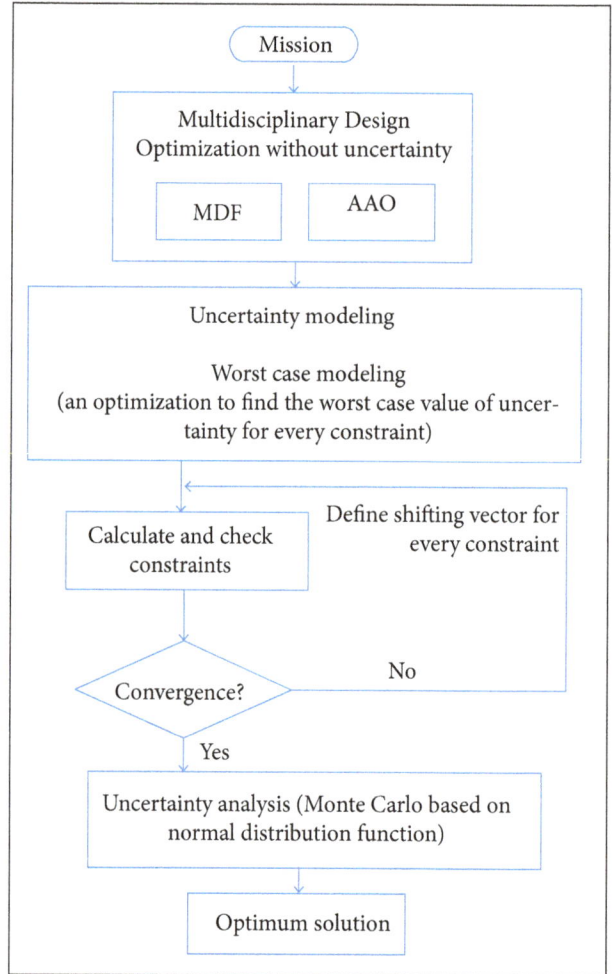

Figure 3. UAV MDO algorithm in the presence of uncertainties.

IMPLEMENTATION OF UMDO METHODS AND THEIR COMPARISON

The results of the MDO algorithm in MDF structure and decoupled method are provided for Global Hawk UAV in the rest of the paper, and parameters such as total mass, runtime code, and the percentage of success have been compared.

To do so, mission information of this UAV was considered as design code input and the output extracted. Then, outputs were compared to real UAV information. Code inputs for Global Hawk redesign are: 575 km/h of velocity cruise, 28 h of flight endurance, maximum altitude of 18 km, and payload mass of 1,360 kg.

In Table 4, by the percentage of success, we wonder what is the percentage of likelihood of fulfillment of all constraints (as explained in previous seasons) in the presence of uncertainty for the final design. As seen in this table, result differences in MDF method with existing information from Global Hawk are between 13 – 23% in mass and less than 9% in geometrical specifications. The following points could be referred to as the reasons for these differences:

- Lack of sufficient information about Global Hawk flight scenario.
- Lack of information about airfoil and lifting instruments at the time of takeoff and landing.
- Lack of information about accurate stall speed and some of functional parameters.
- Lack of information about payload geometric specifications, sub-systems layout, and fuel.
- Lack of consideration of uncertainties.

Convergence procedure in MDF frameworks is shown in Fig. 4. As can be seen, the success rate of optimal response from MDF algorithm (without uncertainty) is 51 and 100% for decoupled method. The run time of this code in decoupled method is 1,600 s more than the MDF algorithm

(without uncertainty). The total mass obtained from the decoupled method is 1,974 kg more than that of the MDF algorithm.

The total mass obtained from the optimal design algorithm in the presence of uncertainties in decoupled method is 14.1 t. This means that, to compensate the failure probability in multidisciplinary optimal algorithm without uncertainties, the mass is increased in order to increase the success chance from 51 to 100%, considering that the uncertainties have also resulted in less difference between the obtained responses and the real values. Regarding the obtained result, although designing without uncertainties in a more optimal way, it is not reliable.

In Figs. 5 and 6, the results of the Monte Carlo analysis are presented for thousands of performances. In Fig. 7, it is observed the 3-D view; in Fig. 8, the presented charts resulted from redesigned UAV simulation of motion.

Table 4. Outputs of design code for Global Hawk redesign in presence of uncertainty.

Section	Value		
	Real value	MDF	Decoupling approach
Wing span (m)	39.9	39	39
Wing area (m²)	63.02	68.2	68.25
Body diameter (m)	1.42	1.28	1.3
Body length (m)	14.5	14.5	14.5
Empty mass (kg)	5,868	5,229	5,538
Takeoff mass (kg)	14,628	12,159	14,133
Propellant mass (kg)	7,400	5,570	7,235
Time of optimization(s)	-	21,507	23,107
Percentage of success	-	51	100

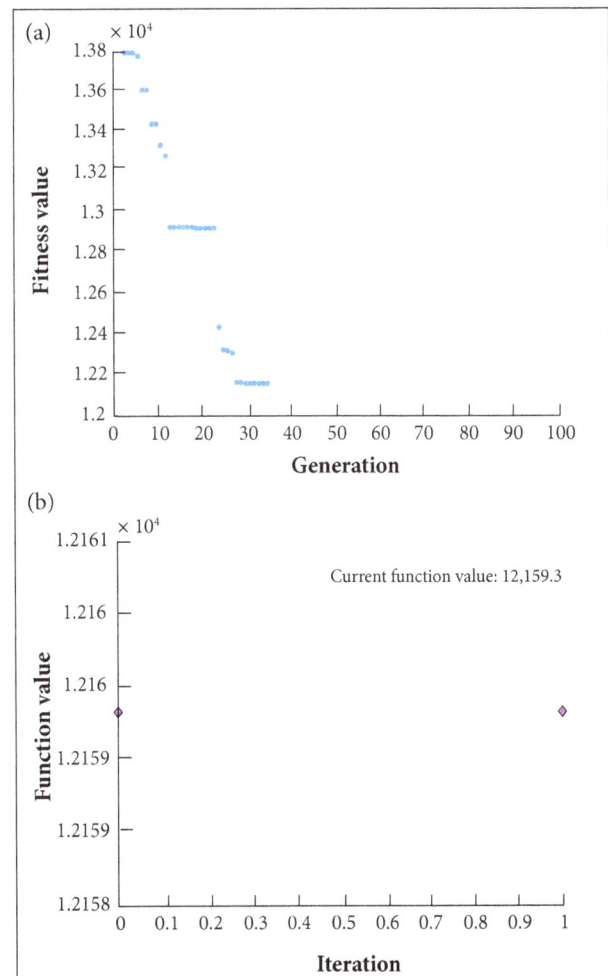

Figure 4. (a) Convergence trend in GA algorithm ; (b) SQP.

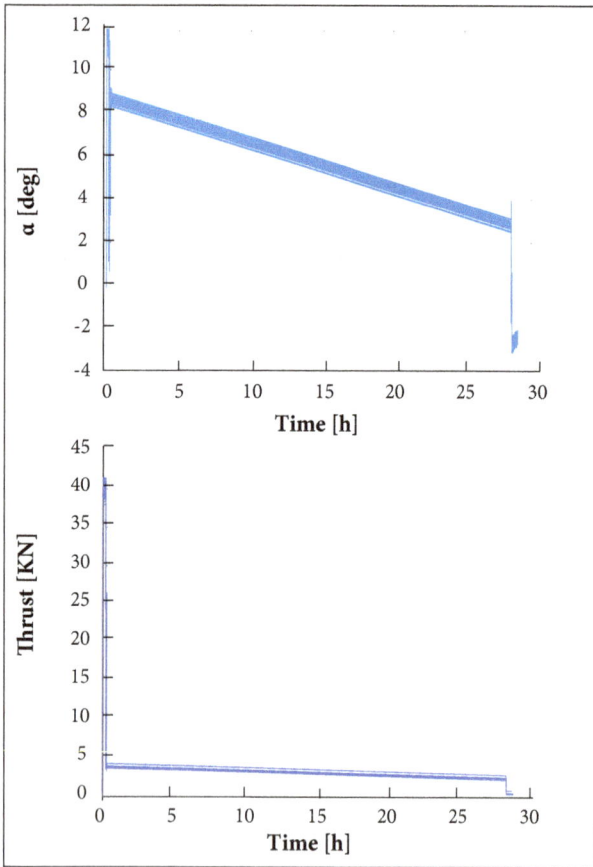

Figure 5. Monte Carlo analysis: angle of attack and thrust *versus* time.

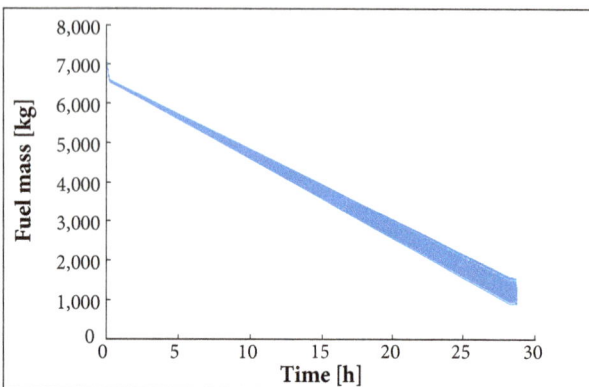

Figure 6. Monte Carlo analysis: fuel consumption *versus* time.

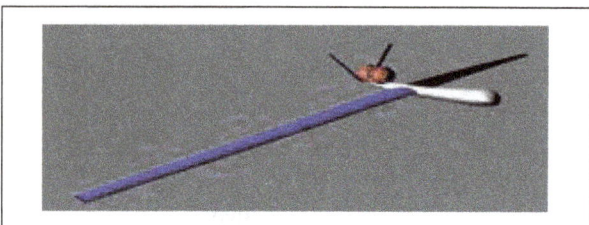

Figure 7. 3-D view of the redesigned UAV.

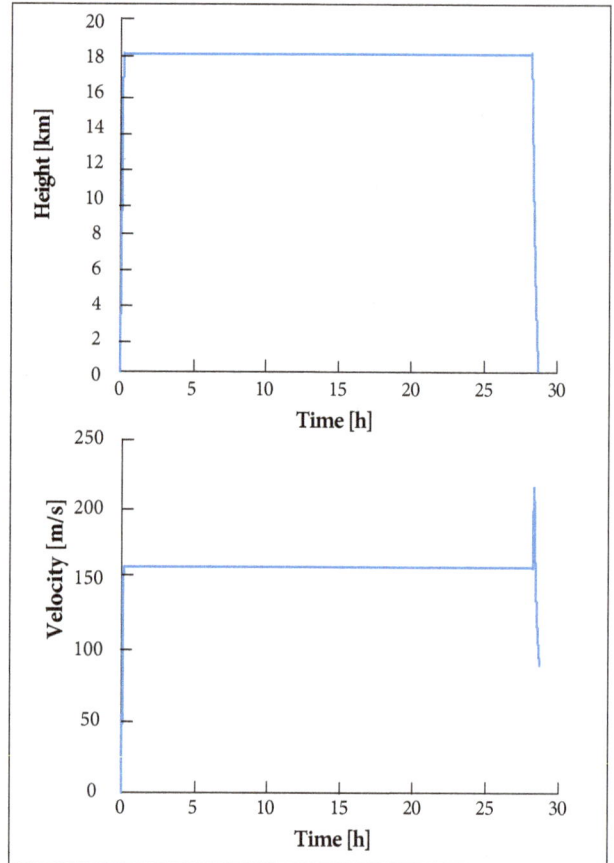

Figure 8. Height and velocity *versus* time.

CONCLUSION

The present study introduced MDO and UMDO frameworks in UAVs, implemented and compared as a case study. To this aim, sub-systems and disciplines involved in design were modeled, and a proposed algorithm for conceptual design of UAV was developed based on multidisciplinary optimization method in MDF framework, using genetic algorithm optimization method and SQP. This algorithm includes 11 common parameters, 15 uncommon parameters, 11 coupled parameters (a total of 37) and 33 constraints.

Regarding the results achieved from MDO in the presence of uncertainties for operational UAV, the results are explained as:

- The comparison of the code outputs, results of the movement simulation, and Monte Carlo analysis for multidisciplinary optimal design code output in the presence of uncertainties indicates the correctness and reliability of results of the proposed algorithm.

- To compensate the failure probability in multidisciplinary optimal algorithm without uncertainties, the mass is increased in order to increase the success chance to 100%.

The advantages of the provided approach are: independence from probability mathematics, appropriate when there is insufficient data to approximate the uncertainties variables or develop a good estimate of the probability density function shape, appropriate speed to calculate the best reliable response, and proper success rate in the presence of uncertainties.

Totally, regarding the presence of the uncertainties in the real world, it is better to consider the mass increase penalty in order to increase the reliability of the design. With the help of this algorithm, we can, according to the defined operation for the UAV and the level of uncertainties of the proposed models in the design, achieve a reliable and optimal answer in the conceptual design phase with the least time and highest accuracy.

AUTHOR'S CONTRIBUTION

Hosseini M and Nosratollahi M, conceived the idea and co-wrote the main text; Hosseini M, prepared the tables and figures. All authors discussed the results and commented on the manuscript.

REFERENCES

Ahn J, Kwon JH (2006) An efficient strategy for reliability-based multidisciplinary design optimization using BLISS. Structural and Multidisciplinary Optimization 31(5):363-372. doi: 10.1007/s00158-005-0565-6

Ahn J, Lee S, Kim JS. (2002). A robust approach to pre-concept design of UCAV considering survivability. Proceedings of the 9th AIAA/ISSMO Symposium on Multidisciplinary Analysis and Optimization; Atlanta, USA.

Buonanno MA (2005) A method for aircraft concept exploration using multicriteria interactive genetic algorithms (PhD thesis). Georgia Institute of Technology.

Çavuş N (2009) Multidisciplinary and multiobjective design optimization of an Unmanned Combat Aerial Vehicle (UCAV) (Master's thesis). Ankara: Middle East Technical University.

Choi SM, Nguyen NV, Kim WS, Kim S, Lee JW, Byun YH (2010) Multidisciplinary Unmanned Combat Air Vehicle (UCAV) system design using multi-fidelity models. Proceedings of the 48th AIAA Aerospace Sciences Meeting Including the New Horizons Forum and Aerospace Exposition; Orlando, USA.

Du L, Choi KK, Youn BD (2006) Inverse possibility analysis method for possibility-based design optimization. AIAA J 44(11):2682-2690. doi: 10.2514/1.16546

Du X, Chen W (2004) Sequential optimization and reliability assessment method for efficient probabilistic design. J Mech Des 126(2):225-233. doi: 10.1115/1.1649968

Eisler CA (2003) Multidisciplinary optimization of conceptual aircraft design (Master's thesis). Ottawa: Carleton University.

Geethaikrishnan C (2003) Multidisciplinary design optimization strategy in multi-stage launch vehicle conceptual design (PhD thesis). Bombay: Indian Institute of Technology.

Giunta A (1997) Aircraft multidisciplinary design optimization using design of experiments theory and response surface modeling methods (PhD thesis). Blacksburg: Virginia Polytechnic Institute and State University.

González LF, Srinivas K, Périaux J, Whitney EJ (2005) Multidisciplinary design optimization of Unmanned Aerial Vehicles (UAV) using multi-criteria evolutionary algorithms. Proceedings of the 6th World Congress of Structural and Multidisciplinary Optimization; Rio de Janeiro, Brazil.

Goraj Z (2005) Design challenges associated with development of a new generation UAV. Aircr Eng Aerosp Tec 77(5):361-368. doi: 10.1108/00022660510617086

Hajela P (2002) Soft computing in multidisciplinary aerospace design - new directions for research. Progr Aero Sci 38(1):1-21. doi: 10.1016/S0376-0421(01)00015-X

Hendrich TJM (2011) Multidisciplinary design optimization in the conceptual design phase creating a conceptual design of the blended wing-body with the BLISS optimization strategy (Master's thesis). Delft: Delft University of Technology.

Hu J, Qiu Z (2010) Non-probabilistic convex models and interval analysis method for dynamic response of a beam with bounded uncertaint. Appl Math Model 34(3):725-734. doi: 10.1016/j.apm.2009.06.013

Iqbal LU (2009) Multidisciplinary design and optimization (MDO) methodology for the aircraft conceptual design (PhD thesis). West Lafayette: Purdue University.

Jaeger L, Gogu C, Segonds S, Bes C (2013) Aircraft multidisciplinary design optimization under both model and design variables uncertainty. J Aircraft 50(2):528-538. doi: 10.2514/1.C031914

Lee DS, Gonzalez LF, Srinivas K, Auld DJ, Wong KC (2007) Aerodynamic shape optimisation of Unmanned Aerial Vehicles using hierarchical asynchronous parallel evolutionary algorithms. Int J Comput Intell Res 3(3):229-250.

Lee JW, Nguyen NV, Choi SM, Kim WS, Jeon KS, Byun YH (2009) Multidisciplinary Unmanned Combat Air Vehicle-UCAV design optimization using variable complexity modeling. Proceedings of the 9th AIAA Aviation Technology, Integration, and Operations Conference (ATIO); Hilton Head, USA.

Mattos BS, Secco NR (2013) An airplane calculator featuring a high-fidelity methodology for tailplane sizing. J Aerosp Technol Manag 5(4):371-386. doi: 10.5028/jatm.v5i4.254

Morris SJ, Kroo I (1990) Aircraft design optimization with dynamic performance constraints. J Aircraft 27(12):1060-1067. doi: 10.2514/3.45982

Neufeld D (2010) Multidisciplinary aircraft conceptual design optimization considering fidelity uncertainties (PhD thesis). Toronto: Ryerson University.

Olds JR (1993) Multidisciplinary design techniques applied to conceptual aerospace vehicle design (PhD thesis). Raleigh: North Carolina State University.

Perez RE, Liu HHT, Behdinan K (2004) Evaluation of multidisciplinary optimization approaches for aircraft conceptual design. Proceedings of the 10th AIAA/ISSMO Multidisciplinary Analysis and Optimization Conference; Albany, USA.

Rao SS (1992) Reliability-based design. New York: McGraw-Hill.

Rao SS, Cao L (2002) Optimum design of mechanical systems involving interval parameters. J Mech Des 124(3):465-472. doi: 10.1115/1.1479691

Raymer DP (2002) Enhancing aircraft conceptual design using multidisciplinary optimization (PhD thesis). Stockholm: Royal Institute of Technology.

Rowell LF, Braun RD, Olds JR, Unal R (1999) Multidisciplinary conceptual design optimization of space transportation systems. J Aircraft 36(1):218-226. doi: 10.2514/2.2428

Schueller G, Jensen H (2008) Computational methods in optimization considering uncertainties - an overview. Comput Meth Appl Mech Eng 198(1):2-13. doi: 10.1016/j.cma.2008.05.004

Smith, Mahadevan S (2003) Probabilistic methods for aerospace system conceptual design. J Spacecraft Rockets 40(3):411-418. doi: 10.2514/2.3961

Sóbester A, Keane AJ (2006) Multidisciplinary design optimization of UAV airframes. Proceedings of the 47th AIAA/ASME/ASCE/AHS/ASC Structures, Structural Dynamics, and Materials Conference; Newport, USA.

Sobieszczanski-Sobieski J, Agte JS, Sandusky RR (2000) Bilevel integrated system synthesis. AIAA J 38(1):164-172. doi: 10.2514/2.937

Tekinlap O, Cavus N (2012) Multiobjective conceptual design of an Unmanned Combat Air Vehicle. Proceedings of the 12th AIAA Aviation Technology, Integration, and Operations (ATIO) Conference and 14th AIAA/ISSMO Multidisciplinary Analysis and Optimization Conference; Indianapolis, USA.

Tianyuan H, Xiongqing Y (2009) Aerodynamic/stealthy/structural multidisciplinary design optimization of Unmanned Combat Air Vehicle. Chin J Aeronaut 22(4):380-386. doi: 10.1016/S1000-9361(08)60114-4

Tu J, Choi KK, Park YH (1999) A new study on reliability-based design optimization. J Mech Des 121(4):557-564. doi: 10.1115/1.2829499

Vittal S, Hajela P (2003) Probabilistic design using empirical distributions. Proceedings of the 44th AIAA/ASME/ASCE/AHS/ASC Structures, Structural Dynamics, and Materials Conference; Norfolk, USA.

Yao W, Chen X, Luo W, van Tooren M, Guo J (2011) Review of uncertainty-based multidisciplinary design optimization methods for aerospace vehicles. Progr Aero Sci 47(6):450-479. doi: 10.1016/j.paerosci.2011.05.001

Yu X, Choi KK, Chang KH (1997) A mixed design approach for probabilistic structural durability. Struct Optim 14(2):81-90. doi: 10.1007/BF01812509

Zill T , Böhnke D, Nagel B (2011) Preliminary aircraft design in a collaborative multidisciplinary design environment. Proceedings of the 11th AIAA Aviation Technology, Integration, and Operations (ATIO) Conference; Virginia Beach, USA.

19

Evaluation of Conceptual Midcourse Guidance Laws for Long-Range Exoatmospheric Interceptors

Mohsen Dehghani Mohammad-abadi[1], Seyed Hamid Jalali-Naini[1]

ABSTRACT: This paper presents a comprehensive study on the performance analysis of 8 conceptual guidance laws for exoatmospheric interception of ballistic missiles. The problem is to find the effective thrust direction of interceptor for interception of short-to-super range ballistic missiles. The zero-effort miss and the generalized required velocity concept are utilized for interception of moving targets. By comparison of the 8 conceptual guidance laws, the thrust direction is suggested to be in the direction of generalized velocity-to-be-gained, or constant velocity-to-be-gained direction, rather than to be in the direction along zero-effort miss, or that of linear optimal solution for long-to-super range interception. Even for short coasting ranges, the generalized velocity-to-be-gained may be utilized because of reasonable computational burden for required velocity rather than the numerical computation for zero-effort miss or linear optimal solution with the same miss distance error. In addition, the fuel consumption of the suggested direction has less sensitivity due to estimation error in intercept time. The guidance law based on constant velocity-to-be-gained direction and the optimal solution are suitable for satellites launch vehicles and space missions.

KEYWORDS: Exoatmospheric midcourse guidance, Effective thrust direction, Zero-effort miss, Velocity-to-be-gained, Long-range interceptor, Anti-ballistic guidance.

INTRODUCTION

Exoatmospheric intercept guidance improvements are of high interest in anti-ballistic air defense systems. The main subjects in this area are focused on midcourse and terminal phases of flight for anti-ballistic interceptors. The design considerations for the midcourse guidance are different from the terminal phase one. In the midcourse phase, the on-board trajectory optimization and trajectory shaping are the main issues, whereas the noise contamination and hit probability against very-high speed targets are the key issues for a terminal guidance law (Zarchan 2012).

The literature on exoatmospheric intercept guidance laws can be categorized into intercept guidance laws against moving targets and guidance laws for space missions including ballistic missiles. Since the concepts and guidance algorithms of ballistic missiles are similar to space vehicle guidance laws, we put them in the same category. The early literature on the subject of optimal 2-point guidance for interception of moving targets is based on zero-effort miss (ZEM) in flat-Earth model (Bryson and Ho 1975). In this case, the acceleration command in the optimal energy problems is obtained proportional to ZEM vector. Precisely speaking, the commanded acceleration of optimal energy guidance laws with final constraints in linear systems is obtained in the form of the predicted error vector pre-multiplied by a gain matrix. In the case that the final position vector is only constrained, the solution simplifies to ZEM vector pre-multiplied by a time-varying gain matrix (Rusnak and Meir 1991). In a special case, if the airframe and control systems are assumed to be identical for 3 axes, the matrix gain

1.Tarbiat Modares University – Faculty of Mechanical Engineering – Aerospace Group – Tehran/Tehran – Iran.

Author for correspondence: Seyed Hamid Jalali-Naini | Tarbiat Modares University – Faculty of Mechanical Engineering – Jalal Al Ahmad Street, No 7 | P.O. box14115-111 – Tehran/Tehran – Iran | E-mail: shjalalinaini@modares.ac.ir

simplifies to a scalar, *i.e.* the commanded acceleration becomes proportional to ZEM (Rusnak and Meir 1991). Two classes of explicit guidance laws based on ZEM have been developed with different assumptions for interceptor dynamics and types of target maneuvers (Jalali-Naini 2004).

In spherical-Earth model, in spite of the assumption of a perfect control system, the optimal maneuver is not obtained in the direction of ZEM because of the non-linear nature of the gravitational acceleration. Most literature on exoatmospheric intercept problems utilized the ZEM vector as an effective direction for thrust vectoring of the interceptor (Massoumnia 1995; Feng *et al.* 2009; Li *et al.* 2013). The ZEM can be approximated in an inverse-square gravity field (Li *et al.* 2013; Mohammad-abadi and Jalali-Naini 2016) or numerically computed on-board with a reasonable integration time step as in predictive guidance scheme (Zarchan 2012). As mentioned before, even ZEM is computed exactly in the spherical-Earth model; the acceleration command along the ZEM is not, mathematically, an optimal solution.

On the other hand, guidance laws for space missions are based on the concept of required velocity and velocity-to-be-gained (Battin 1999; Martin 1965, 1966). At a first glance, the concepts of the 2 guidance categories seem to be different. The concepts of required velocity and velocity-to-be-gained can also be utilized or generalized for interception of moving targets (Jalali-Naini and Pourtakdoust 2005; Chen *et al.* 2010). The velocity-to-be-gained vector becomes proportional to ZEM when the gravitational acceleration is assumed to be constant. In a linearized inverse square gravity field, the velocity-to-be-gained vector is obtained in the form of ZEM pre-multiplied by a time-varying gain matrix (Jalali-Naini and Pourtakdoust 2007). Both ZEM and required velocity can be calculated for a linearized gravity field. Several solutions have been presented using a linearized gravity with different assumptions as treated by Newman (1996) and Deihoul and Massoumnia (2003) for interception of ballistic missiles. The ZEM was also obtained for a linear gravity considering control system dynamics and target maneuvers (Jalali-Naini 2008).

Several anti-ballistic guidance schemes were presented based on ZEM, as mentioned earlier. In these guidance schemes, the corrective maneuver is applied in the direction proportional to ZEM, but the guidance gain is modified, manipulated, and/or theoretically or empirically designed. For space missions, Battin (1999) introduced a guidance scheme in order to keep the direction of velocity-to-be-gained constant, and Sokkappa

(1966) obtained an approximate optimal guidance assuming Q-matrix to be constant. Circi (2004) compared Battin's formula with the numerical optimal solution for a satellite launch vehicle. For short-range anti-ballistic interception, guidance laws based on ZEM perform well whereas for long-to-super range interception, the maneuvering direction needs to be modified to the direction of velocity-to-be-gained or possibly an optimal one. The question is: what direction should be utilized for what ranges. The present study focuses on quantifying the answer to this question, based on accuracy and some implementation issues. Fortunately, several efficient algorithms are available for on-board computation of required velocity and Q-matrix (Zarchan 2012; Arora *et al.* 2015; Ahn *et al.* 2015).

There is another type of guidance laws for space missions or interception in exoatmosphere, referred to as General Energy Management (GEM) for solid rocket motors without cut-off capability (Zarchan 2012). In this class of guidance schemes, the maneuvering direction is somewhat deviated from a desired direction, ZEM or velocity-to-be-gained, in order to manage the wasting of extra fuel of the rocket so as the space vehicle reaches the required velocity at burnout. Since our study focuses on optimal energy guidance, GEM-type guidance schemes are beyond its scope.

In this article, the performance of the midcourse phase of exoatmospheric interceptors is compared for 8 conceptual guidance schemes. It is assumed that this midcourse phase is followed by a coasting phase. In other words, the interceptor is due to reach near the target position, coasting ballistically, where a small kinetic kill vehicle (KKV) is due to separate in order to intercept its target with minimum effort.

BASIC FORMULATION

The governing equations of motion for a particle P (interceptor or target) with a given acceleration vector, $\mathbf{a}_p(t)$, are given by:

$$\dot{\mathbf{r}}_p = \mathbf{v}_p \tag{1a}$$

$$\dot{\mathbf{v}}_p = \mathbf{a}_p(t) \tag{1b}$$

where: $\mathbf{r}_p$, $\mathbf{v}_p$, and $\mathbf{a}_p$ denote position, velocity, and acceleration vectors at current time t in an inertial reference, respectively; the subscript p also represents the particle P.

The final position (at final time t_f) is obtained by integrating twice with respect to the time as follows:

$$\mathbf{r}_p(t_f) = \mathbf{r}_p(t) + t_{go}\mathbf{v}_p(t) + \int_t^{t_f}\int_t^{\xi} \mathbf{a}_p(\tau)d\tau d\xi \qquad (2)$$

where: τ and ξ are dummy indices for time. By converting the preceding double integral to the single one, we have:

$$\mathbf{r}_p(t_f) = \mathbf{r}_p(t) + t_{go}\mathbf{v}_p(t) + \int_t^{t_f}(t_f - \xi)\mathbf{a}_p(\xi)d\xi \qquad (3)$$

where: $t_{go} = t_f - t$ is the time-to-go until the final time.

In an exoatmospheric free-flight motion, we have $\mathbf{a}_p = \mathbf{G}_p$, where $\mathbf{G}_p$ is the gravitational acceleration, i.e. $\mathbf{G}_p = -\mu\mathbf{r}_p/|\mathbf{r}_p|^3$ for a spherical-Earth model, and μ is the Earth's gravitational constant. Therefore, Eq. 3 may be written in the following form:

$$\mathbf{r}_p(t_f) = \mathbf{r}_p(t) + t_{go}\mathbf{v}_p(t) + \int_t^{t_f}(t_f - \xi)\mathbf{G}_p[\mathbf{r}_p(\xi)]d\xi \qquad (4)$$

The preceding equation simplifies for a special case of constant gravity, that is,

$$\mathbf{r}_p(t_f) = \mathbf{r}_p + t_{go}\mathbf{v}_p + \frac{1}{2}\mathbf{G}_p t_{go}^2 \qquad (5)$$

The 3-D intercept geometry with respect to an inertial reference ($Oxyz$) is shown in Fig. 1, in which the interceptor I, having velocity $\mathbf{v}_I$, is pursuing its target T, with velocity $\mathbf{v}_T$. The interceptor and target position vectors are denoted by $\mathbf{r}_I$ and $\mathbf{r}_T$, respectively. The relative position $\mathbf{r}$ and velocity $\mathbf{v}$ for the interception problem are defined as:

$$\mathbf{r} = \mathbf{r}_T - \mathbf{r}_I \qquad (6a)$$

$$\mathbf{v} = \mathbf{v}_T - \mathbf{v}_I \qquad (6b)$$

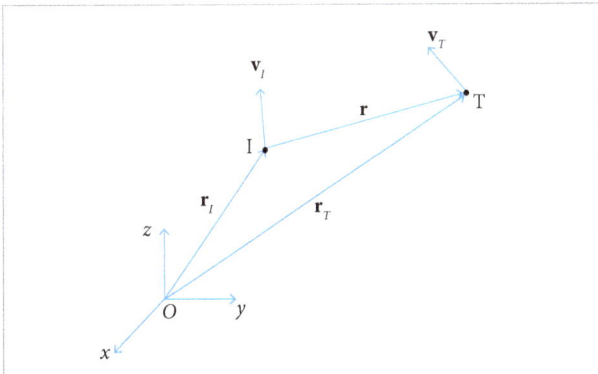

Figure 1. Engagement geometry.

The relative equations of motion are then given by:

$$\dot{\mathbf{r}} = \mathbf{v} \qquad (7a)$$

$$\dot{\mathbf{v}} = \mathbf{a}_T - \mathbf{a}_I \qquad (7b)$$

where: $\mathbf{a}_T$ and $\mathbf{a}_I$ are the respective target and interceptor accelerations in inertial reference. Using Eqs. 3 and 6a, the final relative position is written as:

$$\mathbf{r}(t_f) = \mathbf{r}_T(t_f) - \mathbf{r}_I(t_f) = \mathbf{r}_T(t) + t_{go}\mathbf{v}_T(t) + \qquad (8)$$
$$+ \int_t^{t_f}(t_f - \xi)\mathbf{a}_T(\xi)d\xi - \mathbf{r}_I(t) - t_{go}\mathbf{v}_I(t) - \int_t^{t_f}(t_f - \xi)\mathbf{a}_I(\xi)d\xi$$

Therefore, the relative formulation for miss distance is expressed as follows:

$$\mathbf{r}(t_f) = \mathbf{r}(t) + t_{go}\mathbf{v}(t) + \int_t^{t_f}(t_f - \xi)[\mathbf{a}_T(\xi) - \mathbf{a}_I(\xi)]d\xi \qquad (9)$$

In an exoatmospheric free-flight motion, we have $\mathbf{a}_T = \mathbf{G}_T$ and $\mathbf{a}_I = \mathbf{G}_I$, and the substitution yields:

$$\mathbf{r}(t_f) = \mathbf{r}(t) + t_{go}\mathbf{v}(t) + \qquad (10)$$
$$+ \int_t^{t_f}(t_f - \xi)\{\mathbf{G}[\mathbf{r}_T(\xi)] - \mathbf{G}[\mathbf{r}_I(\xi)]\}d\xi$$

The solution of the preceding equation is not straightforward. One approach to the problem is the linearization of its non-linear term, i.e. gravitational acceleration, that is,

$$\mathbf{G}_I = \mathbf{G}_T + \mathbf{E}(t)\mathbf{r} \qquad (11)$$

where:

$$\mathbf{E}(t) = -\frac{\partial \mathbf{G}}{\partial \mathbf{r}_I}\bigg|_{\text{reference point}} , \quad \mathbf{G} = \mathbf{G}_T - \mathbf{G}_I \qquad (12)$$

Therefore, the linearized state-space form for an exoatmospheric interceptor having thrust acceleration, $\mathbf{a}_{th}$ ($\mathbf{a}_I = \mathbf{G}_I + \mathbf{a}_{th}$), is obtained as (Newman 1996):

$$\frac{d}{dt}\begin{bmatrix}\mathbf{r}\\\mathbf{v}\end{bmatrix} = \begin{bmatrix}\mathbf{0} & \mathbf{I}\\-\mathbf{E}(t) & \mathbf{0}\end{bmatrix}\begin{bmatrix}\mathbf{r}\\\mathbf{v}\end{bmatrix} + \begin{bmatrix}\mathbf{0}\\-\mathbf{a}_{th}\end{bmatrix} \qquad (13)$$

where: $\mathbf{I}$ is a 3×3 identity matrix, and

$$\mathbf{E} = \frac{\mu}{R^3}\left[\mathbf{I} - 3\frac{\mathbf{R}\mathbf{R}^T}{R^2}\right], \quad R = |\mathbf{R}| \qquad (14)$$

Several assumptions can be made for the parameter $\mathbf{R}$ such as $\mathbf{R} = \mathbf{r}_I$ (Newman 1996) and $\mathbf{R} = \mathbf{r}_T$, $\mathbf{R} = (\mathbf{r}_I + \mathbf{r}_T)/2$, $\mathbf{R} = \mathbf{r}_T(t_f)$ or $\mathbf{R} = \mathbf{r}_I + \mathbf{r}_T + \mathbf{r}_T(t_f))/3$ as treated by Deihoul (2003). The author claimed that the last relation for $\mathbf{R}$ gives better results, so it is used in our comparison study.

The solution of the homogenous differential equation ($\ddot{\mathbf{r}} + \mathbf{Er} = \mathbf{0}$) is obtained in terms of the state-transition matrix, $\mathbf{\Phi}(t, t_0)$, as treated by Newman (1996) when $\mathbf{E}$ is assumed to be a constant matrix, that is,

$$\mathbf{r}(t) = \mathbf{\Phi}_{11}(t, t_0)\mathbf{r}(t_0) + \mathbf{\Phi}_{12}(t, t_0)\mathbf{v}(t_0) \tag{15}$$

where: $\mathbf{\Phi}_{ij}$ is a 3×3 submatrix ($i, j = 1, 2$) in the following partitioned matrix form:

$$\mathbf{\Phi}(t, t_0) = \begin{bmatrix} \mathbf{\Phi}_{11} & \mathbf{\Phi}_{12} \\ \mathbf{\Phi}_{21} & \mathbf{\Phi}_{22} \end{bmatrix} \tag{16}$$

and

$$\mathbf{\Phi}_{11} = \mathbf{\Phi}_{22} = \mathbf{I} - \frac{1}{2!}\mathbf{E}(t - t_0)^2 + \frac{1}{4!}\mathbf{E}^2(t - t_0)^4 - \dots \tag{17}$$

$$\mathbf{\Phi}_{12} = \mathbf{I}(t - t_0) - \frac{1}{3!}\mathbf{E}(t - t_0)^3 + \frac{1}{5!}\mathbf{E}^2(t - t_0)^5 - \dots \tag{18}$$

$$\mathbf{\Phi}_{21} = -\mathbf{E}(t - t_0) + \frac{1}{3!}\mathbf{E}^2(t - t_0)^3 - \frac{1}{5!}\mathbf{E}^3(t - t_0)^5 + \dots \tag{19}$$

Therefore, the linearized relation of the miss distance vector in terms of the current states is given by:

$$\mathbf{r}(t_f) = \mathbf{\Phi}_{11}(t_f, t)\mathbf{r} + \mathbf{\Phi}_{12}(t_f, t)\mathbf{v} \tag{20}$$

ZERO-EFFORT MISS

The ZEM at the current time, $\mathbf{ZEM}(t)$, is the distance that the interceptor would miss its target position if the interceptor made no corrective maneuver ($\mathbf{a}_{th}$) after the time t (Zarchan 2012), that is,

$$\mathbf{ZEM}(t) = \mathbf{r}_I^*(t_f) - \mathbf{r}_I(t_f)\big|_{\mathbf{a}_{th}(\xi)=\mathbf{0} \text{ for } \xi \geq t} \tag{21}$$

where: $\mathbf{r}_I^*(t_f)$ is the desired final position of the interceptor.

For an exoatmospheric interceptor ($\mathbf{a}_I = \mathbf{G}_I + \mathbf{a}_{th}$), we have:

$$\mathbf{r}_I(t_f)\big|_{\mathbf{a}_{th}(\xi)=\mathbf{0} \text{ for } \xi \geq t} = \mathbf{r}_I + t_{go}\mathbf{v}_I + \\ + \int_t^{t_f} (t_f - \xi)\mathbf{G}[\mathbf{r}_I(\xi)\big|_{\mathbf{a}_{th}=\mathbf{0}}]d\xi \tag{22}$$

The substitution of Eq. 22 into Eq. 21 yields:

$$\mathbf{ZEM}(t) = \mathbf{r}_I^*(t_f) - \mathbf{r}_I - t_{go}\mathbf{v}_I - \int_t^{t_f} (t_f - \xi) \\ \mathbf{G}[\mathbf{r}_I(\xi)\big|_{\mathbf{a}_{th}=\mathbf{0}}]d\xi \tag{23}$$

For a free-falling target, i.e. $\mathbf{a}_T = \mathbf{G}_T$, using Eq. 10, the ZEM relation can also be expressed in relative coordinates as follows:

$$\mathbf{ZEM}(t) = \mathbf{r} + t_{go}\mathbf{v} + \int_t^{t_f} (t_f - \xi) \\ \{\mathbf{G}[\mathbf{r}_T(\xi)] - \mathbf{G}[\mathbf{r}_I(\xi)\big|_{\mathbf{a}_{th}=\mathbf{0}}]\}d\xi \tag{24}$$

where: $\mathbf{r}_I^*(t_f) = \mathbf{r}_T(t_f)$.

The preceding relation simplifies for a special case of constant gravity or for the case that the interceptor is assumed to be near its target as treated by Massoumnia (1995), that is,

$$\mathbf{ZEM}(t) = \mathbf{r} + t_{go}\mathbf{v} \tag{25}$$

Using Eq. 20, the linearized ZEM relation is given by:

$$\mathbf{ZEM}_{\text{Lin}}(t) = \mathbf{\Phi}_{11}(t_f - t)\mathbf{r} + \mathbf{\Phi}_{12}(t_f - t)\mathbf{v} \tag{26}$$

Two different definitions of ZEM are utilized in the guidance theory. In basic definition, the ZEM is defined as a miss distance vector without further control effort. The intercept time is not imposed to the intercept problem and it is the time of the nearest distance between an interceptor and its target without further control effort. The second definition is based on a specified final time and comes from linear optimal guidance laws with the assumption of a fixed final time. The ZEM vector for the basic definition is, here, denoted by $\mathbf{ZEM}_{\text{min}}$ whereas it is denoted by $\mathbf{ZEM}$ for the second definition. The final time, the time of the nearest distance denoted by $t_{f_{ZEM}}^*$, for the basic definition is obtained by $\partial|\mathbf{ZEM}|/\partial t_f = 0$. For example, in a special case of constant gravity, from Eq. 25, we have:

$$\frac{\partial|\mathbf{ZEM}|}{\partial t_{go}} = 0 \Rightarrow t_{go_{ZEM}}^* = -\frac{\mathbf{r} \cdot \mathbf{v}}{|\mathbf{v}|^2} \tag{27}$$

where: $t_{go_{ZEM}}^* = t_{f_{ZEM}}^* - t$.

It is worth noting that the component of ZEM perpendicular to the interceptor-target line-of-sight (LOS), $\mathbf{ZEM}_{\text{PLOS}}$, may be replaced for $\mathbf{ZEM}$ in a guidance formulation as treated by Zarchan (2012).

GENERALIZED REQUIRED VELOCITY

The required velocity, $\mathbf{v}_R$, for Lambert's problem is defined as an instantaneous velocity, required to satisfy the final position constraint in a specified final time (Battin 1999). This concept is well-known in space missions and surface-to-surface applications. The implementation of guidance laws based on the required velocity may be implicit or explicit.

The required velocity concept may be generalized for an intercept problem against a moving target in the endoatmosphere considering interceptor dynamics. The interceptor desired velocity $\mathbf{v}^*$ is the velocity that makes ZEM equal to zero. This desired velocity is referred to as generalized required velocity (Jalali-Naini and Pourtakdoust 2005). The interceptor dynamics is, here, assumed to be perfect, the interceptor moves in the exoatmosphere, and a moving target is considered. For brevity, we use the term "required velocity" instead of "generalized required velocity".

For example, the required velocity for the case of constant gravity is obtained from Eq. 23 as:

$$\mathbf{v}_R = \frac{\mathbf{r}_T(t_f) - \mathbf{r}}{t_{go}} - \frac{1}{2}\mathbf{G}t_{go} \qquad (28)$$

Therefore, the relation between the required velocity and ZEM for the case of constant gravity is simply obtained as:

$$\mathbf{ZEM} = \mathbf{v}_g t_{go} \qquad (29)$$

where: $\mathbf{v}_g$ is referred to as the velocity-to-be-gained ($\mathbf{v}_g = \mathbf{v}_R - \mathbf{v}_I$).

For a spherical-Earth model, the required velocity causes Eq. 23 equal to zero, that is,

$$0 = \mathbf{r}_T(t_f) - \mathbf{r}_I - t_{go}\mathbf{v}_R - \\ - \int_t^{t_f}(t_f - \xi)\mathbf{G}\{\mathbf{r}_I[\mathbf{r}_I(t), \mathbf{v}_R(t), \xi]\}d\xi \qquad (30)$$

The preceding relation may be solved approximately for the guidance problem, which is beyond the scope of the present study.

CONCEPTUAL GUIDANCE LAWS

It is necessary to determine the effective direction of interceptor thrust vector for short-to-super range exoatmospheric intercept problem. The thrust direction is determined by a conceptual guidance law (GL). After calculation of the effective thrust direction, a steering law is needed to convert the errors into commended body rates. The interceptor is assumed to be non-throttleable with thrust cutoff capability. The conceptual guidance laws are, here, categorized in 5 main classes, namely, guidance laws based on ZEM, guidance laws based on linearized ZEM, guidance laws based on generalized velocity-to-be-gained, guidance laws based on constant direction for velocity-to-be-gained, and optimal solution.

GUIDANCE LAWS BASED ON ZERO-EFFORT MISS

In this case, the thrust acceleration is assumed to be applied perfectly in the direction of ZEM. Three guidance schemes may be considered regarding to 2 definitions of ZEM and also the component of ZEM perpendicular to LOS as follows:

$$\mathbf{a}_{th} = |\mathbf{a}_{th}|\,\mathbf{e}_{\mathrm{ZEM}} \qquad \text{for } |\mathbf{ZEM}| > \varepsilon \qquad (31)$$

$$\mathbf{a}_{th} = |\mathbf{a}_{th}|\,\mathbf{e}_{\mathrm{ZEM}_{min}} \qquad \text{for } |\mathbf{ZEM}_{min}| > \varepsilon \qquad (32)$$

$$\mathbf{a}_{th} = |\mathbf{a}_{th}|\,\mathbf{e}_{\mathrm{ZEM}_{PLOS}} \qquad \text{for } |\mathbf{ZEM}_{PLOS}| > \varepsilon \qquad (33)$$

where: $\mathbf{e}_{\mathrm{ZEM}}$, $\mathbf{e}_{\mathrm{ZEM}_{min}}$ and $\mathbf{e}_{\mathrm{ZEM}_{PLOS}}$ are the unit vectors along $\mathbf{ZEM}$, $\mathbf{ZEM}_{min}$, and $\mathbf{ZEM}_{PLOS}$, respectively; ε is an allowable miss distance, determined from practical considerations.

In each guidance law, the powered phase of flight is terminated when its corresponding $|\mathbf{ZEM}|$ becomes equal or less than ε and then it is followed by a coasting phase until intercept. The component of $\mathbf{ZEM}$ normal to LOS is calculated by:

$$\mathbf{ZEM}_{PLOS} = \mathbf{ZEM} - (\mathbf{ZEM}\cdot\mathbf{e}_r)\mathbf{e}_r \qquad (34)$$

where: $\mathbf{e}_r = \mathbf{r}/r$ is the unit vector along LOS ($r = |\mathbf{r}|$).

To calculate the time-to-go until intercept in $\mathbf{ZEM}_{PLOS}$ relation, the component of ZEM along LOS is imposed to be zero. For a special case of a free-falling target in a flat-Earth model, the relation $t_{go} = -r/\dot{r}$ zeros out the LOS component of ZEM.

GUIDANCE LAWS BASED ON LINEARIZED ZERO-EFFORT MISS

Here, conceptual guidance laws are given using linearized relations for the problem. The first is similar to Eq. 31, but with linearized relation for ZEM, that is,

$$\mathbf{a}_{th} = |\mathbf{a}_{th}|\,\mathbf{e}_{\mathrm{ZEM}_{Lin}} \qquad \text{for } |\mathbf{ZEM}_{Lin}| > \varepsilon \qquad (35)$$

where: $\mathbf{e}_{ZEM_{Lin}}$ is the unit vector along $\mathbf{ZEM}_{Lin}$.

The second conceptual guidance is based on the linear optimal guidance law (OGL) obtained by Deihoul and Massoumnia (2003). Their linearized OGL may be expressed in the following form:

$$\mathbf{U}_{Lin} = \mathbf{M}(t_{bgo}, t_{go}, r_T, r_I)\mathbf{ZEM}_{Lin} \tag{36}$$

where: $\mathbf{U}_{Lin}$ is the optimal thrust vector for a throttleable rocket motor; $\mathbf{M}$ is a 3×3 matrix that causes the thrust vector to deviate from $\mathbf{ZEM}_{Lin}$ direction and change its magnitude as well; t_{bgo} is the time-to-go until burnout.

The relation of matrix $\mathbf{M}$ is available in Deihoul and Massoumnia (2003). The thrust direction in the second conceptual guidance is applied in the direction of $\mathbf{U}_{Lin}$ for non-throttleable rocket motors as follows:

$$\mathbf{a}_{th} = |\mathbf{a}_{th}| \, \mathbf{e}_{U_{Lin}} \quad \text{for } |\mathbf{ZEM}_{Lin}| > \varepsilon \tag{37}$$

To compare the performances of the guidance schemes precisely, a third relation based on the exact calculation of ZEM in a spherical-Earth model is written as follows:

$$\mathbf{a}_{th} = |\mathbf{a}_{th}| \, \mathbf{e}_U \quad \text{for } |\mathbf{ZEM}| > \varepsilon \tag{38}$$

where: $\mathbf{e}_U$ is the unit vector along $\mathbf{U}$ calculated by:

$$\mathbf{U} = \mathbf{M}(t_{bgo}, t_{go}, r_T, r_I)\mathbf{ZEM} \tag{39}$$

where $\mathbf{ZEM}$ is computed numerically.

GUIDANCE LAW BASED ON VELOCITY-TO-BE-GAINED

In space missions a class of guidance laws is based on required velocity with the desired thrust acceleration along the velocity-to-be-gained. The required velocity concept and velocity-to-be-gained can be generalized for interception of a moving target, as mentioned before. The conceptual guidance law is then given by:

$$\mathbf{a}_{th} = |\mathbf{a}_{th}| \, \mathbf{e}_{v_g} \quad \text{for } |\mathbf{v}_g| > \varepsilon_v \tag{40}$$

where: $\mathbf{e}_{v_g}$ is the unit vector along the velocity-to-be-gained; ε_v is an allowable velocity-to-be-gained error.

It is worth noting that the preceding conceptual guidance law is equivalent to Eq. 31 for constant gravity assumption as is evident from Eq. 29.

GUIDANCE BASED ON CONSTANT DIRECTION OF VELOCITY-TO-BE-GAINED

An effective direction of thrust acceleration in space missions is obtained by satisfying the relation $\dot{\mathbf{v}}_g \times \mathbf{v}_g = \mathbf{0}$ as follows (Battin 1999):

$$\mathbf{a}_{th} = \mathbf{b} + (q - \mathbf{b} \cdot \mathbf{e}_{v_g})\mathbf{e}_{v_g} \tag{41}$$

where:

$$q = \sqrt{|\mathbf{a}_{th}|^2 - |\mathbf{b}|^2 + (\mathbf{b} \cdot \mathbf{e}_{v_g})^2} \tag{42}$$

$$\mathbf{b} = -\mathbf{Q}\mathbf{v}_g, \quad \mathbf{Q} = \frac{\partial \mathbf{v}_R}{\partial \mathbf{r}_I} \tag{43}$$

This conceptual guidance law causes the direction of velocity-to-be-gained to be fixed in inertial space.

RESULTS AND DISCUSSION

To compare the performance of conceptual guidance laws, a nonlinear flight simulation is utilized. The interceptor and its target are taken as particles in vertical planar motion with perfect dynamics, i.e. the thrust acceleration is assumed to be exactly in the desired direction of computed thrust direction, without any error or delay.

First, guidance laws (Eqs. 31 – 33) with different ZEM definitions, i.e. ZEM, $\mathbf{ZEM}_{min}$, and also the normal component of ZEM, $\mathbf{ZEM}_{PLOS}$, are compared in a flat-Earth model with constant gravity. The interceptor is located at origin (0,0) with $a_{th} = 50$ m/s² ($a_{th} = |\mathbf{a}_{th}|$). The fuel consumption, $\Delta V = |\mathbf{a}_{th}| \, t_{co}$, of the mentioned conceptual guidance laws are shown in Figs. 2 – 4, where t_{co} is the thrust cut-off time, applied when $|\mathbf{ZEM}| \le \varepsilon$. In these figures, the solid lines indicate the fuel consumption when the thrust vector is applied along $\mathbf{ZEM}$ direction. The circle and square signs indicate the values of fuel consumption when the thrust direction is applied along $\mathbf{ZEM}_{min}$ and $\mathbf{ZEM}_{PLOS}$, respectively, for their corresponding resulted final times. In Fig. 2, the fuel consumption is depicted versus predetermined final time for 2 cases of an initial 0 velocity, $v_I(0) = 0$, and a vertical velocity of $v_{I_z}(0) = 1$ km/s for a stationary target at $\mathbf{r}_T = [500 \quad 0]^T$km. As seen in Fig. 2, the fuel consumption depends on the value of the final time when the thrust vector is applied along ZEM.

Also, the minimum fuel consumption is not occurred necessarily when ZEM_{min} is utilized; however, it may occur for a special case. It should be noted that an appropriate final time (or time-to-go until intercept) is estimated in practice.

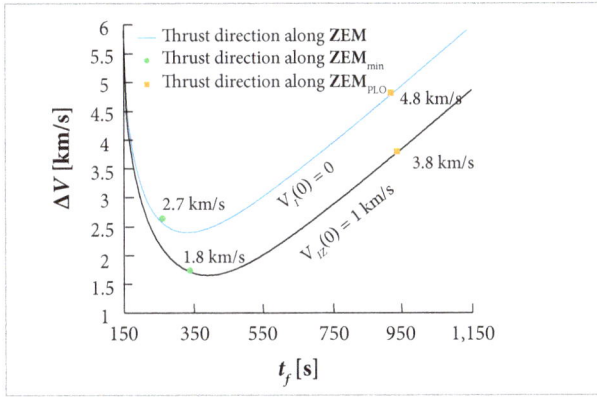

Figure 2. Fuel consumption versus final time for interception of a stationary target in flat-Earth model.

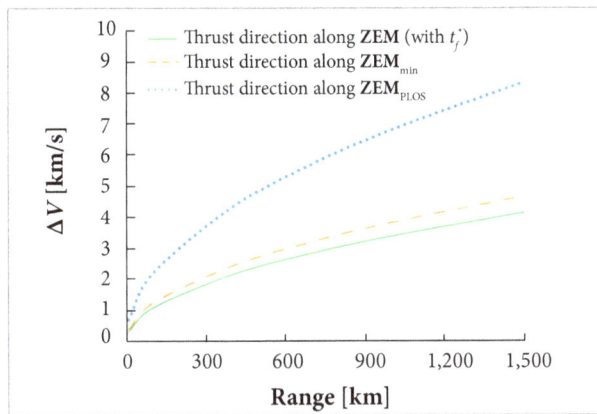

Figure 3. Fuel consumption versus range for interception of a stationary target in flat-Earth model.

In order to investigate more precisely, the fuel consumption is drawn *versus* range in Fig. 3 for the 3 mentioned guidance schemes when the interceptor launches from rest. Other initial values and parameters are similar to Fig. 2. As shown in Figs. 2 and 3, the fuel consumption is highly increased using guidance law (Eq. 33) based on ZEM_{PLOS}. In Fig. 3, the minimum fuel consumption for the case of thrust direction along ZEM is computed by setting the optimum value, t_f^*, for intercept time. In other words, t_f^* is the intercept time for minimum ΔV when the thrust acceleration is imposed along ZEM. The value of ε is, here, chosen as 1 m due to numerical errors.

In the next step, the performances of the conceptual guidance laws are compared in Figs. 4a and 4b for a free-falling target at an initial altitude of 100 km, having a minimum required velocity to hit the origin in a flat-Earth model. Initial values and parameters are similar to Fig. 2, except target position. The target range is taken 100 and 500 km for Figs. 4a and 4b, respectively. A vertical solid line has been drawn for each of Figs. 4a and 4b, showing the maximum possible final time (t_{fmax}), *i.e.* the time of hitting origin (0,0) by the free-falling target. The region at the right-hand side of this vertical line is not acceptable, because the interception of a free-falling target occurs behind the origin in negative altitudes. This is an important difference between the two cases of stationary and free-falling targets.

According to our analysis for a flat-Earth model, the capture criteria for the conceptual guidance law based on ZEM_{PLOS} are highly restricted comparing to the conceptual guidance law based on ZEM, at least, for non-throttleable rockets.

We are now to study the performance of conceptual guidance laws (Eqs. 31, 32, and 40) for a spherical-Earth model in Figs. 5 – 14 against stationary targets. The interceptor fires

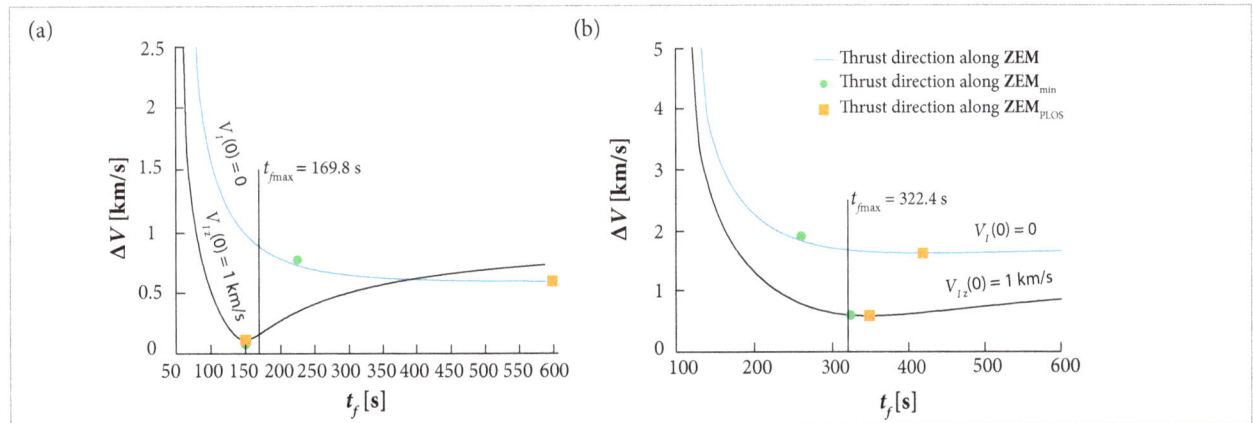

Figure 4. Fuel consumption versus final time for interception of a free-falling target in flat-Earth model. (a) initial range = 100 km; (b) initial range = 500 km.

from position $\mathbf{r}_{I_0} = \begin{bmatrix} 6400 & 0 \end{bmatrix}^T$ km with a velocity of 1 km/s along near vertical horizon (89°) and $a_{th} = 50$ m/s². The initial distance of target from the earth center is taken $r_T = 6{,}500$ km.

First, the fuel consumptions of these guidance laws are plotted in Figs. 5a; 5b and 5c for a target at range angles of 10; 40 and 70°, respectively. As seen in the figures, if the thrust direction is applied along the direction of $\mathbf{v}_g$, the fuel consumption is considerably reduced for long-range applications. To investigate the trajectory of the interceptor based on ZEM direction, four scenarios with different final times are selected. These points, namely, S_1; S_2; S_3 and S_4 are assigned in Fig. 5a for different specified final times when the range angle is 10°.

The typical interceptor trajectories of the mentioned scenarios are illustrated in Figs. 6a, 6b, 6c, and 6d for the points S_1, S_2, S_3, and S_4, respectively, and compared to interceptor trajectories based on $\mathbf{v}_g$ direction with the same final times. The interceptor trajectories based on ZEM and $\mathbf{v}_g$ are viewed by solid and dashed lines, respectively. The thrust cut-off time, t_{co}, is also observed for each trajectory in the figures. In the case of ZEM-based trajectory, increasing total flight time causes an extra revolution of trajectory around the earth center, as shown in Fig. 6d. The maximum limit of total flight time to avoid this phenomena is assigned with S_4 in Fig. 5a and with S_4' in Figs. 5b and 5c with different range angles. The typical interceptor trajectories with final times larger than that of S_4' in Figs. 5b and 5c, are similar to Fig. 6d for ZEM-based trajectories. Moreover, Fig. 7 shows that the fuel consumption of $\mathbf{v}_g$-based guidance law is less sensitive to the estimation of total flight time. In this figure, the total flight time is considered about the total flight time for

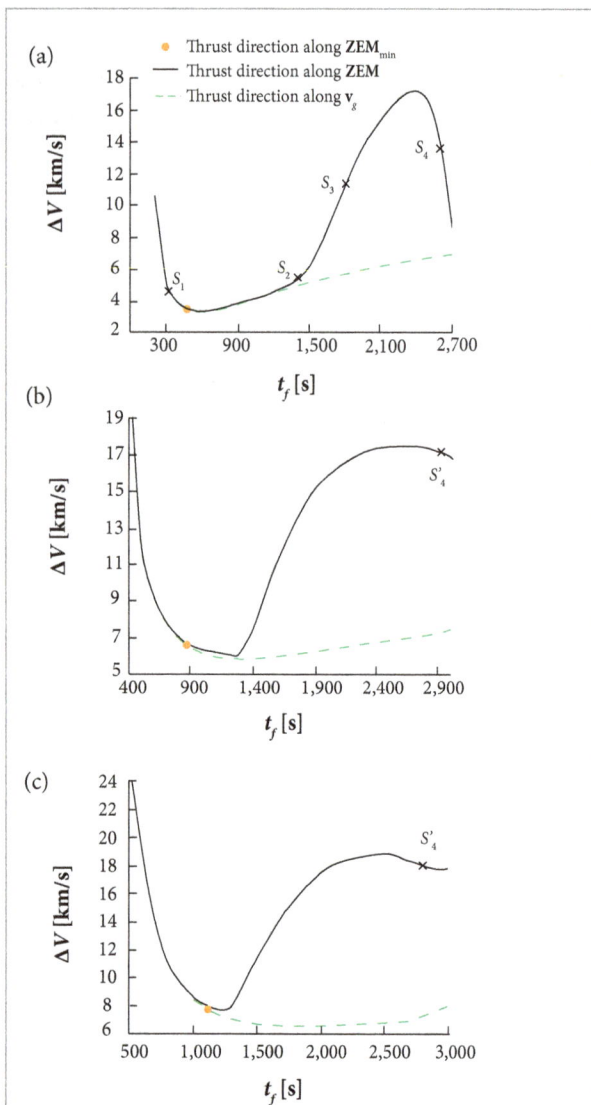

Figure 5. Fuel consumption versus final time for interception of a stationary target in spherical-Earth model. (a) range angle = 10°; (b) range angle = 40°; (c) range angle = 70°.

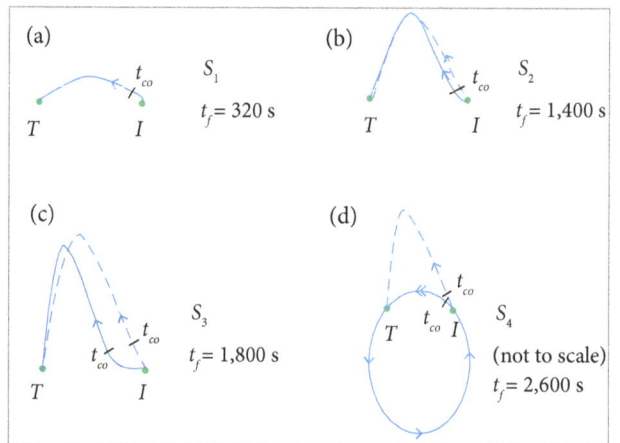

Figure 6. Typical interceptor trajectories for initial range angle = 10° and different values of total flight times (solid line: thrust vector along ZEM direction; dashed line: thrust vector along $\mathbf{v}_g$ direction).

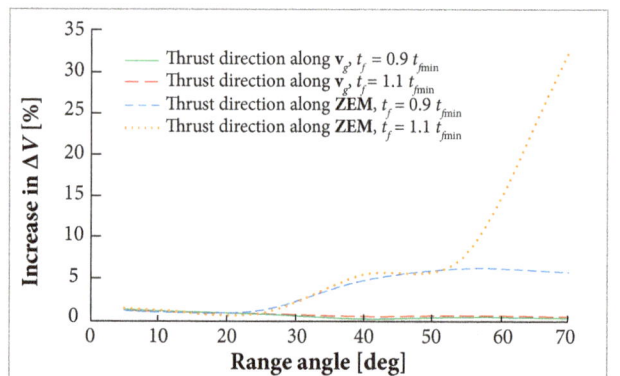

Figure 7. Increase in ΔV for guidance laws based on $\mathbf{v}_g$ and ZEM.

minimum ΔV for each guidance scheme, and it is denoted by $t_{f_{min}}$, which is obtained for each range.

The total rotation of the thrust direction of conceptual guidance laws (Eqs. 31, 32, and 40) can be seen in Figs. 8a, 8b and 8c for different values of thrust acceleration of 50 and 100 m/s² when the range angle is 70°. The final times are chosen

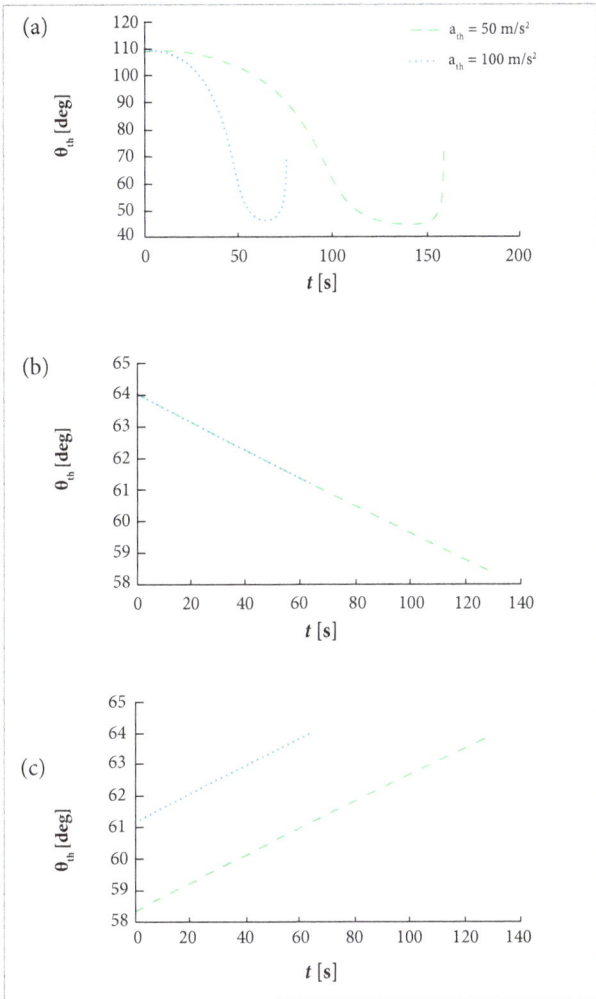

Figure 8. Thrust angle profiles with respect to the equatorial plane for interception of a stationary target for minimum energy orbit with initial range angle = 10°. (a) thrust vector along ZEM direction; (b) thrust vector along $\mathbf{v}_g$ direction; (c) constant $\mathbf{v}_g$ direction.

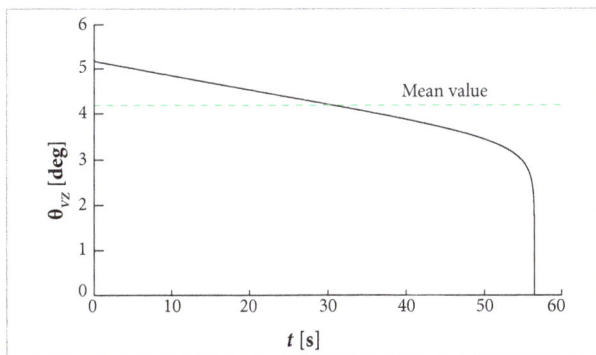

Figure 9. History of the angle between $\mathbf{v}_g$ and ZEM for thrust vector along ZEM direction (minimum energy orbit with a range angle of 10°).

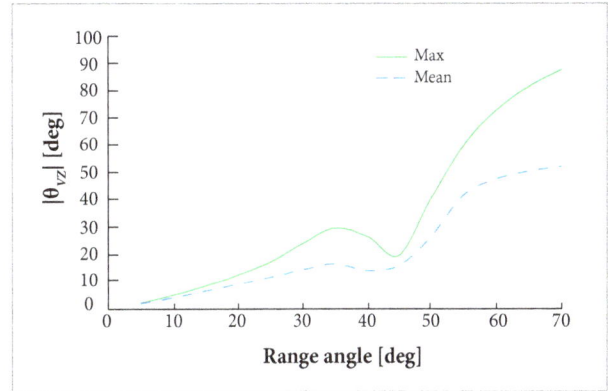

Figure 10. Max and mean of the absolute value of the angle between $\mathbf{v}_g$ and ZEM versus range angle for thrust vector along ZEM direction (minimum energy orbit).

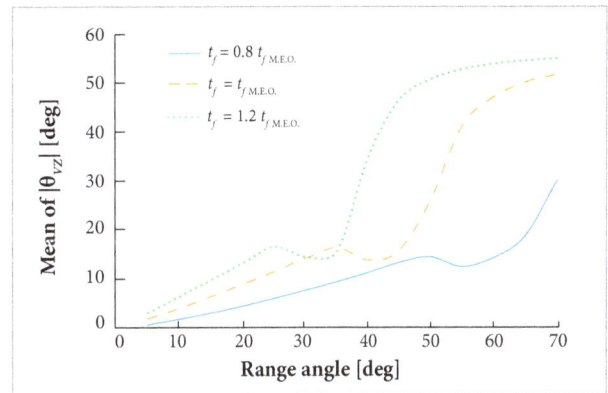

Figure 11. Mean of the absolute value of the angle between $\mathbf{v}_g$ and ZEM versus range angle for thrust vector along ZEM direction with different total final times.

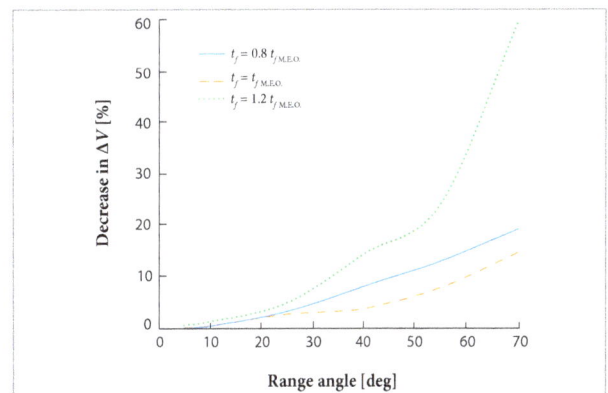

Figure 12. Decrease in ΔV for $\mathbf{v}_g$- based guidance with respect to ZEM-based guidance for minimum and non-minimum energy orbit.

for minimum energy orbit. Other parameters and initial values are similar to Fig. 5. In these figures, θ_{th} is the angle of the thrust acceleration with respect to the equatorial plane. As can be seen in these figures, the guidance law based on **ZEM** has a larger total rotation of the thrust vector than those of the two guidance laws based on $\mathbf{v}_g$. Also, it is observed that the rate of change of θ_{th} is nearly constant for the two guidance laws based on $\mathbf{v}_g$. This property phenomena can be used for implementation of the guidance laws based on $\mathbf{v}_g$.

An important question comes from the implementation point of view: what is the typical value of the angle between **ZEM** and $\mathbf{v}_g$? If this value is large enough to overwhelm the control system tracking error and estimation error of required velocity in the presence of target tracking error, the performance study of the guidance law can go ahead for this purpose.

First, the angle between $\mathbf{v}_g$ and **ZEM**, denoted by θ_{VZ}, is depicted in Fig. 9 *versus* time for an interceptor when its thrust acceleration is applied along **ZEM** for a range angle of 10°. The maximum and the mean of $|\theta_{VZ}|$, $\int_0^{t_{co}} |\theta_{VZ}|\, dt/t_{co}$, are observed by solid and dashed lines, respectively. The maximum and mean

values of $|\theta_{VZ}|$ are shown in Fig. 10 *versus* range angle. The final times in Figs. 9 and 10 are chosen for minimum energy orbit, *i.e.* minimizing the required velocity. To investigate more precisely, the mean value of $|\theta_{VZ}|$ is plotted in Fig. 11 *versus* range angle for a deviation of ±20% with respect to the final time of minimum energy orbit. As expected, the values of θ_{VZ} are large enough to overcome noisy measurements and control system tracking error for medium-to-super range applications. The effect of this deviation on ΔV can be viewed in Fig. 12 where the comparison is made with respect to the fuel consumption of the ZEM-based guidance law.

The effect of target altitude is investigated in Fig. 13. First, the fuel consumption is plotted *versus* target radial position for a range angle of 10° for the minimum energy orbit. As can also be seen in Fig. 5a, there is little difference between the fuel consumptions of $\mathbf{v}_g$- and **ZEM**-based guidance laws. This difference is increased by increasing the range angle. For example, the difference is shown in Fig. 13b for the range angle of 40°. Moreover, the maximum value of the angle between $\mathbf{v}_g$ and **ZEM** versus target radial position can be viewed in Figs. 14a and 14b for minimum and non-minimum energy orbits,

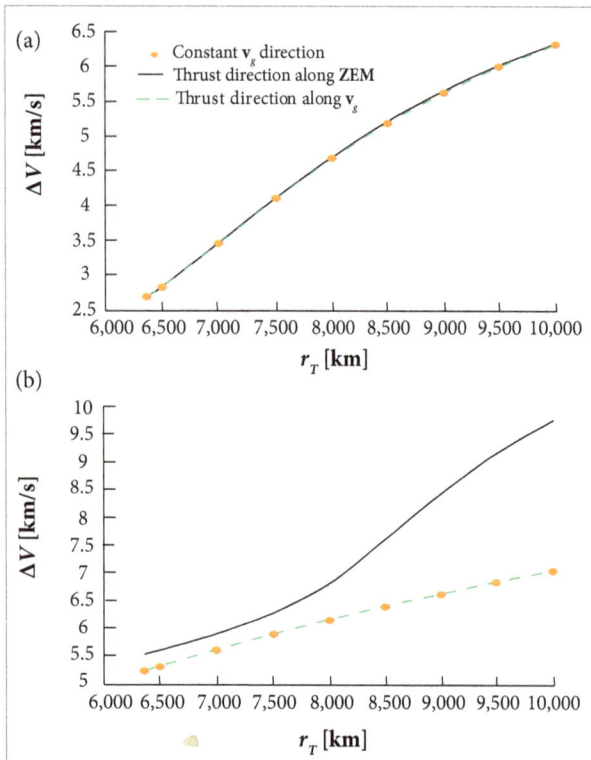

Figure 13. Fuel consumption versus target radial position for interception of a stationary target in spherical-Earth model. (a) range angle = 10°; (b) range angle = 40°.

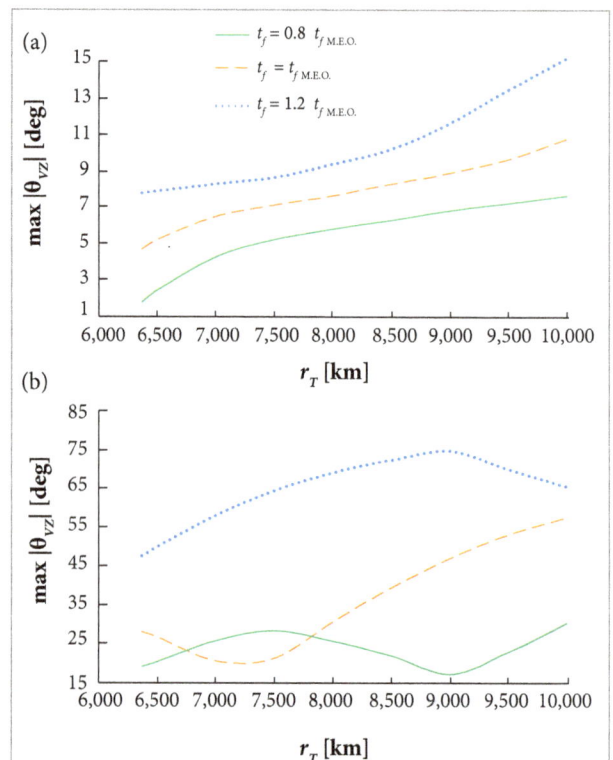

Figure 14. Max of the absolute value of the angle between $\mathbf{v}_g$ and ZEM versus target radial position for thrust vector along ZEM direction. (a) range angle = 10°; (b) range angle = 40°.

e.g. $t_f = (1 \pm 0.2)t_{fM.E.O.}$, respectively, where M.E.O. is the mean energy orbit.

After a preliminary study of conceptual guidance laws for flat-Earth model and also stationary targets in spherical-Earth model, we focus on free-falling targets in the case of spherical-Earth model. The interceptor initial position, velocity, and acceleration due to thrust is taken similar to the studied case of spherical-Earth model with stationary targets.

First, the guidance law based on linearized ZEM, Eq. 26, is compared to the guidance based on the direction computed by linear optimal guidance law (Eq. 36) as obtained by Deihoul and Massuomnia (2003). The miss distance and fuel consumption of the two guidance schemes are plotted in Figs. 15 and 16 *versus* final time for a target at a range angle of 40°, having the required velocity of minimum energy to hit the initial position of the interceptor. The initial distance of target from the Earth center is taken $r_T = 6,500$ km. These 2 guidance laws produce nearly the same results; however, the little differences in the results cannot appear properly in Figs. 15 and 16 because of the

scale of these figures. These results are obtained by setting an optimized value of $t_b = 60$ s in Eq. 37 for the gain matrix $\mathbf{M}$. This analysis turns out the gain matrix $\mathbf{M}$ do not give a significant improvement on the performance of the ZEM-based guidance schemes for an initial range angle less than 40°.

To study more precisely, the initial distance of the target from the earth center increases to 7,500 km. First, the value of t_b is chosen 100 s by the performance analysis based on Fig. 17 for an initial range angle of 90° with 3 different values of total final times. In the next step, the fuel consumptions of four conceptual guidance laws, *i.e.* guidance laws based on $\mathbf{ZEM}_{min}$, based on $\mathbf{ZEM}$, based on the direction of linear optimal problem, and based on velocity-to-be-gained are compared together. For this purpose, the fuel consumption is depicted *versus* final time in Figs. 18a; 18b and 18c for 3 different initial range angles of 60°; 90° and 120°, respectively. The value of t_b is optimized for each range angle. Other parameters and initial values are taken as before. For a typical range comparison, the interceptor travels 29.7° (3,301 km) for an initial range angle of 90° (10,002 km) as plotted in Fig. 19 for a total flight time of 1,600 s. The thrust direction along the velocity-to-be-gained produces less fuel consumption among the other mentioned guidance laws as shown in the Fig. 18. For short range applications, the angle between ZEM and velocity-to-be-gained is negligible as shown in Fig. 20. The angle between the 2 directions increases by increasing the initial range angle. The effect of matrix gain of $\mathbf{M}$ is also increased by increasing the range angle, but for these ranges the direction of the velocity-to-be-gained produces a considerable decrease on fuel consumption. Besides, the estimation error of time-to-go for guidance laws based on velocity-to-be-gained has less sensitivity than that of ZEM-based guidance schemes.

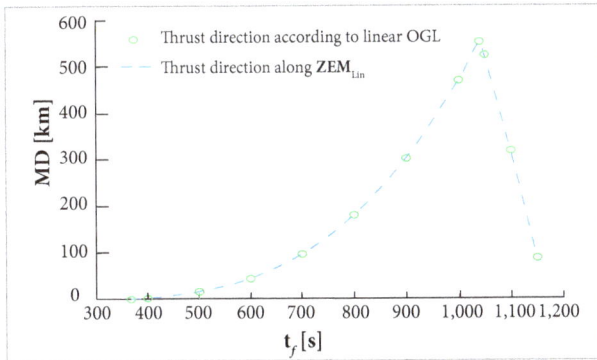

Figure 15. Miss distance (MD) versus final time for interception of a free-falling target in spherical-Earth model using linearized ZEM relation (initial range angle = 40°).

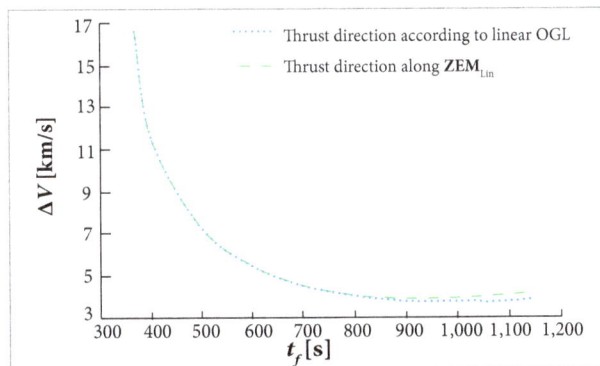

Figure 16. Fuel consumption versus final time for interception of a free-falling target in spherical-Earth model (initial range angle = 40°).

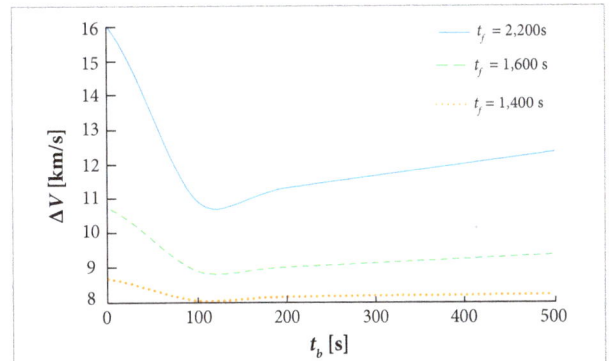

Figure 17. Fuel consumption versus burnout time for interception of a free-falling target in spherical-Earth model, when thrust vector is along linearized ZEM direction (initial range angle = 90°).

The angle of thrust direction with respect to the equatorial plane is plotted in Fig. 21 *versus* time for 3 guidance laws. As can be seen, there is a little difference between the velocity-to-be-gained direction and the direction of guidance laws based on constant $\mathbf{v}_g$ direction. The maximum difference for an initial range angle of 70° is about 5.26°. In addition, the rate of change of thrust direction for guidance laws based on velocity-to-be-gained are very small comparing to that of the guidance laws based on ZEM. To investigate more precisely the difference between the thrust angles under the 2 conceptual guidance laws based on velocity-to-be-gained, θ_{BV}, is shown in Fig. 22 in three forms, *i.e.* max of $|\theta_{BV}|$, mean of $|\theta_{BV}|$, and mean of

θ_{BV} for a minimum energy orbit and a stationary target. Initial values for interceptor and its target are the same as Figs. 5 – 7. As seen in Fig. 22, the mean value of θ_{BV} is 1.11° for an initial

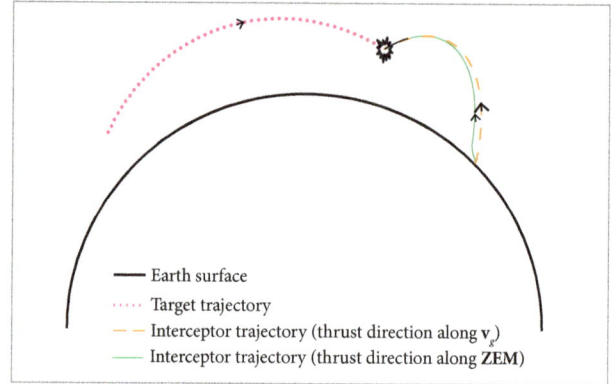

Figure 19. Interceptor and its target trajectories for initial range angle of 90° and total flight time of 1,600 s.

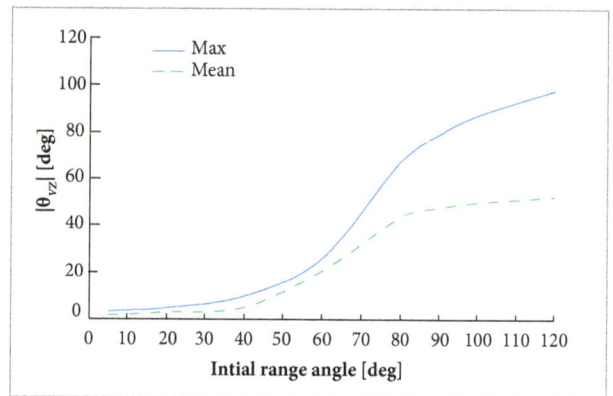

Figure 20. Maximum and mean of the absolute value of the angle between $\mathbf{v}_g$ and ZEM versus initial range angle for interception of a free-falling target (thrust vector along ZEM direction, minimum energy orbit).

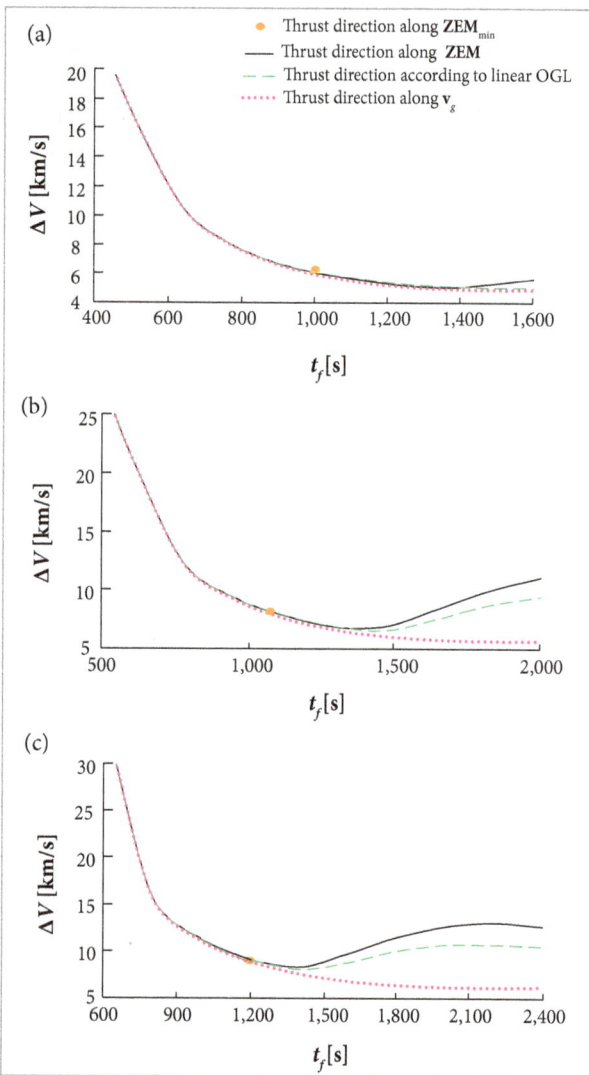

Figure 18. Compassion of fuel consumption of four conceptual guidance laws for interception of a free-falling target in spherical-Earth model. (a) initial range angle = 60°; (b) initial range angle = 90°; (c) initial range angle = 120°.

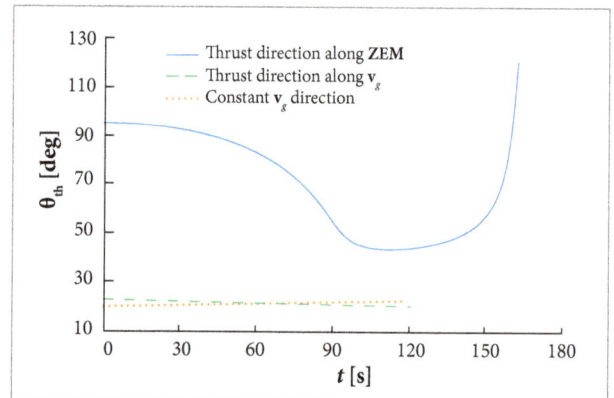

Figure 21. Comparison of thrust angle with respect to the equatorial plane for 3 conceptual guidance laws (initial range angle = 70°; a_{th} = 50 m/s²).

range angle of 70°. Their fuel consumptions are compared in Fig. 23 versus initial range angle for minimum and non-minimum energy orbit. For example, the guidance based on the constant $\mathbf{v}_g$ direction causes a decrease of 0.32% in fuel consumption for minimum energy orbit with respect to the guidance law based on $\mathbf{v}_g$ direction for a range angle of 180°. As seen in Fig. 22, below the initial range angle of 10.3°, max $|\theta_{BV}|$ is less than 1°. However, improving the performance of a guidance law is always of interest by a modified formulation without any additional hardware or extra cost. Fortunately, several iterative and approximate methods are available in literature to calculate the required velocity and Q-matrix for Lambert's problem.

For space missions, Sokkappa (1966) developed a near optimal guidance in closed-loop for throttleable spacecraft assuming constant Q-matrix; however, the Q-matrix was updated for onboard computation. Our simulation results show that the maximum difference between the thrust angles of Sokkappa's

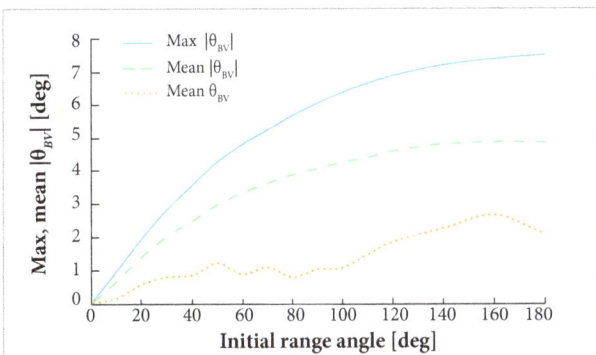

Figure 22. Maximum and mean of the absolute value, and mean value of the angle between the thrust direction along $\mathbf{v}_g$ and constant $\mathbf{v}_g$ direction versus range angle for interception of a stationary target for minimum energy orbit.

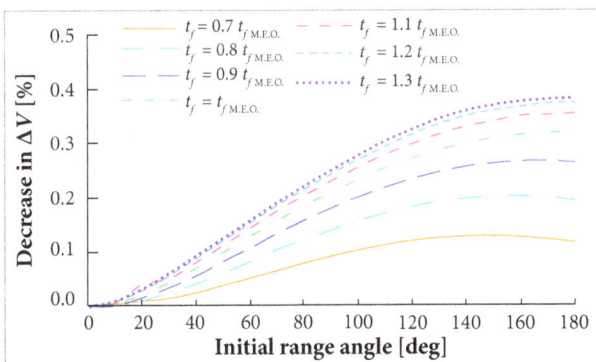

Figure 23. The percentage decrease in ΔV for the guidance law based on constant $\mathbf{v}_g$ direction with respect to $\mathbf{v}_g$ direction for minimum and non-minimum energy orbit.

solution and guidance based on constant $\mathbf{v}_g$ direction is less than 0.5° when the initial range angle is less than 180° (the initial values and parameters are similar to Fig. 22). In addition, Sokkappa compared his optimal solution with the guidance law based on constant $\mathbf{v}_g$ direction for a case of injection from an earth orbit of 100 nautical miles to pass through an inertial point of 180,000 nautical miles. The guidance law based on constant $\mathbf{v}_g$ direction produces ΔV = 3,328.6 m/s, whereas Sokkappa's solution has a 34.2 m/s decrease in ΔV. Also, the performance of guidance law based on constant $\mathbf{v}_g$ direction and numerical optimal solution was compared by Circi (2004) for a satellite launch vehicle. The optimum solution has a decrease of 39 kg for a payload mass of 1,734 kg under the guidance law based on constant $\mathbf{v}_g$ direction for a perigee of 150 km.

It is worth noting that the guidance laws based on velocity-to-be-gained are applicable provided that a required velocity can be defined. For example, a required velocity cannot be defined for a fixed-final-time problem when final position and velocity vectors are both constrained. The impact angle of the conceptual guidance schemes is not considered in this investigation. It may be accomplished using appropriate choice of final time.

CONCLUSIONS

This study suggests the effective thrust direction of an exoatmospheric interceptor for interception of short-to-super range moving targets with final position constraint. This has accomplished using a comprehensive study on conceptual guidance laws with stationary, moving, and free-falling targets. The first class of guidance law is based on ZEM. Three guidance schemes are considered in the first class, regarding to 2 definitions of ZEM and the component of ZEM perpendicular to line-of-sight. The capture criteria of the guidance scheme based on the perpendicular component of ZEM are highly restricted for non-throttleable rockets. The second class of conceptual guidance laws are based on linearized formulation, *i.e.* linearized ZEM and linear optimal control theory for throttleable rockets when its computed thrust direction is applied to non-throttleable rockets. The third class of guidance laws is based on the generalized required velocity and generalized velocity-to-be-gained. Two guidance schemes are considered in the third class. In the first scheme, the thrust acceleration is applied along the velocity-to-be-gained vector, whereas the second scheme tries to avoid the rotation of the velocity-to-be-gained in an inertial space.

The suggested direction of thrust acceleration is along the generalized velocity-to-be-gained, defined based on the generalized required velocity for interception of moving targets. For short-range application, the same results have been achieved when the thrust acceleration applied in the direction of ZEM or in the direction of the generalized velocity-to-be-gained. Increasing the range angle, the difference in performance is appeared. For long-range interception, the suggested thrust direction requires less amount of fuel rather than conceptual guidance laws based on ZEM or linearized formulations. The guidance performance is not improved as expected using a ZEM vector multiplied by a gain matrix to deviate the thrust direction from ZEM vector, as obtained in linearized optimal solutions.

If the intercept time is chosen a bit larger than the minimum energy orbit due to tactical consideration, *e.g.* salvo firing, adjustment of impact angle, etc., the suggested direction will have a significant fuel savings rather than guidance laws based on ZEM. Moreover, the performance of the suggested conceptual guidance law has less sensitive to the estimation error of final time. The guidance scheme based on constant velocity-to-be-gained direction may improve negligibly the fuel performance of the interceptor in the presence of noise for suborbital interception. Finally, the optimal solution does not give a better performance when the target position and velocity are contaminated by noise for suborbital interception such as a ballistic target; however, it improves the performance for a satellite launch vehicle and possibly for interception of orbital targets.

AUTHOR'S CONTRIBUTION

Mohamed-abadi MD performed the numerical solutions and prepared the figures. The idea, assumptions, classifications, and framework belong to Jalali-Naini SH, who wrote the manuscript. Both authors discussed the results and commented on the manuscript.

REFERENCES

Ahn J, Bang J, Lee SI (2015) Acceleration of zero-revolution Lambert's algorithms using table-based initialization. J Guid Control Dynam 38(2):335-342. doi: 10.2514/1.G000764

Arora N, Russell RP, Strange NJ, Ottesen D (2015) Partial derivatives of the solution to the Lambert boundary value problem. J Guid Control Dynam 38(9):1563-1572. doi: 10.2514/1. G001030

Battin RH (1999) An introduction to the mathematics and methods of astrodynamics. Revised edition. Reston: American Institute of Aeronautics and Astronautics.

Bryson AE, Ho YC (1975) Applied optimal control. New York: Hemisphere.

Chen FL, Xiao Y, Chen W (2010) Guidance based on velocity-to-be-gained surface for super-range exoatmospheric intercept. Acta Aeronautica et Astronautica Sinica 31(2):342-349. In Chinese.

Circi C (2004) Hybrid methods and Q-guidance for rocket performance optimization. Proc IME G J Aero Eng 218(5):353-359. doi: 10.1243/0954410042467040

Deihoul AR (2003) Anti ballistic optimal midcourse guidance law (PhD thesis). Tehran: Sharif University of Technology. In Persian.

Deihoul AR, Massoumnia MA (2003) A near optimal midcourse guidance law based on spherical gravity. Scientia Iranica 10(4):436-442.

Feng C, Yelun X, Wanchun C (2009) Guidance based on zero effort miss for super-range exoatmospheric intercept. Acta Aeronautica et Astronautica Sinica 30(9):1583-1589. In Chinese.

Jalali-Naini SH (2004) Modern explicit guidance law for high-order dynamics. J Guid Control Dynam 27(5):918-922. doi: 10.2514/1.5902

Jalali-Naini SH (2008) Generalization of zero-effort miss equations in atmospheric guidance laws with application to midcourse flight (PhD thesis). Tehran: Sharif University of Technology. In Persian.

Jalali-Naini SH, Pourtakdoust SH (2005) Modern midcourse guidance laws in the endoatmosphere. Proceedings of the AIAA Guidance, Navigation and Control Conference and Exhibit; San Francisco, USA.

Jalali-Naini SH, Pourtakdoust SH (2007) A unified approach to intercept guidance laws. Proceedings of the 6th Iranian Aerospace Society Conference; Tehran, Iran.

Li LG, Jing WX, Gao CS (2013) Design of midcourse trajectory for tactical ballistic missile intercept on the basis of zero effort miss. App Mech Mater 397-400:536-545. doi: 10.4028/www. scientific.net/AMM. 397-400.536

Martin FH (1965) Closed-loop near-optimum steering for a class of space missions (PhD thesis). Cambridge: Massachusetts Institute of Technology.

Martin FH (1966) Closed-loop near-optimum steering for a class of space missions. AIAA J 4(11):1920-1927. doi: 10.2514/3.3819

Massoumnia MA (1995) Optimal midcourse guidance law for fixed-interval propulsive maneuvers. J Guid Control Dynam 18(3):465-470. doi: 10.2514/3.21410

Mohammad-abadi MD, Jalali-Naini SH (2016) Approximate solution of zero-effort-miss under gravitational acceleration inversely proportional to the cubic distance. Modares Mech Eng 16(4):135-144. In Persian.

Newman B (1996) Strategic intercept midcourse guidance using modified zero effort miss steering. J Guid Control Dynam 19(1):107-112. doi: 10.2514/3.21586

Rusnak I, Meir L (1991) Modern guidance law for high-order autopilot. J Guid Control Dynam 14(5):1056-1058. doi: 10.2514/3.20749

Sokkappa BG (1966) On optimal steering to achieve required velocity. Proceedings of the 16th International Astronautical Congress; Athens, Greece.

Zarchan P (2012) Tactical and strategic missile guidance. 6th edition. Reston: American Institute of Aeronautics and Astronautics.

Permissions

All chapters in this book were first published in JATM, by Departamento de Ciência e Tecnologia Aeroespacial; hereby published with permission under the Creative Commons Attribution License or equivalent. Every chapter published in this book has been scrutinized by our experts. Their significance has been extensively debated. The topics covered herein carry significant findings which will fuel the growth of the discipline. They may even be implemented as practical applications or may be referred to as a beginning point for another development.

The contributors of this book come from diverse backgrounds, making this book a truly international effort. This book will bring forth new frontiers with its revolutionizing research information and detailed analysis of the nascent developments around the world.

We would like to thank all the contributing authors for lending their expertise to make the book truly unique. They have played a crucial role in the development of this book. Without their invaluable contributions this book wouldn't have been possible. They have made vital efforts to compile up to date information on the varied aspects of this subject to make this book a valuable addition to the collection of many professionals and students.

This book was conceptualized with the vision of imparting up-to-date information and advanced data in this field. To ensure the same, a matchless editorial board was set up. Every individual on the board went through rigorous rounds of assessment to prove their worth. After which they invested a large part of their time researching and compiling the most relevant data for our readers.

The editorial board has been involved in producing this book since its inception. They have spent rigorous hours researching and exploring the diverse topics which have resulted in the successful publishing of this book. They have passed on their knowledge of decades through this book. To expedite this challenging task, the publisher supported the team at every step. A small team of assistant editors was also appointed to further simplify the editing procedure and attain best results for the readers.

Apart from the editorial board, the designing team has also invested a significant amount of their time in understanding the subject and creating the most relevant covers. They scrutinized every image to scout for the most suitable representation of the subject and create an appropriate cover for the book.

The publishing team has been an ardent support to the editorial, designing and production team. Their endless efforts to recruit the best for this project, has resulted in the accomplishment of this book. They are a veteran in the field of academics and their pool of knowledge is as vast as their experience in printing. Their expertise and guidance has proved useful at every step. Their uncompromising quality standards have made this book an exceptional effort. Their encouragement from time to time has been an inspiration for everyone.

The publisher and the editorial board hope that this book will prove to be a valuable piece of knowledge for researchers, students, practitioners and scholars across the globe.

List of Contributors

Diana Carolina Morón Hernández, Freddy Alexander Díaz González, Juan Sebastian Triana Correa and Pablo Roberto Pinzón Cabrera
Universidad Sergio Arboleda – Escuela de Ciencias Exactas e Ingeniería - Bogotá – Colombia

Vahid Tahmasbi and Seyed Mohammad Hossein Karimian
Amirkabir University of Technology – Aerospace Engineering Department – Tehran – Iran

Valdemir Carrara
Instituto Nacional de Pesquisas Espaciais – Engenharia e Tecnologias Espaciais – Divisão de Mecânica Espacial e Controle – São José dos Campos/SP – Brazil
Departamento de Ciência e Tecnologia Aeroespacial – Instituto Tecnológico de Aeronáutica – Divisão de Engenharia Aeronáutica – São José dos Campos/ SP – Brazil

Rafael Barbosa Januzi
Universidade Federal de São Paulo - Instituto de Ciência e Tecnologia - Departamento de Ciência da Computação – São José dos Campos/SP – Brazil

Daniel Hideaki Makita
Universidade Federal de São Paulo - Instituto de Ciência e Tecnologia - Departamento de Engenharia da Computação – São José dos Campos/SP – Brazil

Luis Felipe de Paula Santos and Lidia Shibuya Sato
Departamento de Ciência e Tecnologia Aeroespacial – Instituto Tecnológico de Aeronáutica – Divisão de Engenharia Eletrônica e Computação – São José dos Campos/SP – Brazil

Cleber Souza Corrêa, Gerson Luiz Camillo, Vinicius Milanez Couto, Gilberto Fisch and Felipe do Nascimento Correa
Departamento de Ciência e Tecnologia Aeroespacial – Instituto de Aeronáutica e Espaço – Divisão de Ciências Atmosféricas – São José dos Campos/SP – Brazil.

Fabricio Harter
Universidade Federal de Pelotas – Faculdade de Meteorologia – Pelotas/RS – Brazil

Henrique Oliveira da Mata
Departamento de Ciência e Tecnologia Aeroespacial – Centro de Lançamento de Alcântara – Seção de Segurança de Vôo – Alcântara/MA – Brazil

João Batista Pessoa Falcão Filho, Ana Cristina Avelar, Leonardo Motta Maia de Oliveira Carvalho and João Luiz F. Azevedo
Departamento de Ciência e Tecnologia Aeroespacial – Instituto de Aeronáutica e Espaço – Divisão de Aerodinâmica – São José dos Campos/SP – Brazil

Marcos Galante Boato and Ezio Castejon Garcia
Departamento de Ciência e Tecnologia Aeroespacial – Instituto Tecnológico de Aeronáutica – Divisão de Engenharia Mecânica – São José dos Campos/ SP – Brazil

Marcio Bueno dos Santos and Antonio Fernando Beloto
Instituto Nacional de Pesquisas Espaciais – Laboratório de Integração e Testes – São José dos Campos/SP – Brazil

Ludmila Resende Vargas, Anne Karoline Poli and Rita de Cássia Lazzarini Dutra
Departamento de Ciência e Tecnologia Aeroespacial – Instituto Tecnológico de Aeronáutica – Programa de Pós-Graduação em Ciência e Tecnologia Espaciais – São José dos Campos/SP – Brazil

Camila Brito deSouza
Departamento de Ciência e Tecnologia Aeroespacial – Instituto de Aeronáutica e Espaço – Divisão de Materiais – São José dosCampos/SP – Brazil
Universidade Federal de São Paulo – Instituto de Ciência e Tecnologia - Campus Parque Tecnológico – São José dos Campos/SP – Brazil

Maurício Ribeiro Baldan
Departamento de Ciência e Tecnologia Aeroespacial – Instituto Tecnológico de Aeronáutica – Programa de Pós-Graduação em Ciência e Tecnologia Espaciais – São José dos Campos/SP – Brazil
Instituto Nacional de Pesquisas Espaciais – Laboratório Associado de Sensores – São José dos Campos/SP – Brazil.

Emerson Sarmento Gonçalves
Departamento de Ciência e Tecnologia Aeroespacial
– Instituto Tecnológico de Aeronáutica – Programa
de Pós-Graduação em Ciência e Tecnologia Espaciais
– São José dos Campos/SP – Brazil
Departamento de Ciência e Tecnologia Aeroespacial
– Instituto de Aeronáutica e Espaço – Divisão de
Materiais – São José dos Campos/SP – Brazil

Xiaowei Shao, Zehao Zhang, Jihe Wang and Dexin Zhang
Shanghai Jiao Tong University – School of
Aeronautics and Astronautics – Distributed
Spacecraft System Technology Laboratory –
Shanghai – China

Rouzbeh Moradi and Alireza Alikhani
Ministry of Science, Research and Technology
– Aerospace Research Institute – Astronautics
Department – Tehran/Tehran – Iran

Mohsen Fathi Jegarkandi
Sharif University of Technology– Engineering
College – Department of Aerospace Engineering –
Tehran/Tehran – Iran

Mostafa Zakeri and Mehran Nosratollahi
Space Research Institute – System Engineering
Department – Space System Design Laboratory –
Tehran/Tehran – Iran

Alireza Novinzade
Khaje Nasir Toosi University of Technology –
Faculty of Aerospace – Systems Division – Tehran/
Tehran – Iran

Daniel Schuch
Departamento de Ciência e Tecnologia Aeroespacial
– Instituto Tecnológico de Aeronáutica – Programa
de Pós-Graduação em Ciências e Tecnologias
Espaciais – São José dos Campos/SP – Brazil

Gigliola Salerno
Centro Universitário FEI – Departamento de
Engenharia de Materiais – São Bernardo do
Campo/SP – Brazil

Stefano Mariani and Alberto Corigliano
Politecnico di Milano – Dipartimento di Ingegneria
Civile e Ambientale – Milano/Milano – Italy

Tharcius Augusto Pivetta
Departamento de Ciência e Tecnologia Aeroespacial
– Instituto Tecnológico de Aeronáutica – São José
dos Campos/SP – Brazil

Glauco da Silva
Departamento de Ciência e Tecnologia Aeroespacial
– Instituto Tecnológico de Aeronáutica – São José
dos Campos/SP – Brazil
Departamento de Ciência e Tecnologia Aeroespacial
– Instituto de Aeronáutica e Espaço – São José dos
Campos/SP – Brazil

Carlos Henrique Netto Lahoz
Departamento de Ciência e Tecnologia Aeroespacial
– Instituto Tecnológico de Aeronáutica – São José
dos Campos/SP – Brazil
Departamento de Ciência e Tecnologia Aeroespacial
– Instituto de Aeronáutica e Espaço – São José dos
Campos/SP – Brazil
Massachusetts Institute of Technology – Department
of Aeronautics and Astronautics – Cambridge/MA
– USA

João Batista Camargo Júnior
Universidade de São Paulo – Escola Politécnica –
Departamento de Engenharia de Computação e
Sistemas Digitais – São Paulo/SP – Brazil

Hossein Bonyan Khamseh
Universidade Federal de Minas Gerais – Escola de
Engenharia Elétrica – Departamento de Engenharia
Elétrica – Belo Horizonte/MG – Brazil

Luciana de Barros
Departamento de Ciência e Tecnologia Aeroespacial
– Instituto de Aeronáutica e Espaço – Divisão de
Propulsão Espacial – São José dos Campos/SP –
Brazil
Departamento de Ciência e Tecnologia Aeroespacial
– Instituto Tecnológico de Aeronáutica –
Departamento de Química – São José dos Campos/
SP – Brazil

Afonso Paulo Monteiro Pinheiro and Josemar da Encarnação Câmara
Departamento de Ciência e Tecnologia Aeroespacial
– Instituto de Aeronáutica e Espaço – Divisão de
Propulsão Espacial – São José dos Campos/SP –
Brazil

Koshun Iha
Departamento de Ciência e Tecnologia Aeroespacial – Instituto Tecnológico de Aeronáutica – Departamento de Química – São José dos Campos/SP – Brazil

Ruibin Xue
Beijing Institute of Technology – School of Aerospace Engineering – Key Laboratory of Dynamics and Control of Flight Vehicle – Beijing – China

Gaohua Cai
Beijing Aerospace Automatic Control Institute – Beijing – China

Luiz Felipe Cannaval Sbegue
Universidade de São Paulo – Escola de Engenharia de Lorena – Departamento de Engenharia de Materiais – Lorena/SP – Brazil

Luciene Dias Villar
Departamento de Ciência e Tecnologia Aeroespacial – Instituto de Aeronáutica e Espaço – Divisão de Química – São José dos Campos/SP – Brazil

Majid Hosseini, Mehran Nosratollahi and Hossein Sadati
Malek Ashtar University of Technology – Faculty of Aerospace – Department of Aerospace – Tehran/Tehran – Iran

Mohsen Dehghani Mohammad-abadi and Seyed Hamid Jalali-Naini
Tarbiat Modares University – Faculty of Mechanical Engineering – Aerospace Group – Tehran/Tehran – Iran

Index

www.ingramcontent.com/pod-product-compliance
Lightning Source LLC
Chambersburg PA
CBHW082015190326
41458CB00010B/3192